PRAISE FOR

ARTS OF LIVING ON A DAMAGED PLANET

"What an inventive, fascinating book about landscapes in the Anthropocene! Between these book covers, rightside-up, upside-down, a concatenation of social science and natural science, artwork and natural science, ghosts of departed species and traces of our own human shrines to memory. . . . Not a horror-filled glimpse at destruction but also not a hymn to romantic wilderness. Here, guided by a remarkable and remarkably diverse set of guides, we enter into our planetary environments as they stand, sometimes battered, sometimes resilient, always riveting in their human—and nonhuman—richness. *Arts of Living on a Damaged Planet* is truly a book for our time."

—PETER GALISON, Harvard University

"Facing the perfect storm strangely named the Anthropocene, this book calls its readers to acknowledge and give praise to the many entangled arts of living that made this planet livable and that are now unraveling. Grandiose guilt will not do; we need to learn to notice what we were blind to, a humble but difficult art. The unique welding of scholarship and affect achieved by the texts assembled tells us that learning this art also means allowing oneself to be touched and induced to think and imagine by what touches us."

—ISABELLE STENGERS, author of *Cosmopolitics I* and *II*

"*Arts of Living on a Damaged Planet* exposes us to the active remnants of gigantic past human errors—the ghosts—that affect the daily lives of millions of people and their co-occurring other-than-human life forms. Challenging us to look at life in new and excitingly different ways, each part of this two-sided volume is informative, fascinating, and a source of stimulation to new thoughts and activisms. I have no doubt I will return to it many times."

—MICHAEL G. HADFIELD, University of Hawai'i at Mānoa

ARTS OF LIVING

ON A DAMAGED PLANET

ARTS OF LIVING ON A DAMAGED PLANET

GHOSTS OF THE ANTHROPOCENE

Anna Tsing
Heather Swanson
Elaine Gan
Nils Bubandt

Editors

University of Minnesota Press
MINNEAPOLIS • LONDON

The University of Minnesota Press gratefully acknowledges financial assistance for the publication of this book from the Aarhus University Research Fund.

Illustrations by Jesse Lopez, jesselopez.com.

Published by the University of Minnesota Press
111 Third Avenue South, Suite 290
Minneapolis, MN 55401-2520
http://www.upress.umn.edu

ISBN 978-1-5179-0236-0 (hc)
ISBN 978-1-5179-0237-7 (pb)
A Cataloging-in-Publication record for this book is available from the Library of Congress.

Printed in the United States of America on acid-free paper

22 10 9 8

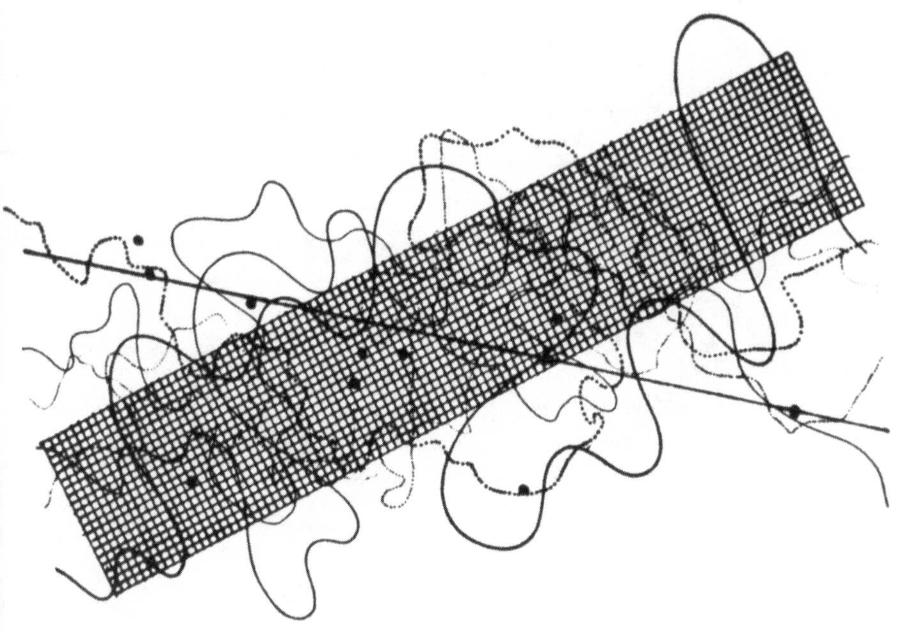

John Cage, *Fontana Mix*, 1958. Copyright by Henmar Press, Inc., New York. Reproduced by kind permission of Peters Edition Limited, London. Courtesy of John Cage Trust.

Instead of a sequence of notes to be played as prescribed by a composer, Cage offers a finite system of elements that can be layered into infinite compositions. Hand-drawn on sheets of paper and clear acetate, Cage's wandering curved lines, randomly placed dots, rectangular grid, and single straight line all intersect when layered, becoming musical scores of indeterminacy.

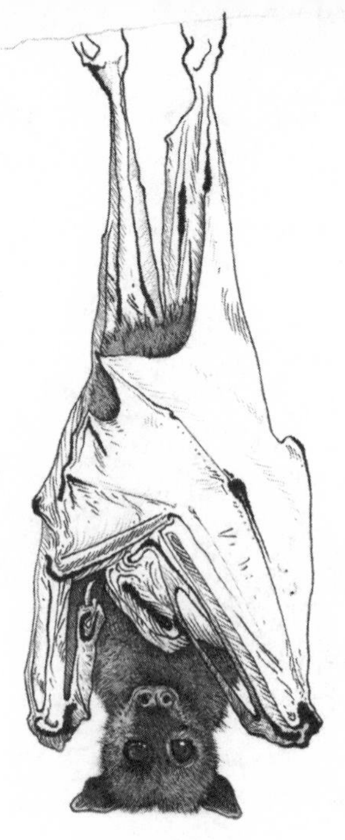

CONTENTS

GHOSTS ON A DAMAGED PLANET

Acknowledgments *Gix*

Introduction: Haunted Landscapes of the Anthropocene *G1*
ELAINE GAN, ANNA TSING, HEATHER SWANSON, AND NILS BUBANDT

1 A Garden or a Grave? The Canyonic Landscape of the Tijuana–San Diego Region *G17*
LESLEY STERN

IN THE MIDST OF DAMAGE *G31*

2 Marie Curie's Fingerprint: Nuclear Spelunking in the Chernobyl Zone *G33*
KATE BROWN

3 Shimmer: When All You Love Is Being Trashed *G51*
DEBORAH BIRD ROSE

FOOTPRINTS OF THE DEAD *G65*

4 Future Megafaunas: A Historical Perspective on the Potential for a Wilder Anthropocene *G67*
JENS-CHRISTIAN SVENNING

5 Ladders, Trees, Complexity, and Other Metaphors in Evolutionary Thinking *G87*
ANDREAS HEJNOL

6 No Small Matter: Mushroom Clouds, Ecologies of Nothingness, and Strange Topologies of Spacetimemattering *G103*
KAREN BARAD

7 Haunted Geologies: Spirits, Stones, and the Necropolitics of the Anthropocene *G121*
NILS BUBANDT

WHAT REMAINS *G143*

8 Ghostly Forms and Forest Histories *G145*
ANDREW S. MATHEWS

9 Establishing New Worlds: The Lichens of Petersham *G157*
ANNE PRINGLE

Coda: Concept and Chronotope *G169*
MARY LOUISE PRATT

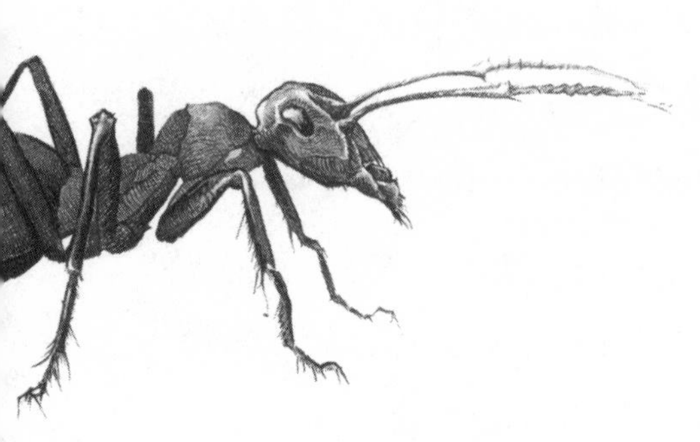

ACKNOWLEDGMENTS

ONE KEY INSPIRATION FOR THIS BOOK was the conference Anthropocene: Arts of Living on a Damaged Planet, which the coeditors organized at the University of California, Santa Cruz (UCSC), on May 8–10, 2014. Its speakers and discussants—including the contributors as well as Bettina Aptheker, Nora Bateson, James Clifford, Chris Connery, William Cronon, Margaret Fitzsimmons, Maya Peterson, Eric Porter, Jenny Reardon, Jessica Weir, and Thomas Wentzer—infused the discussions with passion and powerful arguments. Zachary Caple, Rachel Cypher, Emilie Dionne, Rosa Ficek, Micha Rahder, Kirsten Rudestam, and Craig Schuetze led walk-and-talks and helped with many details. Nadia Peralta and Elena Staley mobilized undergraduates. Gene A. Felice II showed his marvelous "Phytoplankton Confessional." Rune Flikke, Carolyn Hadfield, Frida Hastrup, and Knut Nustad traveled from afar as invited guests. Karen Fowler and Molly Gloss joined us for discussions with Ursula K. Le Guin. Katy Overstreet gracefully managed logistics, while John Weber and Rachel Nelson of the Institute of Arts and Sciences, as well as Faye Crosby of Cowell College, were helpful in many ways—including in creating casual settings for impromptu exchanges. Courtney Mahaney of the Institute for Humanities Research made our web page. Fred Deakin offered advice and support throughout. Patricia Alvarez, Egill Bjarnason, Rebecca Ora, Benjamin Schultz-Figueroa, Tristan Carkeet, and Katrine Duus Therkelsen worked on video documentation. Enosh Baker fed us artfully. A second conference in Aarhus on October 30–31, 2014, continued these discussions. Contributors were joined there by biologist Michael Thomas-Poulsen and anthropologist Annemarie Mol.

Both conferences received major funding from the Danish National Research Foundation. The Santa Cruz conference was also funded by the University of California, Santa Cruz Foundation, the UCSC Council of Provosts, the UCSC Office of Research, Cowell College, the UCSC Social Sciences Division, the UCSC Institute of the Arts and Sciences, the UCSC Living Writers Series, the UC Presidential Chair in Feminist Critical Race and Ethnic Studies, the Institute for Humanities Research, the UCSC Anthropology Department, Literary Guillotine, and Bookshop Santa Cruz. The book has benefited from a subvention from Aarhus University Research Foundation.

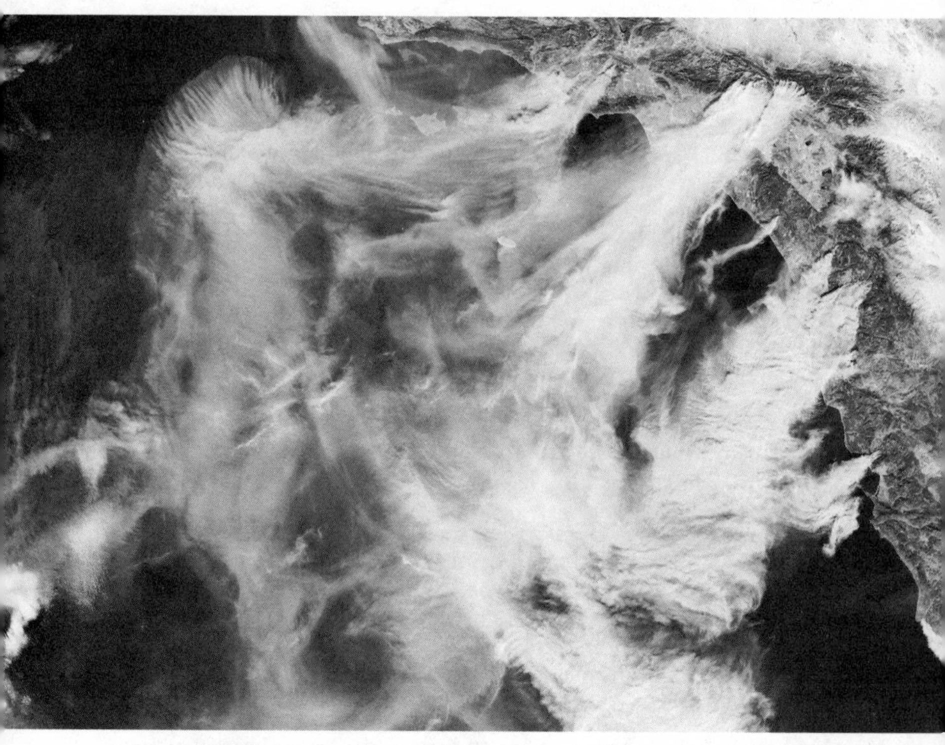

Santa Ana winds in Southern California, courtesy of NASA Earth Observatory.

INTRODUCTION

HAUNTED LANDSCAPES OF THE ANTHROPOCENE

Elaine Gan, Anna Tsing, Heather Swanson, Nils Bubandt

What Kinds of Human Disturbance Can Life on Earth Bear?

The winds of the Anthropocene carry ghosts—the vestiges and signs of past ways of life still charged in the present. This book offers stories of those winds as they blow over haunted landscapes. Our ghosts are the traces of more-than-human histories through which ecologies are made and unmade.

"Anthropocene" is the proposed name for a geologic epoch in which humans have become the major force determining the continuing livability of the earth. The word tells a big story: living arrangements that took millions of years to put into place are being undone in the blink of an eye. The hubris of conquerors and corporations makes it uncertain what we can bequeath to our next generations, human and not human. The enormity of our dilemma leaves scientists, writers, artists, and scholars in shock. How can we best use our research to stem the tide of ruination? In this half of our volume, we approach this problem by showing readers how to pay better attention to overlaid arrangements of human and nonhuman living spaces, which we call "landscapes." Our hope is that such attention will allow us to stand up to the constant barrage of messages asking us to *forget*—that is, to allow a few private owners and public officials with their eyes focused on short-term gains to pretend that environmental devastation does not exist.

We also face a barrage of messages that tell us to keep moving forward, to get the newer model, to have more babies, to get bigger. There is a lot of pressure to grow.

We do not think this work is simple. It requires moving beyond the disciplinary prejudices into which each scholar is trained, to instead take a generous view of what varied knowledge practices might offer. In this spirit, we begin with a literary essay that offers the fine description necessary to pay attention to ruins, but later move to a scientific report on the very long history of human-caused extinctions and an anthropological guide on how to read landscape history in the shapes of trees. These and much more open up the curiosity about life on earth that we will need to limit the destruction we call Anthropocene and protect the Holocene entanglements that we need to survive.

Our era of human destruction has trained our eyes only on the immediate promises of power and profits. This refusal of the past, and even the present, will condemn us to continue fouling our own nests. How can we get back to the pasts we need to see the present more clearly? We call this return to multiple pasts, human and not human, "ghosts." Every landscape is haunted by past ways of life. We see this clearly in the presence of plants whose animal seed-dispersers are no longer with us. Some plants have seeds so big that only big animals can carry them to new places to germinate. When these animals became extinct, their plants could continue without them, but they have been unable to disperse their seeds very well. Their distribution is curtailed; their population dwindles. This is an example of what we are calling haunting.

Anthropogenic landscapes are also haunted by imagined futures. We are willing to turn things into rubble, destroy atmospheres, sell out companion species in exchange for dreamworlds of progress.

Haunting is quite properly eerie: the presence of the past often can be felt only indirectly, and so we extend our senses beyond their comfort zones. Human-made radiocesium has this uncanny quality: it travels in water and soil; it gets inside plants and animals; we cannot see it even as we learn to find its traces. It disturbs us in its indeterminacy; this is a quality of ghosts.

As anthropologists, we imagine our talk of ghosts in kinship with communities around the world, Western and non-Western, who offer nonsecular descriptions of the landscape and its hauntings. Rather than an a priori distinction between modern and nonmodern, however, we open our analysis to practical ways of learning what is out there: the past and the present around us. This book is not about cosmologies but rather about on-the-ground observations, and from varied historical diffractions and points of view. Snake spirits and radioactive clouds share our attention as each draws us closer to the haunted quality of ruined landscapes.

Our use of the term "Anthropocene" does not imagine a homogeneous human race. We write in dialogue with those who remind readers of unequal relations among humans, industrial ecologies, and human insignificance in the web of life by writing instead of Capitalocene, Plantationocene, or Chthulucene (see Haraway, this volume). Our use of "Anthropocene" intends to join the conversation—but not to accept the worst uses of the term, from green capitalism to technopositivist hubris.

As we introduce the chapters, we want to show you both their practical gifts for reading landscapes and their work in grasping that which is hard to grasp—the spookiness of the past in the present. In this introduction we offer the wind as a figure for this uncanniness. Winds are hard to pin down, and yet material; they might convey some of our sense of haunting. Each paragraph in grey italics introduces an article from our volume through its haunting qualities. (Bold phrases are key themes in direct quotation from the essays.) We have included pieces from the "Monsters" half of the book along with "Ghosts" since the sections tell intertwined stories. Although our analytic frames deserve some separation, monsters and ghosts cannot be segregated. Meanwhile, we also use sentences in italics to index crosscurrents among our multiple authorial voices.

The Santa Ana winds pour into Los Laureles Canyon along the Tijuana–San Diego border. The wind is hot and dry, and it carries ghosts. ***Tires are everywhere in this canyon,*** *writes Lesley Stern. Garbage dumps, bulldozed mesas, and steel-fenced borders mix with invasive plants and native gardens in weedy shantytowns. Here is the debris of capitalist waste, the unspectacular afterlives of discarded things. Some tires are repurposed as building materials. Others lie around, dumped by careless dealers. Traffic from the United States flows southbound into Mexico unchecked; not so for reverse traffic. The canyon remembers* ***the movement of things,*** *including unlikely tomatoes growing through toxic sewers and cracked cement. Traces of past, present, and future mix in gardens that sprout from the graves of a violently uneven modernity.*

Like every landscape, Los Laureles Canyon is haunted by its human and nonhuman histories.

The transformation of the formerly biodiverse estuaries and canyons of the U.S.–Mexico border illustrates the predicament of industrial modernity: condominiums line one side, while waste piles on the other. Directives to close the border ask us to shut our eyes to continuing transfers of wealth and waste. Ghosts accumulate on both sides of the border from the residues of violence.

As life-enhancing entanglements disappear from our landscapes, ghosts take their place. Some scientists argue that the rate of biological extinction is now several hundred times beyond its historical levels. We might lose a majority of all species by the end of the twenty-first century.[1] The problem is not just the loss of individual species but of assemblages, some of which we may not even know about, some of which will not recover. Mass extinction could ensue from cascading effects. In an entangled world where bodies are tumbled into bodies (see our Monsters), extinction is a multispecies event. The extinction of a critical number of species would mean the destruction of long-evolving coordinations and interdependencies. While we gain plastic gyres and parking lots, we lose rainforests and coral reefs.

How much longer will we agree to step aside in silence as masters of the universe turn us into property, write our contracts, rape our bodies, sell our histories? How much longer will you and I choose extinction?

We live at the cusp of an extinction event comparable in scale to the Cretaceous–Paleogene (K–Pg) extinction event 65 million years ago that killed off the dinosaurs along with some 75 percent of all life forms on the planet.[2] The difference is that the current event, the "sixth extinction," will not be caused by an asteroid from outer space crashing into earth. The extinction event currently taking shape on the horizon of our shared future is the product of modern industry. How shall we retain the productive horror of our civilization—and yet refuse its inevitability? One method is to notice that the "we" is not homogeneous: some have been considered more disposable than others.

More than fourteen thousand kilometers lie between the Arctic tundra and the tip of South America. American red knots make that great migration each spring on the belly of the wind. They make a critical stop at Delaware Bay, where they feast on the eggs of horseshoe crabs that are emerging from the ocean on a single day in synchronized reproduction.

Human overharvesting of horseshoe crabs, however, has threatened the food supply of these migrating birds. As a result, a multispecies coordination that has taken place over millions of years is suddenly in danger of extinction. Will they leave only ghosts? asks biologist Peter Funch.

How many kinds of time—from *longue durée* evolutionary rapprochements to the quick boom and bust of investment capital—are wrapped up in these encounters? Minor forms of space and time merge with great ones. An extinction is a local event as well as a global one. Extinction is a breakdown of coordinations that has unintended and reverberating effects.

Some earth systems scientists describe the Anthropocene as the "Great Acceleration," the sharp rise in the destructive environmental effects of human industry since the second half of the twentieth century.[3] The massive increase in carbon dioxide, methane, and nitrate emissions into the atmosphere from industrialized agriculture, mineral extraction, petroleum-driven production, and globalized shipping/transportation networks has outpaced all other rhythms of life. Yet the Great Acceleration is best understood through immersion in many small and situated rhythms. Big stories take their form from seemingly minor contingencies, asymmetrical encounters, and moments of indeterminacy. Landscapes show us.

Imagine walking through Monti Pisani in Italy, where pines and abandoned chestnut orchards mingle. Andrew Mathews offers tactics for noticing ***the longue durée*** *of human disturbance as he shows us* ***form, texture, color,*** *a process of* ***constant speculation*** *as pattern. Ghosts become tangible through the form of ancient chestnut stools. Centuries of grafting, cultivation, trade, taxation, and disease are inscribed onto their structure and shape. The landscape emerges from ghostly entanglements: the many histories of life and death that have made these trees, this place.*

Extinction Leaves Traces

To track the histories that make multispecies livability possible, it is not enough to watch lively bodies. Instead, we must wander through landscapes, where assemblages of the dead gather together with the living. In their juxtapositions, we see livability anew. Many great animals that roamed the world in the Ice Age, for example, are now extinct. Their traces are still with us. Northern trees that grow back when cut down, such as oaks, may have evolved that ability in times when elephants trampled them. The ghosts of lost animals haunt these plants, even as the plants live on as our companions in the present.

Giant cave bears, straight-tusked elephants, and spotted hyenas once made their lives in Europe. The ground sloth, the mastodon, the shrub-ox: these were animals of North America. Unprecedented numbers of megafauna species became extinct during the late Quaternary period. Their disappearance from Eurasia, Australia, and the Americas is closely linked to the arrival of modern humans in these continents. As biologist Jens-Christian Svenning argues, their loss is almost certainly anthropogenic.

As humans reshape the landscape, we forget what was there before. Ecologists call this forgetting the "shifting baseline syndrome." Our newly shaped and ruined landscapes become the new reality. Admiring one landscape and its biological entanglements often entails forgetting many others. Forgetting, in itself, remakes landscapes, as we privilege some assemblages over others. Yet ghosts remind us. Ghosts point to our forgetting, showing us how living landscapes are imbued with earlier tracks and traces.

The native American flowers that are now missing from the Great Meadows of the University of California campus in Santa Cruz are ghosts to ecologist Ingrid Parker. Remembering missing flowers alerts her to the amnesia that distorts our perception of landscapes. Today, the Great Meadows are places of beauty and leisure, protected by law as natural havens. But the meadows are recent products of human disturbance. Almost devoid of the native plants that used to grow there, they are grasslands of colonially introduced species. The lifeworlds of indigenous flowers and the Native Americans that lived with them are specters in these grasslands.

Ghosts remind us that we live in an impossible present—a time of rupture, a world haunted with the threat of extinction. Deep histories tumble in unruly graves that are bulldozed into gardens of Progress. Yet *Arts of Living on a Damaged Planet* is also a book of weeds—the small, partial, and wild stories of more-than-human attempts to stay alive. Ghosts, too, are weeds that whisper tales of the many pasts and yet-to-comes that surround us. Considered through ghosts and weeds, worlds have ended many times before. Endings come with the death of a leaf, the death of a city, the death of a friendship, the death of small promises and small stories. The landscapes grown from such endings are our disaster as well as our weedy hope.

Modernist Futures Have Made the Anthropocene

Bad deaths generate their own variety of ghosts. Across mainland Southeast Asia, "green" ghosts arise from deaths in war and in childbirth; these deaths occur before their proper time. How much more, then, does the violence of settler colonialism and capitalist expansion give rise to the ghosts of bad death, death out of time? Here is the terrain of what anthropologist Deborah Bird Rose calls "double death," that is, extinction, which extinguishes times yet to come.

Rose has argued that white Australian settlers brought with them a particular, and peculiar, kind of time.[4] They looked straight ahead to the *future,* a singular path of optimism and salvation informing their dreams and deeds. This *future* is a characteristic feature of commitments to modernity, that complex of symbolic and material projects for separating "nature" and "culture." Moving toward this *future* requires ruthless ambition—and the willingness to participate in great projects of destruction while ignoring extinction as collateral damage. The settlers looked straight ahead as they destroyed native peoples and ecologies. The terrain carved out by this *future* is suffused with bad death ghosts.

Aleksandr Kupny grew up in the hopes of this future, and he is not afraid of ghosts. Kate Brown lets him lead us into the sarcophagus of the destroyed Chernobyl reactor, where he delightedly takes pictures of the wreckage. The ghosts are everywhere. "After forty years in radioactive fields, he said, he can sense decaying atoms." Everyone had warned him that the radioactivity would kill him, but he paid no heed, even after other friends in his community died. "The first few times we went below," Kupny said, "I recorded my dose and wrote it down, but then Sergei asked me why I did that. 'What good will it do you to know? The less you know, the better you will sleep.'"

What better figure for the promises of modernity? The less you know, the better you will sleep. Meanwhile, our safety net of multispecies interdependencies tears and breaks.

Unintentional Consequences Hit Us with New Force

Industrial engineering creates many unplanned effects; what promoters intend is rarely realized. Instead of building toward a single future, many kinds of time swirl through the worlds shaped by the modern *anthropos*. These are our ghosts.

Sometimes we can see the ghosts of relentless waste and manufactured poverty in the forms of stinking garbage and leaky sewers. But there are also ghosts we cannot see and those we chose to forget. They don't sit still. They leave traces; they disturb our plans. They crack through pavements. They tell us about stretches of ancient time and contemporary layerings of time, collapsed together in landscapes.

In 1945, one technology suddenly changed the whole world: the splitting of an atom. The two atomic bombs that destroyed Hiroshima and Nagasaki, respectively, synchronized the world to radioactivity as winds carried radioisotopes around the world. Physicist and philosopher Karen Barad says these acts of war have scarred bodies and landscapes; every radiated cell is now a ghost of war. Technoscientific war changes what we know as matter, and it calls out for new analytic tools that can move us beyond what is big and small, absent or present, inside or outside. For Barad, ghosts are superpositions of past, present, and future. Radioactivity is eerie, a powerful ghost that resets planetary time. Barad invokes quantum field theory to show us haunted landscapes as ***strange topologies****: "Every bit of spacetimemattering is . . . entangled inside all others."*

The synchronizations put into motion by contemporary technology—not just radioactivity but also global pollution, the movements of capital, climate change, and many more—look different when assessed from the perspective of planetary damage. They show us ghosts, the multiple stories of landscape effects. Whereas Progress trained us to keep moving forward, to look up to an apex at the end of a horizon, ghosts show us multiple unruly temporalities.

Death may not, after all, be the end of life; after death comes the strange life of ghosts. Hélène Cixous suggests that ghosts are uncanny because they disturb the proper separation between life and death; they mark a "between that is tainted with strangeness."[5] *Such strangeness, the uncanny nature of nature, abounds in the Anthropocene, where life persists in the shadow of mass death.*

Ladders Are Not the Only Kind of Time

In Europe, northern Renaissance thinkers came up with a great scheme linking classical, religious, and emergent modern thinking. They claimed that life had evolved from simple to complex. This was a grand and optimistic view that placed humans at the top of the Great Chain of Being, the highest rung of a ladder, where God had once resided. Like the Christian religious thought before it, this scheme assumed that we were all in a single time, on a single trajectory.

The storm of the Anthropocene sweeps us off the ladder into the waves of the more-than-human sea, where biologist Andreas Hejnol shows us tunicates, sponges, and jellies. Terrible and wonderful, we hardly know how to give them names. Take them off the ladder of Progress, Hejnol tells us; let them show us their complex designs. Imagine swimming among them rather than locking them into rungs on a ladder that leads only to ourselves. How many evolutionary gifts do these creatures entangle us in?

Some kinds of lives stretch beyond our ken, and for us, they also offer a ghostly radiance. The lichen that grows on tombstones is one example. Every autumn, mycologist Anne Pringle goes to the Petersham Cemetery near Boston to trace the outline of individual lichens, watching their growth on the gravestones of local residents and dignitaries. They grow slowly, and sometimes some disappear. Some are probably the same individuals as those that first found a place to settle when those dignitaries died centuries ago. For fleeting creatures such as ourselves, lichens are more-than-ghosts of the past and the yet-to-come.

Lichens are symbiotic assemblages of species: filamentous fungi and photosynthetic algae or cyanobacteria. Lichens are themselves a kind of landscape, enlivened by their ghosts. Many filamentous fungi are potentially immortal. This does not mean they cannot be killed; yet, unlike humans, they do not die just from age. Until cut off by injury, they spread in networks of continually renewed filaments. When we notice their tempo, rather than impose ours, they open us to the possibility of a different kind of livability.

Many kinds of time—of bacteria, fungi, algae, humans, and Western colonialism—meet on the gravestones of Petersham. The ghosts of multispecies landscapes disturb our conventional sense of time, where we measure and manage one thing leading to another. Lichens may be alive when we are gone. Lichens are ghosts that haunt us from the past, but they also peer at us from a future without us. These temporal feats alert us that the time of modernity is not the only kind of time, and that our metronomic synchrony is not the only time that matters.

Noticing Attunes Us to Worlds Otherwise

When nineteenth-century Japanese polymath Minakata Kumagusu campaigned to maintain the local shrines that the Meiji government planned to raze, he did so both as a scientist and as a participant in local forms of knowledge. Local shrines were sites of remnant old forest, and Minakata hoped to preserve their biodiversity, including the slime molds and fungi that were subjects of his research. At the same time, he felt that folk knowledge, including stories of strange beings and eerie shadow biologies, was key to his ability to learn about nature. Rather than dismissing folk knowledge, he incorporated approaches from it into his scientific work. Indeed, while generally unacknowledged, vernacular—and even "spooky"—insights have informed some of the most important science all over the world. This is a reason to learn from ghosts, however unfamiliar their forms. Our experiments combine natural history and vernacular legacies, learning from precedents nourished by other times and places.

According to the Javanese villagers who befriended anthropologist Nils Bubandt, an ancient spirit snake lives in the geothermal vent of the mud volcano that recently destroyed their homes and livelihoods. The spirit being gives them gifts in the shape of magical stones. While difficult to find and interpret, the stones have the power to change people's luck. So villagers scour the mudflat where their homes used to be, hoping to find the gift of a better fortune in stones. To those who can divine within them the animal forms that hide within, the stones hold the promise of a better life. In a twist of irony, however, these stones are spewed from the volcano that destroyed their lives, a volcano triggered, perhaps, by oil drilling. In this devastated landscape, stones and spirits, petrochemical industry and magic, enliven each other. It is a landscape where nothing is certain. So while villagers blame the oil company for the devastating eruption of mud, geologists argue over the true and proper cause of the eruption: was it natural or anthropogenic? Is the disaster the work of ***geos*** *or* ***anthropos****? The mud volcano is caught in undecidability. Reading the villagers' search for spirit stones in light of such undecidability urges us to see how spirits also possess geology. In troubled, illegible times, ghosts haunt us in many forms.*

In the midst of disaster, stones bring a gift of hope: of fortune, of insight, of the possibility of living-with. In the Anthropocene, multiple conversations with stones are necessary. After all, the Anthropocene is a geological epoch proposed by geologists, climate chemists, and stratigraphers—scientists used to studying stones, rocks, sediments, and chemical cycles. In the Anthropocene, they suggest, humans have become a geological force. Modern industry is laying down indelible strata on the earth that will remain even after we have vanished from the surface of the planet.

To learn the stories of stones, geologists might use the insights of ethnographers and poets. In her poem "Marrow," writer Ursula Le Guin urges us to listen to stones without forcing our will on them. Might such listening be necessary to know the Anthropocene?

> There was a word inside a stone.
> I tried to pry it clear,
> mallet and chisel, pick and gad,
> until the stone was dropping blood,
> but still I could not hear
> the word the stone had said.
> I threw it down beside the road
> among a thousand stones
> and as I turned away it cried
> the word aloud within my ear
> and the marrow of my bones
> heard, and replied.

Shimmer Still Beckons

Smothered by bad death ghosts, it seems easy to give in to inevitability or to climb belligerently up and forward. But there are other ghostly matters shimmering just below our notice. This book argues that, to survive, we need to relearn multiple forms of curiosity. Curiosity is an attunement to multispecies entanglement, complexity, and the shimmer all around us.

"Shimmer" is a gift, too, of the Yolngu people of Australia, as passed to us by Deborah Bird Rose. Shimmer is the seasonal kiss of mutually thrilling encounters among flying foxes and flowering eucalyptus trees, flying fox people, rain, and rainbows. Flying foxes spread eucalyptus pollen and seeds, allowing the trees to reproduce; they are an animal wind in the trees. Rose describes their coordinations through the Yolngu term *bir'yun*—a shimmering into brilliance. *Bir'yun* attends to temporal patterns that emerge from more-than-human shimmerings and dreamings—pulses of ancestral power, of life riding a wave that is always coming: "*bir'yun* shows us that the world is not composed of gears and cogs but of multifaceted, multispecies relations and pulses."

Landscapes shimmer when they gather rhythms shared across varied forms of life. *Shimmer* describes the coming in and out of focus of multispecies knots, with their cascading effects. Yolngu cosmologies inform us; juxtaposed with the stories made available from many arts and sciences, vernacular and academic, we learn the liveliness of landscapes. Landscapes enact more-than-human rhythms. To follow these rhythms, we need new histories and descriptions, crossing the sciences and humanities.

As artists, we conjure magical figures, weave speculative fictions, animate feral and partial connections. We necessarily stumble. And try again. With every mark, difference haunts and struggles to appear anew.

Postcolonial historian Dipesh Chakrabarty points out that consideration of humanity as a geological force troubles the distinction between natural and human history, forcing us into a new kind of historicity.[6] The deep time of geology, climate, and natural science is collapsing into the historical time of human technology. *Anthropos* has become an overwhelming force that can build and destroy, birth and kill all others on the planet. In the new histories and politics that we must form—and as the contributions to *Arts of Living on a Damaged Planet* demonstrate—we must share space with the ghostly contours of a stone, the radioactivity of a fingerprint, the eggs of a horseshoe crab, a wild bat pollinator, an absent wildflower in a meadow, a lichen on a tombstone, a tomato growing in an abandoned car tire. It is these shared spaces, or what we call haunted landscapes, that relentlessly trouble the narratives of Progress, and urge us to radically imagine worlds that are possible because they are already here.

Anthropocene: a time when survival teeters on a question stirring in the marrow of the Earth's bones. What kinds of human disturbances can life on Earth bear? By showing us Progress and Extinction—life's historical entanglement with death in ruined landscapes—ghosts point the way in this half of *Arts of Living on a Damaged Planet*. Turn the book over and follow monsters.

ELAINE GAN explores the timing of human–plant interactions, specifically around rice cultivation, as technologies of life and death that make geopolitical histories. Raised in the big old cities of Manila and New York, she is an artist and interdisciplinary scholar who loves the sounds and smells of the sea. She is art director of the Aarhus University Research on the Anthropocene project and a fellow of the New York Foundation for the Arts in Architecture and Environmental Structures.

The author of *The Mushroom at the End of the World: On the Possibility of Life in Capitalist Ruins* and *Friction: An Ethnography of Global Connection,* **ANNA TSING** has long traced the violences that capitalist extraction inflicts on more-than-human lives and landscapes, as well as the arts and joys of collaborative living that emerge from unruly encounters. She is professor of anthropology at the University of California, Santa Cruz, a Niels Bohr Professor at Aarhus University, and co-convener, with Nils Bubandt, of Aarhus University Research on the Anthropocene (AURA).

Raised in coastal Oregon's patchwork of clearcuts and second-growth forests, **HEATHER SWANSON** spent her childhood wandering in industrially damaged places. Now an assistant professor at Aarhus University, she is fascinated by how histories are embedded in bodies and landscapes. Her forthcoming book, *Caught in Comparisons,* probes the transformation of northern Japan's salmon populations and the watersheds they inhabit.

As an anthropologist, **NILS BUBANDT** has learned to be equally at home with witches, protesters, and mud volcanoes. A professor at Aarhus University, he is co-convener of Aarhus University Research on the Anthropocene (AURA), with Anna Tsing, and works to animate descriptions of the Anthropocene with the voices of spirits. In his book *Democracy, Corruption, and the Politics of Spirits in Contemporary Indonesia,* he portrays the life of spirits at the heart of modern politics.

Notes

1. Peter Raven, "Foreword," in *Atlas of Population and Environment,* eds. Paul Harrison and Fred Pearce (Berkeley: University of California Press, 2000), x.
2. A. D. Barnosky, N. Matzke, S. Tomiya, G. O. Wogan, B. Swartz, T. B. Quental, C. Marshall, et al., "Has the Earth's Sixth Mass Extinction Already Arrived?" *Nature* 471, no. 7336 (2011): 51–57.
3. Will Steffen, Wendy Broadgate, Lisa Deutsch, Owen Gaffney, and Cornelia Ludwig, "The Trajectory of the Anthropocene: The Great Acceleration," *Anthropocene Review* 2, no. 1 (2015): 81–98.
4. Deborah Bird Rose, *Reports from a Wild Country: Ethics of Decolonization* (Kensington: University of New South Wales Press, 2004).
5. Hélène Cixous, "Fiction and Its Phantoms: A Reading of Freud's *Das Unheimliche* (the 'Uncanny')," *New Literary History* 7, no. 3 (1976): 543.
6. Dipesh Chakrabarty, "The Climate of History: Four Theses," *Critical Inquiry* 35, no. 2 (2009): 197–222.

1
A GARDEN OR A GRAVE?

THE CANYONIC LANDSCAPE OF THE TIJUANA–SAN DIEGO REGION

Lesley Stern

WE ARE IN A NATIVE GARDEN. It is fall, though this year the summer seems never-ending. The garden is scrubby, typical coastal sage. The air is hot and dry and the foliage is brown, spiky, and brittle. In spring, there are many more blooms, including a pretty yellow flower, *Glebionis coronaria,* an avaricious invasive. Behind me is a small building, the Estuary Visitors Center. I look out westward, beyond the edges of the garden, to a different terrain: very flat and wet. I am facing westward toward the ocean, but my view of the ocean is blocked by a row of condominiums and a straggle of palm trees on the horizon. This is irritating for someone grown accustomed to the views of a city that faces the ocean and turns its back on the desert. I walk westward on a pathway to the edge of the scrubby native garden. There is a ridge here where the native garden meets the wetland, a dramatic divide. The vegetation is utterly different: fingers of water wind through reeds, dragonflies dart, occasional waterfowl glide, plovers and terns circle and swoop. The sea flows in and out of the tidal channels here.

The Visitors Center is built on the site of a garbage dump. This native garden bordering on a wetland and estuary is a landscape that has been restored, reclaimed as public land—through long and sustained political battles involving many agencies and alliances—and now it is being preserved. Some battles are won, some are lost, and sometimes the dividing line between victors and vanquished is decidedly indecisive. Take the condominium development: it blocks the view of the ocean and imposes an unnatural edge. But it also prevents the navy, who, after the Second World War, owned all of this southwestern corner of the United States and now retain a small area as a helicopter base, from flying directly over this part of the wetlands. Because of the condos, they have to fly around the estuary.

Ninety-five percent of Californian wetlands have disappeared. This estuary is a small area remaining almost intact, designated one of twenty-five important wetlands in the United States. It is in the Pacific flight path for birds, providing food and shelter. It also protects rare and endangered plant species. Although the boundary of the estuary is clearly demarcated, the urban development along its edges includes myriad nonnatives that continue to invade the reserve through permeable borders.

My interest is in canyons. The question I want to begin with is this: can canyons be considered gardens? Or gardenlike? Can canyons, in some instances, be considered to share some features with gardens—particularly in that region where gardens intersect with public space? So why am I here in a landscape very different to canyons? Because this estuary, which has many features of a public garden, is the place where the Tijuana River, via several canyons, empties into the ocean. The most important of these canyons is called in Mexico Los Laureles and in the United States Goat Canyon. Laureles and the estuary are not only contiguous but also continuous. I cannot see Laureles from here; the mountains of Mexico look like small, gentle hills. But I am here today to hike, through this public parkland, to the point where Laureles meets the estuary. There I will encounter an obstruction: the border fence.

If you were to buy a freeway map of this area, or switch on your smart phone, you would have no idea that the San Diego–Tijuana region is built on canyons; indeed, you would have no idea that it is one continuous area, because maps stop at the border of Mexico. Yet all the time on freeways, we drive through or over canyons. The tops and sides of canyons are sliced off, filled in; bulldozers can be seen creating mesas, cutting across valleys, and building huge retaining walls for the freeway system and housing developments. Thankfully, many canyons remain green because, ironically, the city put infrastructure there: buried, sometimes concretized sewage and power lines.

Oscar Romo, coordinator of the Tijuana Watershed and director of the Estuary Research Center, has a three-dimensional topographical map of the region, of the Tijuana river shed, which stretches over a region of approximately 1,750 square miles on either side of the California-Baja California border. His map is about six feet square and sits on top of a table. We stand and walk around the table, looking down, like birds or a navy helicopter hovering. The map is colored and contoured. You see both continuities and divisions that are impossible to discern in the landscape at ground level. What the segueing from green in the north to khaki in the south registers, for instance, is the flow and diversion of water, different kinds of land use, and a different distribution of resources. Mexican farmers tend to use water as it flows—a system that is less productive but also more economic in terms of water use. On the U.S. side, water is engineered and diverted into the All-American Canal. When water usage is averaged out, each person in the United States uses 225 gallons of water per day; in Mexico, 25.

I walk around the park on the eastern periphery. There are condos painted in uniform desert colors and backyards filled with a wonderful array of plants from all over the world—wonderful from my point of view, that is, but not from the perspective of those trying to protect the estuary from invasives. As I walk, I encounter signs: the navy telling me to keep out and the estuary telling me to keep to the paths. I follow as bidden, resenting such directing and controlling and patrolling of traffic in this quasi-wilderness. Suddenly I am diverted from my path to Mexico and directed toward the ocean. The ribbon of water has widened and forms a pool perhaps twenty feet wide, and the vista opens up as the wetlands stretch out and the water laces and disappears into an expansive ocean. I stop, filled with wonder, then turn my back on the ocean to face inland toward a land that is scarified. From here I can see that several canyons—which is to say, several creeks and tributaries of the Tijuana River—feed the estuary. Laureles is just one. I can, by squinting, clarify the point where Laureles is blocked, crossed by *la linea, la frontera,* the border fence. I can see a busy road on the Mexico side constructed by landfill across the canyon's valley. Tiny toy cars whizz silently across on their way to Playas, the single seaside neighborhood in Tijuana, a city that has turned its back on the ocean, turned toward the desert. I know that at the bottom of the canyon, in the estuary, trash inexorably accumulates, dramatically exacerbated when it rains—tires, plastic bottles, mattresses, and more.

Liquids and solids. The solidity of things and the flow of liquid. But it turns out that things are not so solid and that liquid does not always flow.

I drive inland, meet up with Oscar Romo, and drive across the border with him in a vehicle more fitted for the terrain we are about to encounter. Here, at the border, is another landscape: six lanes of traffic in each direction. This is the most populated border crossing in the world. The cars coming into the United States seem scarcely to be moving, but we breeze through into Mexico, no need for passports. We drive along next to the new border fence, a shining steel knife, solid, unscalable, slicing through the landscape, through several *colonias* or neighborhoods, through steep canyons, through bands of improvised housing.

We reach Laureles and stop at the top of the canyon, in what was a small ranch called La Cueva. Everything is different since I was last here, maybe six months ago. The hillside has been shaved, stripped of vegetation, the topsoil scraped away, and the land marked out in chalk into small parcels of land. "For Sale" signs spike the landscape like cacti. The small ranch has been sold to a developer who will give the new owners a formal certificate of ownership, legally worthless but prized in this canyon, where much of the settlement is informal, where most people are squatting. Despite promises, there will be no infrastructure as part of the deal. There will be no sewage or electricity or city garbage disposal. Now that the land has been shaved, you can see vividly, with the clarity of a hawk, the sites where trash has been deposited and accumulated in piles on top of the hillside. There are two *maquiladoras* (factories) out of site, just over the hill, but their presence is registered in two heaps of waste perching on top of the hillside. The third and largest pile is hospital waste from all over the city of Tijuana, much of it the residue of hospital tourism—the large number of people from the United States visiting Tijuana for cheap and good medical and dental care. A few houses are already being erected using recycled wood, permeated by poisonous chemicals—railroad ties, for instance—sent here from the United States. It is illegal to sell this timber in the United States and costly, because it has to go into expensively constructed, impermeable landfill cells. Sometimes the border is curiously permeable.

In January the rains will come. All the waste and silt will sweep down the six miles of the canyon into the estuary, and it will be far too much for the sediment basins to catch. Sediment basins were built ten years ago. In 2003 they were destroyed by El Niño, which also devastated the salt marsh, now completely gone. The topography only needs to change one inch for the ecosystem to be unviable. El Niño brought three feet of rain in. The basins have been rebuilt, but if made any bigger, they will destroy the wildlife.

We climb back into the truck and start to head down into the already settled part of the canyon, into a *colonia* called San Bernardo. From here a shantytown, comprising eighty-five thousand people, spreads all the way down to the southern tip of Mexico. Tijuana has been a magnet of immigration for people from poorer states in Mexico and from other parts of Latin America. The majority of the people are here as illegal immigrants. Most are squatters (on land owned by the Mexican federal government—twenty meters on each side of the creek is government land) though some—through shady dealings with developers—hold land titles. Others have acquired title, though usually not formalized, through the Mexican *hejido* laws whereby squatters, after a certain amount of time, do have certain rights to ownership. What typically happens in these situations is that after a period of political jostling, infrastructure is developed, if haphazardly and unevenly: paved roads, electricity, and sewage lines.

Water has been brought into the top of San Bernardo, but it is too expensive for most people to access directly. There are no sewer lines. Many of the houses are built perilously close to the creek that runs through the canyon, and most are built on the steep sides of the canyon. Where there is little vegetation, the wind blows, erosion happens, and pollutants are carried in the air. Trash is chucked into the creek and dogs and children play there. When the rains come, floods happen very quickly; great channels erupt in the dirt roads. It is impossible for people living here to get out of the canyon.

This community is poor, but like many *colonias* in Tijuana, the vernacular housing that has developed unhindered by zoning laws and in an ad hoc manner, using recycled materials, is inventive and crazy. It happens from the bottom up: first, you build one room or one story, but with a flat roof bordered by a row of rebar projecting into the sky. Then, when possible, you build a second story, maybe expand sideways. You incorporate found, foraged, and salvaged materials: garage doors and bits of discarded housing materials from San Diego, tiles, bottles, tires for retaining walls. Maybe there will be a shop or workshop on the ground floor and housing above. Teddy Cruz, an architect and guerrilla urbanist, points to these approaches to mixed-use development when he urges us to learn from Tijuana. He calls it trickle-up or time-based building.[1] Mike Davis writes, "In the Do-It-Yourself City, bricolage supplants master planning, and urban design becomes a kind of art brut, generated by populist building practices. If only by default, the masses become the city's true auteurs, and architecture is not so much transcended as retranslated through its dynamic vernacular context."[2]

The old neighborhood near the border and the airport—Colonia Libertad—shows how an informal community can eventually grow into a formalized settlement through the development of vernacular building practices along with juridical interventions and the eventual supply of infrastructure. But this community in Laureles, because of the creek that runs through the canyon and into the estuary, is both more crucially situated and more precarious.

The landscape of this canyon bears little relation to either the kind of arid landscape I experience freeway driving across canyons in San Diego or one of the green canyons that are being restored. It seems like a vicious circle, an example of where the destruction of the natural environment—in this case, a canyon and estuary—is unstoppable, only escalating. Yet this landscape is not separate from the freeways and green canyons. They are all part of one system, part of the larger network of canyons. The "green fingers" about which Appleyard and Lynch were so hopeful in their largely unrealized 1974 planning document *Temporary Paradise?* (following in the footsteps of John Nolen's 1908 report) sometimes turn out to be gray and cemented, or macadamized, or brown and dusty.[3] Churned up by mud and torrents of water.

I am curious about what this man, Oscar Romo, so fierce in his protection of the estuary, so virulent in his guarding against invasives, is doing in the heart of San Bernardo, in what could be considered enemy territory as it were, in the place that causes so many problems for the estuary. What he is doing is building a soccer stadium and community center, planting thousands of trees, negotiating with the Mexican federal government to secure an easement that would create a nature reserve running on both sides of the creek for the length of the canyon. Just as importantly, he is gathering evidence for U.S. judiciary bodies: evidence that the trash that ends up in the estuary originates from the United States. Ingeniously, he has planted microchips in plastic bottles so that the circuit of trash can be traced. Other small devices, planted like baby trees in the canyons, contain sensors that register and communicate the movements of solids. This is basically the same technology that the Border Patrol will deploy when we join *la linea* and drive back into the United States. But Romo and the scientists he works with on this sensing project are more interested in sensing the movement of things than people, in using the evidence "sensed" as political ammunition, to persuade both U.S. and Mexican judiciary bodies that if Laureles is the "problem" for the estuary, it is also the solution.

The movement of things. Things like tires. Let's take tires. Trees and tires and tomatoes all matter here, all contribute to a shaping of the landscape. But it is to tires and tomatoes that I will now turn. Tires are everywhere in this canyon. Tomatoes are not so much in evidence, but on this visit, I find two small tomato plants. It is in the commingling of tires and tomatoes that I begin to find some of the answers to the question of canyons and gardens.

Take a tire, any tire. It is a big, ugly, stinky, almost solid thing, a tire. In all probability manufactured in Mexico. As part of a motor vehicle, you (or I, anyway) tend not to notice tires. They make your journey softer, groovier. On the freeway you see cars whiz by and the movement of the tires is just a whir, a blur. But then you hit a particularly vicious and malevolent nail, lying in wait just for you, and you skid and careen, and then suddenly everything stops. It's just as old man Heidegger predicted: when the tool breaks, you notice its thingness[4]—though the tire in Heideggerian terms is not a thing, lovingly handcrafted; it is a mass-produced and ugly object. Probably not musing much upon Heidegger, you take your redundant tire to the shop to trade it in for a new one. Turns out it costs money to get rid of a tire. There are fees. So maybe you will just dump it somewhere. Or maybe you pay the fees, and that's that, your new tire looks exactly like the old one. Your car is intact, tires and all, a solid thing, a piece of property, an asset that can potentially be converted into liquid capital, used as collateral for a loan, to get a mortgage, say, or go on an overseas trip.

But now the dealer has to pay all the fees and go to the bother and expense of getting rid of the redundant tires. Before they are tossed into the landfill, tires—because they sway—have to be cut up, and because of all the toxic properties they contain, they have to be properly retired, and the costs of this are derived in part from the taxes we pay. It is much easier for the dealers sometimes simply to schlep the tires across the border and dump or sell them there, where they will be patched up and reused, or incorporated into buildings, or just dumped. Massive numbers of tires are dumped.

The lawyers representing various recycling agencies are reluctant to intervene in the circuit of tires. They say this: once a U.S. tire enters Mexico, it is a Mexican tire. And what, I wonder, of the tires that ferry us today between San Diego and Tijuana and back to San Diego?

You can see, in the soccer stadium, that Romo has been learning from Tijuana. The seating and the stage area are built from tires. When he began, with the community, clearing this site, it was filled with trash and tires. So rather than engaging in more schlepping, they used the tires just as people have habitually done, to build walls, particularly retaining walls. In the floods, however, the retaining walls in Laureles are in danger of being swept away. Romo brought his engineering skills to bear on the materials at hand—primarily silt and tires, along with permeable pavers made on site—to construct the stadium as both a seating arena and a water filtration system. It is at once very solid, immoveable, in fact, and permeable, amenable to liquids. Runoff from the dirt road above the stadium, rather than trickling or rushing down into the estuary, will seep into the ground through the permeable pavers and then through layers of silt banked in tires and plastic bottles.

The tires are layered in such a way as to present a scalloped edge, in which hundreds of natives were planted. A politician, in a random and rare act of civil cooperation, came in to clean the area up and yanked all the natives out, believing them to be weeds. Gradual replanting is happening.

Solids and liquids. Obstructions and flows. Steel edges and permeable borders.

There are some—like Hernando De Soto, in *The Mystery of Capital: Why Capitalism Triumphs in the West and Fails Everywhere Else*—who argue that the answer lies in a transformation of the system of property rights.[5] An example of this argument in action was made by Lula, ex-president of Brazil, in his attempt to give property rights to some of Brazil's poorest people where the favelas of Rio de Janeiro, as in Tijuana, crawl dramatically up hillsides. The hope was that this, along with systemic support—financial, social, legal—would provide the means for turning property into a liquid form, into credit, and thus would empower the dispossessed. But insofar as such a proposal remains within the circuit of capital, it does not address the environmental factors that occur in a situation such as that of Laureles, where urban development, albeit of an informal type, collides with what we call "nature" in a potentially explosive manner. It also has the danger of letting the state, and all its local judiciary bodies, off the hook.

Solids and flows. Edible and inedible.

If there are microclimates existing and produced in this canyon, there are too microlandscapes. These are rather different to the vistas that we might see at the top of a canyon or even in a valley looking up and out. These landscapes are small, so small that sometimes you have to lie on your belly in the dust, or crouch down and peer under the "legitimate" landscape, to discern the features that delineate the scene. Of course, the distinction between microclimates and microlandscapes is not always clear-cut.

In Laureles on this visit, I am apprehended by two landscapes so distinctive that I might imagine myself in different worlds. There is a center of attraction in each, and it is the same "thing" that appears, looms up out of the earth, and pulls my gaze downward. In each case, I am exhorted by an invisible force to kneel down, to reach out and touch the round green fruit that grows so fantastically in this unlikely and inhospitable terrain. It is a tomato, a cluster, in fact, of green tomatoes. We have seen no other tomatoes in Laureles. The first tomato plant grows out of a crack in the concrete; the second grows a short distance away in a trickle of water beside the dusty, unpaved road. As we kneel down to scrutinize more closely the plant growing in water, a stink rises up, assaults the senses. This trickle of water is an open sewer. But because we are already down on our knees, we stay there and peer more closely at the verdant green plant life that straggles the edge of the road here. There are a variety of juicy weeds, something that looks temptingly like watercress and a sole celery plant, the leaves of which when crushed release an aroma so much stronger and more enticing than the anemic celery on offer in the supermarkets. In both cases, the small bunch of tomatoes has that translucent jade glow that distinguishes tomatoes at the end of the season, tomatoes that—because they will never ripen into redness—revere their greenness. Peering into this microlandscape, I feel shrunken, insectlike, as though I have left the canyon and entered a jungle of gigantic edible delights. It doesn't take long to return to reality. In reality, this microlandscape is not such an aberration; on the contrary, it is an ecology that in miniature represents the larger landscape of the canyon. The unfiltered sewage provides rich nutrients that enable the trees to grow so strongly. And the trees change the larger ecology, purifying the air, filtering water, reducing temperatures, preventing erosion. But the pathogens and toxicity in that trickle of water are concentrated in the tomatoes and celery. The people who live here are endangered, just as the estuary is.

What about the other tomato plant: how does it emerge out of a barren landscape of concrete? It grows in the patio of the NGO Alter Terra run by Oscar Romo, which functions as a scientific field station and is also where he has a nursery of thousands of trees grown from seed and raised to be planted throughout the canyon. However, there are no vegetables here. No one planted a tomato seed; perhaps some salsa fell from a taco, fell between the cracks. Perhaps a seed migrated from elsewhere in the subsoil and pushed its way up, feeding on the nutrients and filtered water. The concrete here is made of permeable matter, under which is silt. Gradually, bit by bit, the nutrients (at the present time from wastewater) will work with the filtration to create a new layer of microbe- and nutrient-rich topsoil. If I lift my head, I see stacks of these permeable pavers that are destined to be used in projects around the canyon and also to be sold to bring income back into the canyon.

Here in Laureles, people garden sparsely in cans, in tires; succulents, roses, cacti, bougainvillea, dot the landscape. Trees are being planted, not necessarily the trees that would once have grown here, but trees that hold the earth and clean the air and soil and provide an infrastructure for urban development. Vegetables, though, given the toxicity of the soil, are a more tricky proposition. I would never have thought that paved roads might be the key to tomatoes. But so long as the boundary between the road and the vegetable patch is permeable, the potential of yet another landscape unfolds, one in which vegetables and people and nonnative trees cohabit in a reshaped ecology, one in which continuity between the canyon and the estuary is preserved and tomatoes might serve to protect the endangered species in the wetlands.

We make our way down through Laureles, but before reaching the bottom, we get into low gear and climb up to the mesa directly above Laureles and drive through an affluent *colonia* called, for reasons we will soon see, El Mirador. Oscar parks the truck and we walk out to the edge of the cliff. We are standing above the very point where Laureles opens out into the estuary. Looking west and out to sea, we have a magnificent view. But if you drop your eyes, there is another landscape, one terraced by obstructions. The canyon settlement ends at a huge culvert over which cars and buses and trucks whiz past, the toy cars I could see in the distance when I began this voyage. They are on a road built in Mexico transversing (just as in San Diego) the canyons. Just beyond this road are two walls and another road slicing across the landscape: the old, rusty, dilapidated border fence and the new, shiny modernist fence, and in between a wide, beautifully paved road built by Homeland Security after 9/11. To build this road, they appropriated a 150-foot-wide band of land parallel to the fence, demolished a mesa right on the border at the estuary, and moved the earth to cut across the open valley. We look back into the United States at this southernmost tip and see bulldozers at work creating huge craters. These are new sediment basins. The ones they built when they constructed the new wall and road were insufficient to contain the pollution from the 2003 El Niño. Since then, there have been slow political negotiations, Homeland Security brought into dialogue with other bodies and urged to pay less attention and less funds to the wall and more to the estuary and to the shantytown in Laureles. These new sediment basins will, they hope, offer greater protection when the rains come this January. Standing here, looking back at the United States at its most southern tip, my tomato optimism dissipates, seeps away into all these chiseled cracks and gouges in the landscape. But as we stand eating delicious tacos in lively Tijuana before joining, with our passports, the slow, dense mass of cars at the border crossing, it dawns on me that Romo hardly ever uses the word "border." This is perhaps because, imaginatively and strategically, and in quotidian practice, he conceives of the boundary not as one between nations but as one that delineates an environmental zone, the Tijuana River watershed. This is the zone in which a whole network of canyons, the green fingers, feel their way, more often than not in urban spaces.

It is like this: a slow and Sisyphean project, a slow changing of the landscape. You plant natives and a politician rips them out; you negotiate with the council to develop infrastructure in a part of San Bernardo and a developer razes the land at the top of the canyon; you clear tons and tons and tons of trash at the mouth of the estuary and Homeland Security rearranges the valley. All of these developments have immense consequences for already endangered species: birds, plants, and people. So what do you do? You negotiate with the politicians, the developers, Homeland Security, the communities living in the canyon, judiciaries, environmental groups, water authorities, and more. You emphasize edges and boundaries that are permeable; you imagine the canyon not as a threat but as a protector. Through small but incremental gestures, you begin to shape a way to see the landscape differently.

To see it as mutable: solids and liquids. Obstructions and flows. Steel edges and permeable borders.

Writing in the interstices between cultural studies, memoir, and environmental history, **LESLEY STERN** expands the ways we see multispecies worlds. This essay is from her genre-bending book-in-process. Her dreamlike work *The Smoking Book* has been described as "an innovative, hybrid form of writing . . . at once intensely personal and kaleidoscopically international."

Notes

1. TED, "How Architectural Innovations Migrate across Borders," video, 13:14, February 5, 2014, https://www.youtube.com/watch?v=aG-ZeDqG8Zk.
2. Mike Davis, "Learning from Tijuana," *Grand Street* 56 (Spring 1996): 35.
3. Donald Appleyard and Kevin Lynch, *Temporary Paradise? A Look at the Special Landscape of the San Diego Region: A Report to the City of San*

Diego (Cambridge, Mass.: Department of Urban Studies and Planning, Massachusetts Institute of Technology, 1974); John Nolen, *San Diego: A Comprehensive Plan for Its Improvement* (Boston: G. H. Ellis, 1908).

4. Martin Heidegger, *What Is a Thing?,* trans. W. B. Barton Jr. and Vera Deutsch, with an analysis by Eugene T. Gendlin (Chicago: H. Regnery, 1967).
5. Hernando de Soto, *The Mystery of Capital: Why Capitalism Triumphs in the West and Fails Everywhere Else* (New York: Basic Books, 2000).

IN THE MIDST OF DAMAGE

HOW SHOULD WE REACT TO THREATS TO LIFE ON EARTH?

Soldiers are sometimes prepared for war by being asked to enter virtual battles over and over until their hearts no longer race. Their trainers hope to produce humans for whom deadly violence has no emotional charge. For those who reject such a mandate, this example can suggest the exotic forms we humans are able to assume. Our species is capable of many strange reactions to violence and damage. To study culture and history is to immerse oneself in this strangeness.

Historian Kate Brown tells one man's story, and yet it is a familiar one for all of us raised within the contradictions of modernity, so optimistic for the future and yet so cruel in its realization. Aleksandr Kupny crawled under the charred reactor at Chernobyl to see the ruins of the nuclear accident. It was an adventure, a chance to see for himself the power of the atom unleashed. His adventure, however, was not just his own, as Brown explains; it was made from the materials of the greater nuclear adventure, which started from the discovery of radium and has taken us now into a world in which nuclear power—and feral radiation—is everywhere. Kupny imagined himself as a "partisan," a fighter for the nation. In the Cold War, modernization, including nuclear power, was the prize, and every fighter brought us closer. Yet now we rename the very goals the Cold War brought us toward as "Anthropocene," the time of environmental endangerment. Brown takes us to the heart of this irony.

Anthropologist Deborah Bird Rose also traffics in cruel optimism. Describing an attempt to wipe out endangered flying foxes in a small white town in Australia, she writes, "Man is the only animal that shoots other creatures with paintball guns, and when the creatures flap around in terror or fall to the ground injured and in shock, man is the only animal that cheers." But Rose has another lesson she wants to teach. Among Aboriginal people of Northern Australia, flying fox pollination of eucalyptus flowers is part of the "shimmer of life." Shimmer draws us into multispecies worlds. It shows us unexpected coordinations, such as those that bring flying foxes to eucalyptus flowers across a surprisingly wide swath of country. "We are called into recognition. . . . In the midst of terrible destruction, life finds ways to flourish, and . . . the shimmer of life does indeed include us."

2

MARIE CURIE'S FINGERPRINT

NUCLEAR SPELUNKING IN THE CHERNOBYL ZONE

Kate Brown

IT IS HARD TO IMAGINE entering a rabbit hole under the charred Chernobyl No. 4 reactor. Aleksandr Kupny did just that many times. He showed me photos he took during his expeditions underground into the reactor cavern. He would go with a friend on the sly without official permission. The explosion that occurred on April 26, 1986, thirty seconds after the reactor was shut down for an experimental safety test, blew off the building's 4 million pound concrete roof as well as the upper walls and part of the machine room.[1] The fire generated by the eruption burned at greater than two thousand degrees Celsius. The tremendous heat melted iron, steel, cement, machinery, graphite, uranium, and plutonium, turning it all into running lava that poured down through the blown floors of the reactor complex.[2] The lava eventually cooled into stalactites, black, sparkling, and impenetrable. One stalactite is called the "elephant's leg" for its thickness, gray shade, and deep furrows. In the months following the accident, scientists estimated (because it was too hot to measure directly) that the elephant's leg emitted ten thousand roentgens an hour. To translate that measurement biologically, ten thousand roentgens means thirty seconds spent near the leg would cause dizziness and nausea for the rest of the week. Two minutes and cells would hemorrhage, four minutes would lead to diarrhea and fever, while five minutes would deliver to most people a fatal dose.[3]

In 1989, Kupny worked as a health physics technician at Reactor No. 3, which was paired with Reactor No. 4 and still functioned after the accident. He had volunteered to go to Chernobyl, like tens of thousands of Soviet citizens in the late 1980s, because he believed it was his patriotic duty to help out after the catastrophe.[4] He was also intrigued professionally. The smoking Chernobyl plant had radiation levels like nowhere else on earth. Chernobyl, Kupny said, "was the Klondike of radiation fields." He approached the blown reactor not with justifiable dread but with a sense of opportunity. Inside the ruined reactor, he had a chance to measure radiation at levels few others could experience. "I didn't look at the Chernobyl sarcophagus with fear. I saw it as a phenomenon. You can't study something you fear."[5]

As the years went by working as a radiation monitor, Kupny had grown more and more curious about the invisible energy he was paid to measure. He wanted to experience firsthand the tremendous power of splitting a nucleus. As a monitor, Kupny had a pass to the hottest areas at the nuclear power plant, including the sarcophagus, the vast concrete tomb built around the smoldering reactor in the months following the accident. The sarcophagus had two entry points, which were cavelike openings used by workers after the radiation levels had cooled to access the crushed control and machinery rooms.

Kupny had the necessary equipment—hazmat suits, gas masks, flashlights, and radiation detectors. He also possessed a camera. His friend Sergei Koshelev had a video camera. Kupny and Koshelev had no formal permission to take their cameras and headlights on their days off and crawl into the sarcophagus, but they knew the guards and the workers, and no one stopped them. "We went there as partisans," Kupny explained. "We took on the risk ourselves. The fewer people who knew about it, the better."

The risks were considerable, though Kupny is cavalier about them. The caverns under the reactor are former rooms—the control room, a pump house, the turbine halls—but they are no longer in the same places or in the same order they were in when the reactor was operating. The basement of the turbine generator hall is filled with water. Planks thrown over it serve as flooring. Spilled oil, slimy and slippery, seeps everywhere. The men must step lightly around cables, potholes, and ankle-trapping crevices. "Just walking," Koshelev noted, is "a

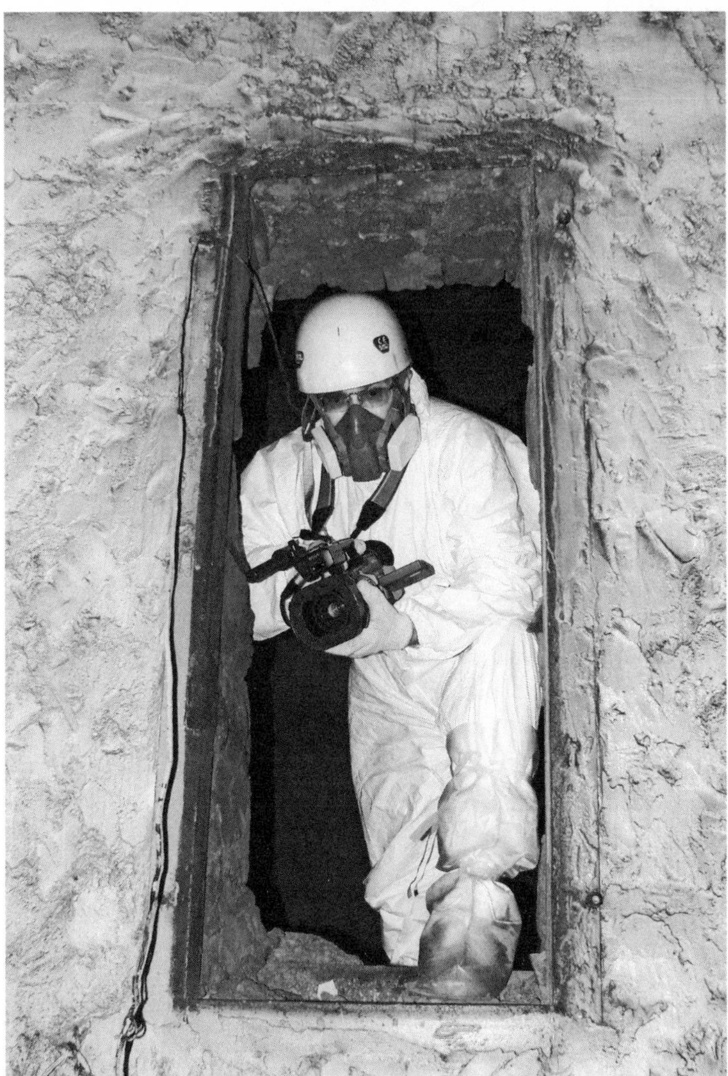

Figure G2.1. Videographer Sergei Koshelev at the entrance to the Chernobyl sarcophagus. Photograph by Aleksandr Kupny.

hazard."[6] In the heavy darkness of the cavern, it is easy to stumble over rubble and cables, to pitch into a ditch, break a leg or arm. Heavy doors could swing shut and jam, trapping them below until help came, if it did. Flashlight batteries are not reliable when exposed to radioactive decay, and they give out, not gradually, but suddenly with no warning. In a pitch of total blackout, the men would then have to feel their way out of the crypt, trying not to panic over the extra time spent exposed. The heavy hazmat suits are hot and make moving cumbersome. The lighter suits made of fabric easily tear, exposing skin and organs to the intense fields of radioactive isotopes. Gamma rays go through both kinds of suits. There are no methods for stopping gamma rays unless the men were to wear seventy-pound lead diving suits, and then they would hardly be able to move at all. All of this means that trip after trip, a body accumulates doses of radioactivity.

They had a half hour, forty minutes max, to stoop, crawl, and wriggle into the underground chambers, take pictures, and get back aboveground before they were overexposed. Despite the risk, the men did not bother with a radiation detector. "The radiation fields do not change much," Kupny remarked. "I knew what spots recorded single digits, and I knew where the numbers spiraled into the tens and hundreds."

I asked Kupny why he did that, why he went down there. Was it the same motivation that drives people to climb Mount Everest, to go places where few have gone before?

Kupny bristled. "I don't do it to put a flag down and beat my chest. I went under to figure out what happened. I see nuclear energy as a natural force. I want to understand this force; the immense power behind the accident. You can start to understand the power by seeing how the reactor fell apart. Unless you go there, you can't understand it."

Kupny is not your usual disaster tourist, one who hopes to snap a photo of ruin and destruction as a metaphor for the larger folly of human societies. For Kupny, crawling into the belly of the burned-out reactor was as close as he could get to entering a mushroom cloud to see how nuclear power works. His life's work has been to visualize and map this scarcely sensible phenomenon. He placed himself before a barrage of hazardous radioactive isotopes because he sought to grasp decaying radioactive isotopes as the elemental force burning at the center of the universe, the energy that gives life and can snatch it away.

It might seem strange to see the Chernobyl accident as a "natural force," but Kupny has a point. When, in the nuclear power plant, Soviet engineers created isotopes not found in nature—strontium, cesium, plutonium—these isotopes, once born, continued living their own "life," following the "natural" properties of radioactive decay, as observed by humans. Like other forces of nature, there is no stopping

Figure G2.2. Sergei Koshelev in the ruin of Reactor No. 4, Chernobyl. Photograph by Aleksandr Kupny.

them until they have run their course. Most artists in the twentieth century concerned with the catastrophic and utopic promises of nuclear energy deployed the powerful visual stimulus of splitting atoms as metaphor. In his photography, however, Kupny does not dwell in the realm of metaphor. Instead, decaying isotopes *are* the raw material of his photography. Kupny is like the geologists of the future who, one thousand years from now, will be able to trace the beginning of the Anthropocene, the epoch in planetary history in which humans dominate, by locating in the earth's strata the first man-made radioactive isotopes, dating from about 1942. Kupny captures on his film what geologists will locate in sedimentary rock and soil, the same decaying particles, still beating strongly thousands of years from now. For Kupny, a man who came of age and lived most of his life as a loyal Soviet citizen, the power of decaying atoms still holds so much promise.

In his fascination with the generative potential of nuclear energy, Kupny is not unusual. He comes from a long line of scientists and technicians who have venerated nuclear power, while silently calculating how best to avoid being destroyed by it.

And that tradition has a particular history. It grew roots at the end of World War II and flourished in the heavily fertilized soil of the Cold War arms race. Without the Cold War, civilian nuclear power reactors like Chernobyl would never have made sense. The technology for civilian nuclear power generation was borrowed from bomb-producing reactors, yet, even with free designs and blueprints, the reactors were not cheap to build, and they were extremely risky to operate. In fact, constructing power reactors to produce slow chain reactions for electricity at a time when oil was flowing cheaply and voluminously from reservoirs in the Middle East makes no sense unless you factor in the Cold War contest, in which each side of the Iron Curtain rushed to assert the supremacy of capitalism over communism (or vice versa). The big bomb producers, in other words, needed a peaceful atom as an antidote to the skin-melting horrors nuclear war presented.

From the first days of the war's end, American propagandists, led by the atom-enthusiast William Laurence of the *New York Times*, denied the fact of radiation-related health effects in Japan and worked to rebrand the destructive "atom" into peaceful "nuclear" power that would generate the "white city" with clean, abundant energy.[7] Promoting nuclear power's benefits for science, medicine, and technology became the chief vehicle for purification of the image of the United States as a nation bent on reducing the globe to an irradiated ruin.

In pursuit of this cause, one of the first two films Hollywood distributors sent to occupied Japan in 1946 was an MGM biopic called *Madame Curie*. The fact that U.S. occupation officials chose to show *Madame Curie*, about the first scientist to grasp the power of radiation and celebrate it, in Japan, the first country to suffer en masse from the radioactive effects of nuclear warfare, says as much about U.S. officials' desire to repurpose the atomic bomb as about their insensitivity to survivors. The film, made in 1943, issues an innocence and naivety that became more difficult to sustain after August 1945. It shows the Curies in their ramshackle lab at the moment of discovery. In a soft focus shot, Pierre Curie looks at the empty space behind Marie's shoulder and evokes the now limitless metaphysical horizon: "If we can prove the secret of this new element, then we can look into the secret of life itself deeper than ever before in the history of the world."

The deep secret at the center of the atom was power, raw and unvarnished. Civilian nuclear reactors became the key to harnessing and utilizing that power. In 1953, after the public started to question the safety of fallout from accelerating nuclear bomb tests, the U.S. National Security Council resolved that "economically competitive nuclear power" must become "a goal of national importance."[8] President Dwight Eisenhower, genuinely alarmed himself by the quick tempo of the nuclear arms race, unveiled the Atoms for Peace program, dedicated to using nuclear power for medicine and to generate electricity "too cheap to meter." In his speech, Eisenhower offered to share American nuclear technology with other countries.[9] With that salvo, a race began between the Cold War superpowers to outproduce each other not only in bombs but also in civilian nuclear power plants. American companies, prodded on by generous federal subsidies, geared up. General Electric led in building up a "nucleonics" division with fourteen thousand employees in the late 1950s. By 1957, U.S. officials had signed forty-nine agreements with countries and had plunked down twenty-nine small research reactors abroad, some in states that had trouble providing general education and medical care to their populations.

Soviet leaders gamely joined the race with the Americans. In 1954, Soviet engineers plugged in the world's first nuclear power-generating reactor in Obninsk, a nuclear research city near Moscow. The reactor produced all of five megawatts for the grid, but the propaganda value of juxtaposing the peaceful Soviet atom against the martial American atom was immeasurable. In the decades to follow, Soviet engineers built reactors outside of major cities and in countries of their East European allies, and in the late 1970s, they started gearing up for a major reactor complex north of Kiev to service the western sections of the country. Soviet society welcomed nuclear power and accepted the idea, blasted from billboards, that the "Soviet atom was a worker, not a soldier."

Selling the idea of nuclear power in Japan, however, was not as seamless. In 1954, an American test in the Bikini Atoll went disastrously wrong and contaminated a Japanese fishing vessel, the *Lucky Dragon,* with a thick coating of black fallout. By the time they reached port in Japan, the ship's sailors were already suffering radiation poisoning. The public panicked on learning that the ship's radioactive catch was making its way through Japanese fish markets. A few months later, the antinuclear film *Godzilla* hit Japanese theaters. Nine million Japanese paid to see the sci-fi horror film depicting a deep-sea prehistoric monster wakened by oceanic nuclear tests.[10] The confused and angry beast pulverized Japanese cities with his radioactive breath in ways that looked much like the atomic bomb-flattened Hiroshima. As protests against nuclear testing erupted in Japan, Japanese and American leaders quickly settled in closed-door meetings on a deal to transfer U.S. nuclear designs and fission products to Japan. Among Japanese leaders, who were eager to secure an independent power source for industrial expansion, the choice between the two competing visions of nuclear power—that of *Madame Curie* or of *Godzilla*—was easy.

Aleksandr Kupny came of age in this intellectual and social current. He grew up in the shadow of civilian reactors. His father was a nuclear engineer who rose to become the director of a large nuclear complex in the Russian Urals and then later in southern Ukraine. Kupny didn't study well in school and never managed to get a university degree, but he knew reactors. He settled into a series of blue-collar jobs as a radiation monitor in reactor cities throughout the USSR. I met him in a leafy café terrace in Slavutych, Ukraine, the city built to house nuclear plant operators and cleanup workers after the abandonment in 1986 of the city of Pripiat next to the Chernobyl power plant. He moved to Slavutych the year it was founded, in 1989, after volunteering to work on the cleanup. Kupny, in short, spent his career in intimate contact with the mysterious and elusive power of radioactive decay. It had been his lifetime study.

I teased him, "How can you study something you can't sense?"

"The most common truism about radiation is that humans cannot sense it," Kupny replied, "but that is not true. With a dosimeter, you can hear radiation. With a camera, you can see it. If the levels are high enough, you can taste it on your tongue. It tastes metallic."

"And feel it?"

Kupny nodded in affirmation. After forty years in radioactive fields, he said, he can sense decaying atoms.

The photos Kupny snapped inside the sarcophagus look like an episode from *Planet of the Apes*. Wrecked machinery and dated equipment lie upended amid peeling paint and rusting, twisted steel. Clouded, frozen control room dials rest amid wires dangling from fuse boxes.

Scattered everywhere are cement blocks tossed on end. Yet the most haunting aspect of Kupny's already-haunting photos is the snowfall of tiny crystalline flakes that float through every scene of silent ruin, lending the photos a deep-sea feel. In one photo, two tiny, streaking comets mysteriously light up the corner of a room that Kupny experienced as cast in total darkness.

The tiny orange flecks and flashing lights are not an aberration in Kupny's film. As Kupny snapped photos, the multitude of photons of radioactive energy swarming around him imposed their image on his film. Imperceptibly, as he pressed his shutter button, the effervescence of the decaying reactor fuel jeweled the atmosphere in the buried chamber, lighting it up like an elaborate, spidery chandelier. These points of light are not representations. They are energy embodied. The specks are none other than cesium, plutonium, and uranium self-portraits.

Figure G2.3. Points of light. Radioactive energy imprinted on an image of ruins of Reactor No. 4. Note the sparks of light in the upper right corner. Photograph by Aleksandr Kupny.

In 1946, David Bradley, a radiation monitor for Operation Crossroads, sliced open a puffer fish he caught in the warm waters of the Bikini Atoll a few days after American generals used the chain of islands and coral reefs as a nuclear testing ground. Bradley had pulled the fish from an irradiated lagoon, and after cutting it open, he placed the fatty tissue down on a photographic plate in a darkroom. He was reproducing an 1896 experiment by Antoine Becquerel, who accidentally burned an image of a copper cross onto a glass plate he had left in a drawer with rock containing a uranium sulfur compound. Inadvertently, Becquerel captured uranium's energy in the form of light illuminating the cross's outline. Bradley sought to have the puffer fish, which he figured absorbed an appreciable dose of radiation, serve the same function of radioactive source as had uranium in Becquerel's test.

He returned to the darkroom several hours later to find that the fish's contaminated bones, organs, and not-yet-digested last meal had burned their images onto the photographic paper to create what art historian Susan Schuppli calls "a new kind of photo-synthesis." Schuppli relates the puffer fish's radioautograph to the silhouettes of residents of Hiroshima and Nagasaki that were etched into concrete and stone after the August 1945 blasts. To leave the world as a shadow meant that the bodies of the victims, not their family members, crafted their own, last memorial. Schuppli believes that radioautographs undercut the idea that humans have a monopoly on image making. They show that matter can write its own history by generating knowledge of itself.[11]

But that can only happen when bodies switch places in the darkroom drawer with specimens of uranium. A radioautograph is the opposite of an x-ray, with radioactive energy emanating out of rather than into a body. The work that goes into retrieving uranium from underground depths, isolating it, refining it, and setting up the conditions for sustained and controlled or explosive and uncontrolled chain reactions that then emanate energy that is then absorbed into bodies is behind both the self-portrait of the puffer fish's skeleton and the raw, pure crystals of light in Kupny's photographs. Marie Curie knew that work intimately. Over the course of five years, she and her husband-collaborator Pierre distilled eight tons of pitchblende down to a few grams of radium. They did so by boiling and condensing the pitchblende, a uranium compound, in huge vats. The indefatigable Marie Curie would stay at the lab late into the night to stir the mixture with poles taller than she. She breathed in the vapors. Her clothes became soaked in it. She heaved buckets, pouring and sometimes spilling compounds on her hands, where burns developed. Over time and through her labor, her body and that of the isotopes in the pitchblende joined.

The film *Madam Curie* made a big impression on Japanese audiences. It triggered a widespread interest in nuclear science and in the scientist herself. After Curie died from aplastic anemia, a blood disease often brought on by exposure to doses of radiation, Japanese collectors purchased some of her papers, which ended up in a Tokyo library. In the early twenty-first century, manga artist Erika Kobayashi went to take a look at one of Marie Curie's notebooks. It was not exceptional, just a sketchpad for her notes and observations. But in the library, Kobayashi pulled out a Geiger counter and held it close to the notebook. She watched the needle rise. Kobayashi was amazed to find that, seventy years after her death, Marie Curie's radioactive fingerprint still registered. Curie never visited Japan, but the touch of her finger emits energy in Japan, the only country in the world to experience the disasters of both martial and peaceful atoms.[12] Without images, we have a hard time imagining disaster. Marie's radioactive fingerprint ended up precisely where nuclear disaster played out in human history in spectacular and visual form. But Marie wasn't thinking about catastrophe and destruction when she left her fingerprints on her notebook. Her dream was to create a source of energy that would heal wounds and relieve human want and misery.

What Curie began, the hopes she invested in achieving a pure source of radiant energy, Kupny carries along. He lives just outside the abandoned Chernobyl Zone, which is a vast wastescape generated by global economies hungry for energy and global political powers craving ideological preeminence. Kupny believes that a lot went wrong at the Chernobyl nuclear power plant, but what he has in mind when he discusses mistakes is not the accident but its aftermath. He argues vehemently that the remaining Chernobyl reactors should never have been shut down in 2002. They were perfectly safe and still had many years in them, he believes. He disagrees with the decision to halt the construction of reactors that were being built when Reactor No. 4 blew. And he wrote a series of articles about how the new French-built stainless steel shelter going up over the aging sarcophagus is wasting money and will not work.

Kupny is convinced that nuclear plants offer an acceptable and minimal health risk. I asked him how much of a dose did he think he received during his spelunking expeditions into the sarcophagus. Kupny said he didn't know.

"The first few times we went below," Kupny said, "I recorded my dose and wrote it down, but then Sergei [Koshelev] asked me why I did that. 'What good will it do you to know? The less you know, the better you will sleep.'"

"I thought about it and stopped recording my dose," Kupny continued. "Any health problem I got, I would question, 'was it because of my dose?' That is how people do it. They know they were exposed and they want to pin every illness or injury they have on that fact." Kupny pointed to his head. "Those who are sick a lot—it comes from the mind. They torture themselves."

Certainly exposure to radiation doesn't seem to have been a problem for Kupny personally. At age sixty-eight, he is a decade past the average male life expectancy in Ukraine. Nor does he look anywhere close to the grave. He appears to be ten years younger, is lithe and agile, with a brisk step and quick mind.

"As it does with food," he told me with a smile, "radiation works on me like a preservative."

Kupny's intellectual influences follow a straight line directly from the nuclear industry where he spent his career. Officers, especially in nuclear navies, where sailors live in close contact with tiny submarine reactors, frequently refer to "hormesis," the idea that exposure to radiation or a toxin in small doses has favorable biological effects. Scientists working for nuclear agencies often focus on psychological stress as the chief health concern for civilian populations exposed to low doses of radiation. Populations show higher rates of disease, these researchers believe, because of worries over exposure, not from the exposures themselves. These ideas are controversial, and a great deal of evidence suggests that small doses of radioactive isotopes and toxins are more, rather than less, dangerous than large doses, and that low-dose exposure to radiation can cause a host of nonmalignant diseases.[13] The concepts of hormesis and radiophobia work well, however, as long as a person is healthy. You can then believe, as Kupny suggested, that those who were anxious brought on their poor health.

Kupny speaks for a lot of people locally who would choose the risk of possible contamination down the road in exchange for jobs and energy independence in the short term. In fact, in 2014, few

Ukrainians thought much about the Chernobyl accident or the Zone it left behind. In Ukraine, in the midst of a civil war, people were worried instead about a drastic drop in the value of the Ukrainian hryvnia, about the uncertainty of the national government, and about renewed incursions from their Russian neighbor. Anxieties about invisible, intangible radiation were luxuries only foreigners could afford. That's what foreigners do: they go to the Chernobyl Zone to imagine disaster. They pay money to do that.

Kupny took the risk of crawling under the sarcophagus for the cause as a "partisan." Partisans fought during World War II and sacrificed to defend the Soviet Union. As a radiation monitor, Kupny was a member of the working class. At a nuclear power plant, radiation monitors do not do heavy lifting like construction crews, plumbers, or janitors, but their jobs are nonetheless "dirty." Radiation monitors responded to emergencies. They were called in for suspected spills or leaking pipes. They inspected barrels, jumped down into trenches, and canvassed the edges of cooling ponds. Monitors were, in other words, on the front lines, patrolling and securing the borders of human contact with radioactive isotopes. After spending a lifetime trying to visualize and map radiation, it makes sense that Kupny fell a bit in love with his subject and wanted to get closer to know more. By calling himself "a partisan," Kupny was also signaling that he was a Soviet man who had faith in the socialist version of high modernism generally and, specifically, in the regenerative force of nuclear power.

Kupny's respect for the Soviet project is not unusual. He is a well-known local writer and artist in Slavutych. His ideas are shared commonly among his fellow citizens for reasons that are not mysterious. In Kupny's lifetime, Soviet socialist modernization delivered a great deal. It put an end to illiteracy, epidemics, and famine, while cleaning up daily life with indoor plumbing, public bathhouses, and central heating. Socialism also enriched life with pensions, paid vacations, affordable housing, free health care and education, and an impressive network of libraries, theaters, concert halls, and clubs. The socialist state lavished this attention foremost on trained, urban, blue-collar workers like Kupny, while outwardly focusing political rhetoric on equality and the ascendance of the working class. The contrasts in 2014 with the past were stark. In postsocialist, post-shock-capitalism society, working classes have been left to chronic poverty and joblessness, while pensions have evaporated, paid vacations are a nostalgic

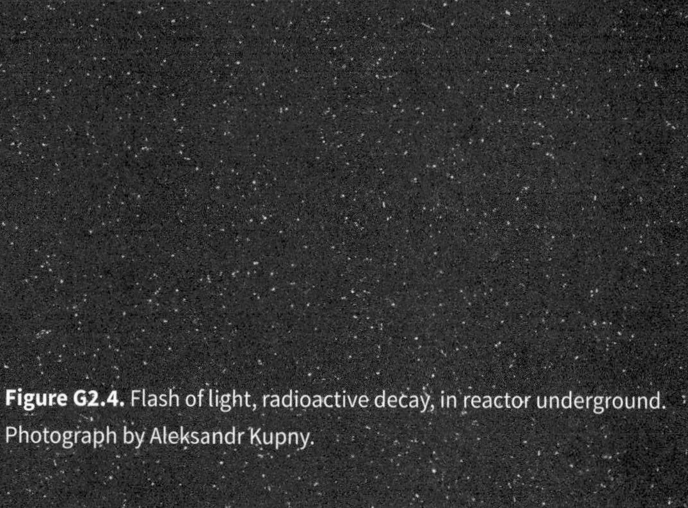

Figure G2.4. Flash of light, radioactive decay, in reactor underground. Photograph by Aleksandr Kupny.

memory, and equality—well, no one thinks about equality anymore. On this panorama, the achievements of the Soviet technocratic planning state look better and better.

Kupny's faith in the socialist version of high modernism, in the ability of man to invent machines to continually improve life, ecology, the animal world, and virtually everything else, is particularly striking because he witnessed firsthand the worst missile the technomodernist experiment tossed at the globe. The Chernobyl disaster took apart faith not only in nuclear safety but in convictions about risk management, disaster relief, and the capacity of states that are powerful and organized enough to produce nuclear reactors but too weak, indecisive, and financially strapped to contain them or clean them up afterward.[14] Yet even after watching Chernobyl unravel, Kupny maintains confidence in human ingenuity, which, as much as it makes a mess of things, goes in search of new technologies to clean up, move on, and, hopefully, improve. I had imagined that kind of optimism was reserved for the young, but not in Kupny's case. Though it would be easy to despair, he does not shrink from hope. Choosing the good atom, the worker atom, over the destructive, soldier atom is one way he maintains that hope.

There is another way that Kupny is a "partisan." Partisans were loyal Soviet subjects. From the first steps into preschool, Soviet citizens were exhorted to act as role models for others in the Soviet Union and around the world. As the fortunate minority to be part of a socialist society, they had a duty to sacrifice themselves to keep socialism going, to see that it was strong and spreading to others less fortunate. Partisans, at the same time, were not regular army. They did not take directions from superior officers or wait for orders. They acted on their own. Sneaking into the sarcophagus like a burglar is a bit like carrying out a raid. He took a portion of radiation, which he didn't measure, much like a factory worker who pockets some metal to make a toy to bring home to his child. This kind of creative, spontaneous labor, for which Kupny, the artisan, takes no payment, announces his freedom.[15] In Marxist terms, he is taking ownership of the means of production, for nothing in a nuclear reactor is so essential to production as the spinning, bouncing, decaying isotopes Kupny brought home on his film. Kupny is self-activated to understand his world, grasp its source of power, and he does so in the context of a society that has turned its back on the accident and on the promises of a society that saw nuclear power as the solution to most basic human problems of food, heat, and medicine. He redirects our attention to the source. He wants us to visualize those flickering, falling leaves of energy lit with autumnal light as a way of seeing his lost world of socialism with all its possibilities and promises.

Prize-winning historian **KATE BROWN** offers landscape biographies through which we might recover the lost histories of modernist wastelands. Her recent book *Plutopia: Nuclear Families, Atomic Cities, and the Great Soviet and American Plutonium Disasters* is the first definitive account of the great plutonium disasters of the United States and the Soviet Union. She is professor of history at the University of Maryland, Baltimore County.

Notes

1. N. P. Baranovskaia, *Chornobyl's'ka trahediia: Narysy z istoriï* (Kiev: Instytut istoriï Ukraïny NAN Ukraïny, 2011), documents 54, 79. In a report of the Secretary of the Kiev Communist Party (Komparty) to the Central Committee of the Communist Party of Ukraine, dated April 26, 1986.
2. Kyle Hill, "Chernobyl's Hot Mess, 'the Elephant's Foot,' Is Still Lethal," *Nautilus,* December 4, 2013, http://nautil.us/blog/chernobyls-hot-mess-the-elephants-foot-is-still-lethal.
3. Ibid.
4. Baranovskaia, *Chornobyl's'ka trahediia,* documents 103, 128, Information of the Communist Party of Ukraine, May 12, 1986.
5. Aleksandr Kupny, interviewed by the author, Slavutych, Ukraine, June 14, 2014.
6. Aleksandr Kupny, *Zhivy poka nas pomniat* (Kupny: Kharkov, 2011), 83.
7. Peter Bacon Hales, *Atomic Spaces: Living on the Manhattan Project* (Champaign: University of Illinois Press, 1999), 346–51; Spencer R. Weart, *The Rise of Nuclear Fear* (Cambridge, Mass.: Harvard University Press, 2012), 81.
8. Weart, *Rise of Nuclear Fear,* 82.
9. Craig Nelson, "'The Energy of a Bright Tomorrow': The Rise of Nuclear Power in Japan," *Origins: Current Events in Historical Perspective* 4, no. 9 (2011), http://origins.osu.edu/article/energy-bright-tomorrow-rise-nuclear-power-japan.
10. Barak Kushner, "Gojira as Japan's First Postwar Media Event," in *In*

Godzilla's Footsteps: Japanese Pop Culture Icons on the Global Stage, ed. William M. Tsutsui and Michiko Ito (New York: Macmillan, 2006), 42.

11. Susan Schuppli, "Dirty Pictures: Toxic Ecologies as Extreme Images," paper presented at the Radioactive Ecologies Conference, Montreal, March 15, 2015; Schuppli, "Radical Contact Prints," in *Camera Atomica,* ed. John O'Brian, 277–91 (London: Black Dog, 2014).
12. Marie Curie's papers in France's Bibliotheque National are stored in lead-lined boxes. Researchers have to sign a waiver to work on her papers. Eoin O'Carroll, "Marie Curie: Why Her Papers Are Still Radioactive," *Christian Science Monitor,* November 7, 2011.
13. The best synthesis of work on nonmalignant health effects from Chernobyl radiation is A. V. Iablokov et al., *Chernobyl: Consequences of the Catastrophe for People and the Environment*, Annals of the New York Academy of Sciences 1181 (Boston: Blackwell on behalf of the New York Academy of Sciences, 2009).
14. For the state's failures in cleaning up nuclear disasters, see Kate Brown, *Plutopia: Nuclear Families, Atomic Cities, and the Great Soviet and American Plutonium Disasters* (Oxford: Oxford University Press, 2013), 282–338.
15. Owen Hatherly, *Landscapes of Communism* (London: Penguin, 2015), 532.

3

SHIMMER

WHEN ALL YOU LOVE IS BEING TRASHED

Deborah Bird Rose

ANGIOSPERMS FORM THE GREAT FAMILY of flowering plants, and their way of life is to entice. They invite, or lure, others through their dazzling brilliance of color, scent, and shape, and they reward their visitors—birds, mammals, insects—with nutrients. In their desire to attract others whom they need for pollination, they have brought forth a worldwide multispecies potlatch. The bling of life owes much to this exuberant, alluring family. In this chapter, I focus on relationships in Australia between angiosperms and one of their important symbiont partners, flying foxes (of the genus *Pteropus*). I aim to explore a matrix of power, desire, and lures and to move across several species and cultures to draw our attention to the brilliant shimmer of the biosphere and the terrible wreckage of life in this era that we are coming to refer to as the Anthropocene.

I best learned about the shimmer of life from Aboriginal people in the Victoria River region of Australia's Northern Territory, a place to which I have been returning for more than thirty years. I use the concept of shimmer to frame this chapter because I believe it is susceptible to a "reciprocal capture" with Western thought.[1] For philosopher Isabelle Stengers, "reciprocal capture" is "an event, the production of new, immanent modes of existence" in which neither entity transcends the other or forces the other to bow down.[2] It is a process of encounter and transformation, not absorption, in which different ways of being and doing find interesting things to do together.

Stengers proposes that possibilities for new modes of existence emerge in acts of reciprocal capture, and it is my hope that an encounter with shimmer may help us better to notice and care for those around us who are in peril. Here, at the edge of extinction, is the place to begin, when the worlds that one loves—including angiosperms and flying foxes—are being trashed.

The extinction crisis is very real. Rates of extinction are perhaps ten thousand times the background rate; as ecologist Steven Harding says, we are hemorrhaging species.[3] When scientists offer us numbers, they are talking about a kind of measurement of verifiable presence or, in a sense, presumptive absence. Such numbers are a proxy to which it is worth paying attention. However, what is actually occurring is more dire than the numbers indicate. There are the functional extinctions, the extinction cascades, the extinction vortexes; these are ways in which, as things start to slip down that death road, other things start going too. Relationships unravel, mutualities falter, dependence becomes a peril rather than a blessing, and whole worlds of knowledge and practice diminish. We are looking at worlds of loss that are much greater than the species extinction numbers suggest. As I will show shortly, shimmer, the ancestral power of life, arises in relationship and encounter, so extinction cascades drag shimmer from the world. The loss is both devastating and barely comprehensible.

Extinction cascades involve failing connectivities. Of the many stories one might tell about multispecies connectivities, the starting point for me is in Aboriginal Australia, where I have been learning about multispecies kinship and connectivity for many years. The stories might be said to begin "in the beginning" with the Dreamings, also known as the creation ancestors. The Dreamings are the creators of much of the biotic life of earth. They are shape-shifters and are the founders of kin groups. Those kin groups include the human and the nonhuman descendants of the ancestors. In the area where I have decades of research experience, flying fox (*P. alecto*) Dreamings are the founding ancestors of many people. Equally, the everyday creatures are major pollinators of many trees in the region. Life flows from ancestors into the present and on into the future, and from the outset it is a multispecies interactive project involving (minimally) flying foxes, angiosperms, and human beings.

I have been adopted into a group of flying fox people, and for me this has been an astonishingly powerful lure, calling me to consider how I might live an ethic of kinship and care within this multispecies family. With it comes a burden: the commitment to bear witness to the shimmering, lively, powerful, interactive worlds that ride the waves of ancestral power. This commitment calls me to engage in forms of scholarship that encourage "passionate immersion" in the lives of both humans and nonhumans.[4]

Brilliance of Motion and Encounter

Shimmer is an Aboriginal aesthetic that helps call us into these multispecies worlds. I use the term "aesthetic" in a nontechnical way to discuss things that appeal to the senses, things that evoke or capture feelings and responses. I want to think of aesthetics, in part, in angiosperm terms, that is, in terms of lures that both entice one's attention and offer rewards.

In his classic essay titled "From Dull to Brilliant," anthropologist Howard Morphy discusses art in the Arnhem Land region of North Australia.[5] His focus is on the Yolngu term *bir'yun,* which translates as "brilliant" or "shimmering." This is an aesthetic that is found in many parts of Australia and is not limited to art. It pervades ritual, dance, and many aspects of life more widely. Morphy's analysis is one of the best for getting at the multispecies processes that concern me. As Morphy describes brilliance (or shimmer—both terms are good), the process of Yolngu painting starts off with a rough blocking out of shapes and then shifts to fine-grain crosshatching. When a painting has just its rough shape, the artists describe it as "dull." The crosshatching shifts the painting to "brilliant," and it is the brilliance of finely detailed work that captures the eye. *Bir'yun* is the shimmer, the brilliance, and, the artists say, it is a kind of motion. Brilliance actually grabs you. Brilliance allows you, or brings you, into the experience of being part of a vibrant and vibrating world. When a painting reaches brilliance, for example, people say that it captures the eye much in the way that the eye is captured by sun glinting on water. It is a capture that is all over the place: water capturing and reflecting the sun, the sun glinting on the water, the eyes of the beholders captured and enraptured, the ephemeral dance of it all. It is equally a lure: creatures long to be grabbed, to experience that beauty, that surprise, that gleaming ephemeral moment of capture.

Bir'yun, or shimmer, or brilliance, is—people say—one's actual capacity to see and experience ancestral power. This is to say that when one is captured by shimmer, one experiences not only the joy of the visual capture but also, and more elegantly as one becomes more knowledgeable, ancestral power as it moves actively across the world.

I learned about *bir'yun* through dancing all night. In contrast to Morphy, I did not work with people who were visual artists; rather, I encountered people who focused on performance, who connected bodies and earth through dance and song.[6] In music, there are multiple different temporal patterns, and it is through those temporal patterns that one starts to experience shimmer. Ethnomusicologists describe these experiences as iridescence.[7] The temporal patterns are a kind of foregrounding and backgrounding, flipping back and forth to the point where the music and the dance become iridescent (or shimmer) with ancestral power.

As the example of brilliance in sun on water indicates, *bir'yun* exceeds human action. "From Dull to Brilliant" can be read as an account of ecology: the earth shimmers. The ecological patterns are manifold, and for the sake of encountering just one example, we might think about pulse. Recall that within the domain of art, a pulse is performed first by bringing the painting from dull to brilliant. Then, at the end of a ritual, the brilliant paintings are rubbed out; they are made dull again. A similar process takes place with dance: first the singing and dancing, which generate iridescence, then the songs and motions that close it down again. At an ecological scale in northern Australia, one of the most obvious patterns is the pulse between wet and dry seasons. The desiccation of the dry season dulls the landscape in many ways (although the country is always beautiful): there is a winding back of fertility, a loss of water, and thus loss of the possibility for sun to glint on water. But then, things begin to move toward brilliant again: the lightning starts to spark things up, the rains start to bring forth shiny green shoots, and rainbows offer their own kind of brilliance. Shimmer comes with the new growth, the everything-coming-new process of shininess and health, and the new generations.[8]

This is the same motion as that of the paintings—from dull to brilliant, and then back to dull, and then back to brilliant. Ecological pulses come from and enable the experience of ancestral power. Indeed, for power to come forth, it must recede. For shimmer to capture the eye, there must be absence of shimmer. To understand how absence brings forth, it must be understood not as lack but as potential. This is where one grasps, afresh, the awful disaster of extinction cascades: not only life and life's shimmer but many of its manifold potentials are eroding.

Human Exceptionalism

Western legacies of mechanism are not a good way to appreciate either dull or brilliant.[9] The term *bir'yun*—which does not distinguish between domains of nature and culture—is characteristic of a lively pulsating world, not a mechanistic one. *Bir'yun* shows us that the world is not composed of gears and cogs but of multifaceted, multispecies relations and pulses. To act as if the world beyond humans is composed of "things" for human use is a catastrophic assault on the diversity, complexity, abundance, and beauty of life.

The legacies of Western mechanism have manifested through repeated assertions of human exceptionalism—that man is the only animal to make tools; that man is the only animal with language, a sense of fairness, generosity, laughter; that man is the only mindful creature. On one hand, all of these claims to exceptionalism have been thoroughly undermined. Other beings also do wonderful and clever things; we are not a unique outlier but rather are part of various continua. On the other hand, however, the term *Anthropocene* reminds us that it is not yet time to jettison a sense of human exceptionalism. Instead, by foregrounding the exceptional *damage* that humans are causing, the Anthropocene shows us the need for radically reworked forms of attention to what marks the human species as different.

One of the great tasks and opportunities for our moment and for the environmental humanities is "to stay with the *human* trouble," to use and tweak Donna Haraway's term.[10] Let us make another recursion across the terrain of our species, this time trying to tell more truthful accounts than those that stress our wondrous superiority. In an ecologically attentive recursion, we find that man is the only animal to voraciously, relentlessly, and viciously wreck the lifeworld of earth. Man is the only animal systematically to torture members of its own species, as well as members of countless other species, and to engage in seemingly endless and often wildly indiscriminate killing. Precisely because human cruelty tends to drop out of our conversations, I want to insist that we linger with it. It is terrible stuff to have to stay with for too long, but those who suffer, whether human or more-than-human, don't have a choice. They have to stay with it, because they are experiencing it. At the very least, we who have not yet been drawn into the vortex of violence are called to recognize it, name it, and resist it; we are called to bear witness and to offer care.

A recent example of violence comes from an event in the northern Australian town of Charters Towers in December 2013. A group of residents had complained for some time about a maternity camp of flying foxes in a municipal park. In their view, the creatures were "pests." And so they organized and conducted an assault. It happened with local government approval; in no way was it a dirty little secret.[11] The assault showed us (yet again) that man is the only animal that attacks defenseless creatures with smoke, water cannons, and firecrackers; that uses helicopters to fly low so as to terrify flying foxes and create downdrafts that break their wing bones. Man is the only animal that shoots other creatures with paintball guns, and when the creatures flap around in terror or fall to the ground injured and in shock, man is the only animal that cheers.

This portrait of human cruelty is as one-sided as were earlier accounts of our wondrous superiority. But when we highlight the pitiless and destructive qualities of humans, we see the desperate need to find ways to recuperate relational and mutually beneficial sides of the story about who we are and of what we are capable.

Ability to Care

One of the most interesting things about humans is our remarkable plasticity as individuals and as a species. While cruelty is indeed one of the great insignia of a distinctly human way, there are other sides of our capacity that help us bring ourselves into fellowship with others. In our multispecies ethnographic research at the edge of extinction, my colleague Thom Van Dooren and I have been working with volunteers, scientists, and other people who are doing reparative and protective work with creatures whose species-futures are in peril.

Flying fox carers are one set of passionate people who work at the edge of extinction and who have opened their lives and homes to others. Although there is no way to know exactly how many of these winged creatures dwelled in Australia prior to British settlement, it is certain that their populations have been radically reduced. Today, two

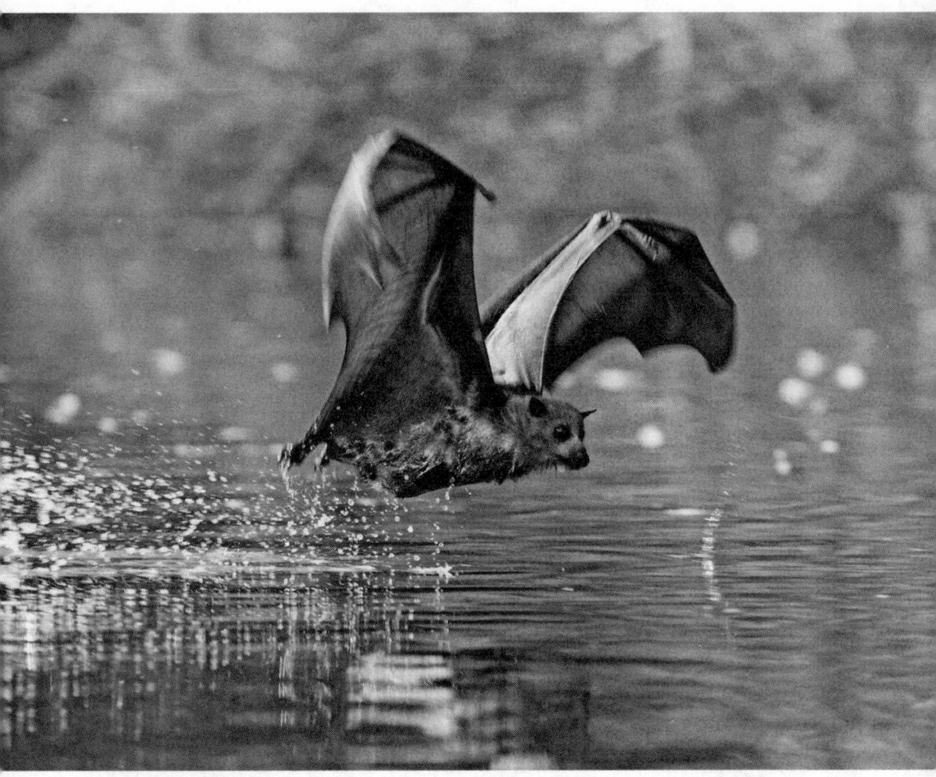

Figure G3.1. Flying fox "bellydipping." Photograph by Nick Edards/Half Light Photographic, 2009.

mainland species of flying foxes are listed as vulnerable to extinction under the Commonwealth Environment Protection and Biodiversity Conservation Act of 1999.[12] Caring for flying foxes can take the form of habitat protection and many other forms of activism and include developing arts and ethics of multispecies conviviality.[13] The most intimate modes of care involve orphans; the foster babies must be made to feel part of a family. They have to be fed and touched regularly. Human intentionality infuses care practice; youngsters will die without tactile, vocal, sociofamilial care.

When childhood is over, flying foxes don't want parents anymore, so care involves preparing them for adult forms of sociality. They go into a soft release program where they learn to interact with other flying foxes and to fly and navigate. Finally, they return to the bush. And what happens then? Some will be lucky and some won't. Many will be injured by anthropogenic hazards and end up back in care. So care is an ethical response involving tenderness, generosity, and compassion, and care is an ongoing assumption of responsibility in the face of continuing violence and peril. Pulses of harm and care offer a peculiarly telling story of the Anthropocene, highlighting multispecies entanglements, conflicting ways of being human, mass death, and, through it all, as I discuss later, a great and joyful desire for life.

This recursion across the terrain of our species acknowledges both human violence and the care that seeks to mitigate the effects of violence. Implicitly, it acknowledges the multispecies utility of modern human inventiveness in the form of antibiotics, infant formula, and bottles, for example. At its heart we see beautiful modes of careful attention, and we see them as relational responses that are lured out of us through encounters with others and that enable us to participate in the shimmer of life.

A Love Story

Care is part of a wider story of desire that takes me back to angiosperms and life's shimmer. Flowers and flying foxes come together every year with beautiful timing and exquisite generosity, giving each other great kisses that bring forth new generations. On one side of this kiss is a cohort of Australian trees—Eucalypts, Corymbias, and other members of the Myrtaceae family. The trees put out bright, showy, perfumed flowers because they need to be pollinated, and in fact, many of them need outcrossed pollination. They do best when their pollen is taken from tree to tree over some distance. On the other side, flying foxes are highly attuned to smell, have excellent night vision that is especially attentive to light colors, and have foraging patterns that keep them moving from tree to tree across wide areas. They often travel fifty kilometers per night, and many travel more than a thousand kilometers per year. Trees produce their most nutritious and abundant pollen and nectar at night, when the flying foxes are abroad. Birds and bees get the sloppy leftovers the next morning.

Figure G3.2. Flying fox in flowers. Photograph by Nick Edards/Half Light Photographic, 2008.

Blossoming takes place sequentially and flying foxes somehow know when trees start to bloom hundreds of kilometers away from where they are, and off they go en masse. Both trees and flying foxes depend on these encounters. The tongue meets the flower and the flower meets the tongue in a kiss of symbiotic mutualism. Trees call out to flying foxes in languages of color and scent, and flying foxes respond with gusto.

One of the great utterances of other species is this beautiful assent to life. We have a good word for it in English. It is the great, expressive, demonstrative "yes." Let us consider the lush, extravagant beauty, flamboyance, and dazzling seductiveness with which Eucalypts say yes. They burst open sequentially, and when they burst, every branch and twig says, "Yes! More! More buds, more flowers, more color, more scent, more pollen, more nectar!" More and more, and all that can be conjured from within the tree to reach out into the world with this great, vivid, multisensorial call "yes!" And for their part, the flying foxes come racing to respond. Their yes includes their long tongues that are perfectly adapted to sucking up nectar and their delicate whiskers that pick up pollen and distribute over 70 percent of it intact. They carry Eucalypt futures on their furry little faces, and across the patchy and increasingly fragmented landscapes of contemporary Australia, the renewal of woodland and forest life hinges on this specific yes. A new generation of trees is carried on the fur and the tongue and on the wings that beat through the night, bringing the animal to the trees and bearing the trees' gifts along to other trees. At the same time, a new generation of flying foxes is nurtured into life with lashings of glorious nectar.

Saying Yes

We humans, too, can be saying yes. There is a fantastically large set of contexts within which to say yes, but to stay with the flying foxes, it is clear that to celebrate the lives of flying foxes is to say yes to Eucalypts and thus to say yes to dry sclerophyll woodlands and to rainforests. It is to say yes to photosynthesis and to say yes to oxygen. Why would one not? We breathe in, we breathe out. In this world of connectivity, we live to celebrate another day and to experience life's shimmer as it comes forth in our lives with all manner of tears, happiness, grief, commitment, love, exuberance, and celebration. Of course, we humans are part of it. The waves of ancestral power that shimmer and grab are also exactly the relationships that bring us forth and sustain us. The kiss of life is an ancestral blessing, alive, brilliant, and pulsating in the world around us and within us.

At the same time, we may also think about what is refused when we turn away from all this abundance. Instead of the kiss of life, we humans too frequently offer a resounding no, and every no also ripples and reverberates across animals and trees, through photosynthesis and oxygen, even into the breath and into the heartbeat and rhythms of life itself. No invades the shimmer of life, unmaking ancestral power. In this time of extinctions, we are going to be asked again and again to take a stand for life, and this means taking a stand for faith in life's meaningfulness. We are called to live within faith that there are patterns beyond our known patterns and that, in the midst of all that we do not know, we also gain knowledge. We are called to acknowledge that in the midst of all we cannot choose, we also make choices. And we are called into recognition: of the shimmer of life's pulses and the great patterns within which the power of life expresses itself. We are therefore called into gratitude for the fact that in the midst of terrible destruction, life finds ways to flourish, and that the shimmer of life does indeed include us.

Through her writing on extinction, ethics, and Aboriginal ecological philosophy, **DEBORAH BIRD ROSE** has been a key figure in enlivening the environmental humanities. Her work, based on long-term research with Aboriginal peoples in Australia, focuses on multispecies communities in this time of climate change. She is professor in the Environmental Humanities Program, University of New South Wales, Sydney, and cofounder of the journal *Environmental Humanities.* Her most recent book is *Wild Dog Dreaming: Love and Extinction.*

Notes

The research with carers was funded by the Australian Research Council. I would like to thank Thom van Dooren, all the carers who shared their time and thoughts with me, and, as ever, the flying fox people of Yarralin and Lingara, who brought me into their world of kinship. A small portion of this essay was previously published in *Forum for World Literature Studies* 6, no. 1 (2014).

1. Isabelle Stengers, *Cosmopolitics I,* trans. Robert Bononno (Minneapolis: University of Minnesota Press, 2010).
2. Ibid., 35.
3. Stephan Harding, "Gaia and Biodiversity," in *Gaia in Turmoil: Climate Change, Biodepletion, and Earth Ethics in an Age of Crisis,* ed. Eileen Crist and H. Bruce Rinker (Cambridge, Mass.: MIT Press, 2010), 107.
4. Anna Tsing, "Arts of Inclusion, or, How to Love a Mushroom," *Australian Humanities Review* 50 (2011): 19.
5. Howard Morphy, "From Dull to Brilliant: The Aesthetics of Spiritual Power among the Yolngu," *Man,* New Series 24, no. 1 (1989): 21–40.
6. Deborah Bird Rose, "Pattern, Connection, Desire: In Honour of Gregory Bateson," *Australian Humanities Review* 35 (June 2005), http://www.australianhumanitiesreview.org/archive/Issue-June-2005/rose.html; Bird Rose, "Dreaming Ecology: Beyond the Between," *Religion and Literature* 40, no. 1 (2008): 109–22.
7. Catherine Ellis, "Time Consciousness of Aboriginal Performers," in *Problems and Solutions: Occasional Essays in Musicology Presented to Alice M. Moyle,* ed. Jamie Kassler and Jill Stubington, 149–85 (Sydney: Hale and Iremonger, 1984).
8. My description here is extremely minimal. The shimmer of ecological pulses arises also in their intersecting and crosscutting ripples, for example, the work of the sun ensures that rain does not take over the earth, and the work of rain ensures that the sun does not take over; their pulse is complementary, and each part arrives as relief from an excess of the other. There are winds, of course, and the play of winds on grasses and on clouds. The effect, if one is attentive, is an iridescence or shimmer sparking up at these larger scales.
9. It might be thought that the disenchantment of earth life brings about a pervasive dullness, so it is important to distinguish between "dull" as a monotone of absence and "dull" as the term is being used in this context, as part of a pulse, a zone of potentiality.
10. Donna Haraway, *When Species Meet* (Minneapolis: University of Minnesota Press, 2008).
11. An excellent account can be seen on YouTube in the form of a documentary by Noel Castley-Wright investigating the animal cruelty applied

to flying fox colonies: "State of Shame—Queensland's Legislated Animal Cruelty," video, 22:13, April 6, 2014, https://www.youtube.com/watch?v=owF5D6k_9-U.

12. Australian Government Department of Environment, "Flying-Foxes: Environment Law," https://www.environment.gov.au/biodiversity/threatened/species/flying-fox-law.

13. Thom Van Dooren and Deborah Bird Rose, "Storied-Places in a Multispecies City," *Humanimalia* 3, no. 2 (2012): 1–27.

FOOTPRINTS OF THE DEAD

WE ARE SURROUNDED BY TRACES OF THE PAST, a veritable garden of ghosts.

Walk down the street with biologist Jens-Christian Svenning, and he will show you creatures from another age—from beetles to roadside trees. Contemporary organisms evolved a very long time ago, in quite different circumstances; landscapes are ephemeral compared to species, Svenning explains in this imaginary walk. The plants and animals we know have had to adjust to today's overwhelming human presence. But there are also *missing* creatures—and particularly the giant animals that died off following human conquests of their territories. The missing megafauna leave gaps in the ecologies with which they once lived: ghostly footprints.

Andreas Hejnol shows us another kind of ghost: the spectral presence of evolutionary time inside the bodies of organisms. Every new species inherits parts of its body plan from earlier organisms. For those who want to admire the diversity of life, the trick is not to imagine this inheritance as teleological progress, the climbing of a ladder toward the sun. Instead, we might appreciate the ghostly presence of ancestors inside us, which makes it possible for us to do whatever we do.

Physicist and feminist theorist Karen Barad takes us into the spectral time of the nuclear age. The clocks stopped when the atom bomb fell on Hiroshima, she reminds us. At this moment, the world performs a strange topology. Small is large, large is small: radiation makes its way into the tiny cells that make up bodies, even as it flies across the globe. The effects are immediate, and also long delayed; in the shadow of radiation, the line between material and immaterial no longer makes sense. Barad's provocation that there is an atom bomb inside each morsel of life shakes up our ability to differentiate scales—and haunts us.

"What necropolitics is this?" asks anthropologist Nils Bubandt, exploring "the life and death effects—intended as well as unintended—of . . . ruination and extinction." Like Hejnol and Barad, Bubandt shows how our nicely planned distinctions fall apart in the presence of ghosts. Are there "natural disasters" anymore in the Anthropocene? he wonders, contemplating an Indonesian mud volcano that might be "natural" and might be "human made." In the fractures between human deeds and deep time, indecidable geologies erupt, and with them ghosts and spirits of the Anthropocene. ●

4

FUTURE MEGAFAUNAS

A HISTORICAL PERSPECTIVE ON THE POTENTIAL FOR A WILDER ANTHROPOCENE

Jens-Christian Svenning

A LONG-TERM ECOLOGICAL AND BIOGEOGRAPHIC perspective is an important art for living on a damaged planet: it helps us both to see what factors have slipped away from our present-day landscapes and ecosystems and to imagine how we might overcome their absences. Ecological history shows us that contemporary ecosystems are in a very unusual state. Today, the biggest wild animal in most places would be a kangaroo or a deer. This is a novel situation on time scales of thousands to millions of years. Wildernesses and other natural areas around the world are haunted by ghosts of giant animals: mastodons, ground sloths, and tapirs in the Appalachian forests; giant wombat-like marsupials in Australia; and elephants, rhinos, and aurochs in Europe. These giants occurred all across the world until recently, and their disappearance—which continues today, as seen in the rhino-poaching crisis in Africa—is tightly linked to us, *Homo sapiens*.

What, then, is the future of wild megafaunas (often defined as animal species larger than or equal to forty-five kilograms body mass) in an epoch, increasingly referred to as the Anthropocene, where human activities have become so dominant that they are now a planetary force?[1] In fact, as much of the world is suffering from ongoing, often rapid defaunation, a pertinent question is perhaps, is there any future?[2] Or is a wild megafauna just a phenomenon of the past? Conversely, could megafaunas play a role in a wilder Anthropocene, where uncontrolled processes and wild agents are allowed and even promoted—as argued in proposals for "rewilding"?[3] These are themes I will consider in this essay, taking a long-term ecological and biogeographic perspective.

Seeing Past Worlds

Why is a long-term paleoecological perspective relevant when discussing the Anthropocene and, in particular, the biodiversity of the Anthropocene? It takes a little historical and evolutionary context to answer that question. Basically, most of the species that we have around us today are extremely old. For example, many endemic plant species in California are more than 2 million years old.[4] Extant species are therefore much older than the Anthropocene, irrespective of which definition of the Anthropocene we use.[5] Most species are also older than even the Holocene, the warm period starting at the end of the last ice age, 11,700 years B.P., and characterized by the near-global occurrence of modern humans *(H. sapiens)* and their activities.[6] The ecological characteristics of present-day species are to a large extent inherited from their ancestors and are therefore also for this reason much older than the Anthropocene. This means that species are adapted to the anthropogenic activities and cultural environments of the Anthropocene and late Holocene in only limited and rather superficial ways. Instead, they are primarily adapted to the conditions that characterized ecosystems *before* their modification by modern humans. If we want to foster conditions that will maintain the earth's biodiversity, it is therefore crucial that we understand the ecosystems in which our current species evolved.

This presents a methodological challenge, however. It is difficult to imagine past ecosystems. Scientists refer to the "shifting baseline syndrome" to describe our tendency to imagine that environmental conditions at the edge of our own memories represent the way the world used to be.[7] This is why a paleoecological perspective is so important: attention to longer histories allows us to appreciate the rich, diverse landscapes that have existed in pasts beyond human memory. These pasts show a potential for biodiversity in many settings that far exceeds what we see today. Importantly, given the rapid environmental changes of the Anthropocene, looking at how environments and their biota have varied enormously in the past can also allow us to gain an understanding of the potential for biodiversity under different climatic conditions.[8]

One of the most striking things one sees through a paleoecological approach is the vast abundance and richness of large animals prior to the arrival of *H. sapiens* (Figure G4.1).

For most of the last 40 million years or so, rich megafaunas were characteristic of all major geographic regions.[9] Large terrestrial animals, evolved after the catastrophic mass extinction that killed off all the nonbird dinosaurs, had become key actors across the planet. But today, rich megafaunas are restricted to a few regions, principally protected areas in southern and eastern Africa and parts of southern Asia.

Consider the megafauna that existed in Europe during the Last Interglacial period, some 125,000 years ago. This time period is relevant, as this is the latest period with a climate similar to today but

Figure G4.1. Large animals (megafauna) were diverse and abundant across all continents, except Antarctica, until the Late Pleistocene. Here is an illustration of the rich preextinction North American megafauna, including several proboscideans, the Columbian mammoth *(Mammuthus columbi)* and American mastodon *(Mammut americanum),* and a diversity of other large herbivores and carnivores. Image by Roman Uchytel, http://prehistoric-fauna.com/.

prior to the arrival of modern humans in Europe. There was another hominid around at that point, namely, the Neanderthal *(H. neanderthalensis),* but no modern humans. During this era, central and western Europe hosted trees, herbaceous plants, invertebrates, and small vertebrates that were largely the same as those that occur in the region today, with minor exceptions, such as a few regionally extinct species, for example, three-way sedge *(Dulichium arundinaceum),* which is now restricted to North America.

In contrast, the megafauna of the Last Interglacial deviated strongly from the megafauna that occurs in the region today.[10] The presently native species, such as red deer *(Cervus elaphus),* roe deer *(Capreolus capreolus),* and wild boar *(Sus scrofa),* were all present and widespread. The same is true for species that now have restricted distributions in the region but were widespread in recent prehistory (earlier in the Holocene), for example, Eurasian elk (moose; *Alces alces*), bison (*Bison* sp.), wolf *(Canis lupus),* lynx *(Lynx lynx),* and brown bear *(Ursus arctos).* Wild horse *(Equus ferus)* and aurochs *(Bos primigenius)*—species that now only or primarily survive in domesticated and feral forms—were likewise widespread. Fallow deer *(Dama dama),* which today is widespread in the region due to human introduction, was restricted to the Middle East and the Balkans in the Holocene but was widespread and common in central and western Europe during the Last Interglacial. However, the diversity went far beyond this.[11] Elephants—in the form of the large, straight-tusked elephant *(Elephas antiquus),* an extinct relative of the Asian elephant *(E. maximus)*—were common. There were also two now-extinct species of rhinoceros (*Stephanorhinus kirchbergensis* and *S. hemitoechus*), relatives of the near-extinct so-called Sumatran rhinoceros *(Dicerorhinus sumatrensis)*; the extinct giant deer *(Megaloceros giganteus)*; the extinct, mostly herbivorous cave bear *(U. spelaeus)*; and a number of large carnivores such as cave lion *(Panthera leo spelaea),* leopard *(P. pardus),* and spotted hyena *(Crocuta crocuta).* In southern parts of Europe, there was also a now extinct species of water buffalo *(Bubalus murrensis)* as well as the common hippo *(Hippopotamus amphibius)*—then also frequent in the Thames region but now restricted to Africa.

It would be hard to find any place with such a rich megafauna today. At temperate latitudes, even the fauna of Yellowstone National Park appears meager in comparison. Only well-protected nature areas in Africa and southern Asia have megafaunas that come close to this diversity of large species, including the so-called megaherbivores (herbivores more than one ton in body weight, with particularly strong effects on vegetation).[12] However, before *H. sapiens,* such faunas were the norm, widely occurring on all continents—and they had been there for tens of millions of years.[13]

This point is well illustrated by considering the biggest of the biggest, the proboscideans. If we look just a little bit back in time, ten

Figure G4.2. Prior to the late Quaternary megafauna extinctions, proboscideans were diverse and abundant on all continents, except Australia and Antarctica, and in most ecosystems, but they are now restricted to three surviving species in small areas of Africa and Asia. Owing to their large size and great strength, and their resulting near-immunity to predation, elephants have a strong ecosystem effect, notably via destroying woody vegetation. Here African savanna elephants *(Loxodonta africana)* are feeding on acacias in Maasai Mara in Kenya. Contemporary savannas in Africa have less woody biomass than South American savannas in large part due to effects of the surviving African large herbivores. Photograph by Jens-Christian Svenning.

thousand to seventy thousand years ago, there were proboscideans, such as elephants, mammoths, and mastodons, on all continents, except Australia and Antarctica, as well as on many islands close to continents.[14] There were more than fifteen species, including today's African and Asian elephants as well as well-known extinct species such as the woolly mammoth *(Mammuthus primigenius)* and the American mastodon *(Mammut americanum)* (Figure G4.2).

They occurred in all types of biomes, from tropical rain forests and savannas to temperate steppes and prairies, from temperate forests to Mediterranean woodlands and the arctic tundra. There were dwarf or dwarfish elephants on the Californian islands, in the Mediterranean, in Japan, and on various islands in Indonesia. Although this richness of large animals now seems strange, for millions of years, their abundance was part of the typical state of our world.

Humans and Megafauna Extinctions

Since the 1960s, scientists have debated why so much of the megafauna has disappeared.[15] A key starting point for understanding this loss is its uniqueness. The disproportionate loss of large species and the resulting megafauna-poor ecosystems in most of the world are unprecedented in post-Mesozoic ecological history, that is, since large mammals evolved some 40 million years ago.[16] It is not that there has not been a lot of extinction going on in this period. There has. However, these extinctions never singled out megafauna. For example, when the ice ages started 2.6 million years ago, there were massive extinctions around the world, affecting plants and animals broadly, for example, trees, mollusks, and some big animals.[17] However, these mass extinctions did not affect the megafauna disproportionately.

The late Quaternary megafauna extinctions were different. While plant, insect, and small animals species stayed more or less the same, the big animals disappeared. There are currently only two credible hypotheses for this. One suggests that the extinctions have been caused by the strong climate variability of the late Quaternary.[18] The other links the extinctions to the expansion of modern humans across the world, primarily invoking hunting as the decisive factor.[19] There is ample evidence that prehistoric humans were capable big-game hunters who could take down even the largest species. Several recent reviews have addressed this issue, pointing to humans as the main driver of large animal loss.[20] In a recent study, we tested the two competing hypotheses using macroecological analyses of a carefully updated global mapping of all extinct large mammals.[21] This study shows that extinction rates have been high in much of the world but also variable in strength, with the strongest extinctions in the New World and Australia. Furthermore, it found that extinction rates were strongly linked to human biogeography, but weakly linked to ice age–warm period climate shifts. Notably, extinctions were uniformly high in places where *H. sapiens* was the first hominid to arrive and uniformly low in the sub-Saharan center of hominid evolution, with intermediate rates in regions colonized by pre-*sapiens* hominids.[22] The arrival of modern humans appears to be the main driving factor in the severe extinction of megafauna in most land areas outside Africa.

What makes Africa unique here? Humans evolved in Africa, so perhaps this explains it. The classical explanation for the relative persistence of megafauna animals in Africa is that the megafaunas here were able to coevolve with and adapt to hominids. A nonexclusive alternative is that megafauna extinctions also happened here, but earlier, due to the activities of pre-*sapiens* humans. There is some evidence for this alternative, notably that some animals, such as saber-toothed cats, went extinct earlier in Africa than elsewhere.[23] Nevertheless, it is also clear that more megafaunas have survived to the present day in Africa than elsewhere. Still a third explanation considers *H. sapiens* as an invasive species, which, during its global expansion, escaped many of the pests and pathogens that had coevolved with it in its area of origin, allowing greater human population increases and bigger impacts in newly invaded regions.[24]

A somewhat underappreciated fact is that megafauna extinctions did not exclusively occur during the coldest part of the Last Glacial or during the climatic shifts during the subsequent warming, but both started earlier in some places and continued later. Hence, there is no general temporal match linking extinctions to periods of strong climatic change or climatic stress, even when looking at extinctions at sites with high temporal resolution.[25] Notably, quite a number of megafauna species survived into the early Holocene in South America, Eurasia, and Africa, with extinctions continuing across millennia and with most of these clearly linked to human impacts.[26] Hence, Pleistocene mammal extinctions and the numerous examples of human-driven historical and ongoing range declines in remaining megafauna species, such as elephants, rhinos, bison, and large carnivores, can be seen as one continuous process of defaunation.[27]

When one looks at these overall extinction patterns, there are at least three important take-home messages for the Anthropocene. The first is that these extinctions are without ecological replacement. There have been a lot of extinctions in the history of the earth, but in most cases, they have been extinctions in which one species has replaced another in its specific ecological niche. In this case, however, megafauna species have disappeared without replacement. Second, we can say that these extinctions are largely or completely anthropogenic. Third, the faunas that we see today, where large species are generally poorly represented, are a highly unusual condition in terms of the evolutionary history of biodiversity.

Ecosystem Effects of Megafauna Extinction

With the increasing realization that the current megafauna-poor conditions are unusual and largely anthropogenic, there has been increasing interest in understanding the ecological consequences of these losses. What have been the roles of the large top carnivores and large herbivores in ecosystems, and what does their absence mean for ecosystems? While most research to date has focused on the ecological consequences of historical or near-historical megafauna losses, an increasing number of researchers are exploring Late Pleistocene and early Holocene extinctions for a longer perspective.[28] These studies provide important fundamental insight into the ecosystem role of megafauna. A seminal study on the ecological role of prehistoric megafauna was published already in 1982, highlighting its likely importance in the dispersal of seeds of many large-seeded or large-fruited plant species in the Americas, with the severe megafauna losses in this region likely causing declines in many of these plant species.[29] Recent studies corroborate this hypothesis. In all contexts, there is now increasing evidence that large animals play key roles, notably via top-down trophic interactions and associated trophic cascades, that is, via effects from the upper levels of food webs penetrating downward and likely even affecting biogeochemical cycling and climate.[30]

One of the most famous examples concerns the ecological consequences of the reintroduction of the wolf *(C. lupus)* to Yellowstone National Park in 1995, after its eradication in 1926. The reintroduction

Figure G4.3. Wolves *(Canis lupus)* were introduced to Yellowstone National Park in the 1990s to restore their top-down ecological effects on the ecosystem, mainly via their predation on large ungulates such as American elk *(Cervus canadensis)*. The reintroduction is reported to have resulted in strong ecological effects, including increased regeneration of willow, poplar, and aspen. Nevertheless, many parts of the Yellowstone landscape are still strongly affected by herbivory pressure, as seen by the abundant woody regeneration inside this enclosure in stark contrast to the near-treeless surrounding landscape. The latter is consistent with increasing evidence for strong megafauna effects on the vegetation prior to the late Quaternary extinctions. Photograph by Jens-Christian Svenning.

reinstated a series of trophic cascades with important ecosystem effects. Wolves began to limit the population size and space use of American elk *(C. canadensis)* and of the competing, but smaller, coyote *(Canis latrans)*. Hereby, the wolf reintroduction had strong indirect consequences for vegetation dynamics (notably increased regeneration of poplar and aspen [*Populus* spp.] and willow [*Salix* spp.]) along with a variety of other direct and indirect effects on the population dynamics of many other species (Figure G4.3).[31]

The exact nature of these dynamics and the importance of wolves have been controversial, illustrating the need for further research.[32] However, similar wolf-driven trophic cascades are reported from other regions[33] and are also reported for many other large carnivores.[34]

When it comes to prehistoric extinctions, we have much less knowledge of their ecosystem impacts, although more and more studies are trying to address this.[35] There is evidence from both Australia and North America that megafauna losses have had strong impacts on vegetation.[36] At a Queensland site, paleoecological evidence indicates that large animal extinctions caused a shift from a system with lots of fauna, little fire, and a mix of fire-sensitive and fire-robust species of plants to the current fire-driven system with many fewer fire-sensitive species.[37] Other studies provide evidence that the severe megafauna losses in South America have resulted in higher tree cover in this region's savannas compared to Africa[38] and that trees adapted for megafauna-mediated seed dispersal are still declining in response to the loss of their dispersers many millennia ago.[39] Megafauna are constant gardeners, one might say, and their extinctions have long-term ecological effects.

As a case, let us consider the role of the formerly rich megafauna in shaping the vegetation within the temperate forest biome. For more than a decade, ecologists have debated whether the traditional view of this biome as one dominated by closed tall-canopy forest is correct.[40] Some researchers have argued that the effects of large wild herbivores were once strong enough to create a more diverse vegetation structure in temperate regions, perhaps in places even a dynamic savanna-like system.[41] The dense temperate forests imagined to have covered much of Europe and eastern North America prior to agriculture could therefore be a sign not of "untouched" wilderness but of vast extinctions.

Although many questions remain open, it is increasingly clear—from a broad range of paleoecological indicators—that preagricultural ecosystems within the temperate forest biome were not uniformly closed forest but were often a mosaic of dense woody vegetation and semi-open and open vegetation, perhaps especially in lowlands and close to water.[42] Importantly, it is clear that before *H. sapiens*'s arrival and the loss of much of the large-herbivore diversity, including the largest species, such as elephants and rhinos, there was much more open and semi-open vegetation as well as denser large-herbivore populations, with a considerable proportion of the landscape having high large-herbivore densities (on the order of 2.5 or more big animals per hectare).[43] We should probably imagine pre-modern human temperate biome ecosystems as complex, dynamic, and species-rich mosaics of densely forested areas, open woodlands, and grass- and heathlands, with highly diverse, often dense populations of large animals with strong effects on vegetation structure.[44]

Toward a Wilder Anthropocene

Today, the Late Pleistocene and Holocene attrition of the earth's megafaunas continues in most of the world.[45] Currently, approximately 21 percent of mammals are threatened with extinction with a bias toward large species, so that approximately 60 percent of herbivores of one hundred kilograms or more in body mass are threatened with extinction.[46] For example, all five species of rhinoceros are highly endangered, with most populations undergoing rapid declines from excessive poaching. However, in some parts of the world, defaunation is becoming replaced by refaunation. In Europe and North America, large carnivores and large herbivores are making comebacks, even recolonizing anthropogenic landscapes that are home to large numbers of people. In Europe, most large mammals are undergoing marked population increases and range expansions.[47] Species like beaver *(Castor fiber)*, red deer *(Cervus elaphus)*, and wild boar *(Sus scrofa)* have exhibited strong reexpansions and are now common in many parts of the European continent. Wolves are reexpanding into western Europe. Other species, such as lynx *(L. lynx)*, Eurasian elk *(A. alces)*, and brown bear *(U. arctos)*, are still more localized but nevertheless clearly expanding. Similarly, in North America, species like the wolf and the cougar *(Puma concolor)* are likewise reexpanding. These refaunation dynamics reflect a variety of societal changes that have led to greater tolerance of these species and formal legal protections. Human population declines in rural areas—a part of increasing urbanization—have been a contributing factor in some areas. In Europe, for example, approximately 1 million hectares of agricultural land are abandoned each year, providing habitat for expanding megafauna species.[48]

In addition to this relatively spontaneous recovery of some megafauna, reintroductions are also becoming more and more common. Until recently, reintroductions have been performed as a *species* conservation measure, to restore particular animals to areas from which they had recently been extirpated. However, more and more frequently, reintroductions are being carried out with ecosystem restoration as the key aim. The increasing realization that large animals are often important for ecosystem function and biodiversity as well as growing recognition of their widespread loss has inspired a new ecological restoration approach: trophic rewilding.[49] This approach uses species introductions to restore top-down trophic interactions and associated trophic cascades to promote self-regulating, biodiverse ecosystems.[50] Importantly, it has been proposed to restore not just recently extirpated species but also species lost during the Late Pleistocene and early Holocene extinctions using functionally similar exotic species as ecological replacements.[51] The introduction of wolves to

Figure G4.4. Beavers *(Castor fiber)* were nearly exterminated in Europe by the early twentieth century but have made a remarkable comeback via spontaneous reexpansions as well as a number of reintroductions; they now number more than three hundred thousand individuals. The beaver was hunted out from Denmark more than a thousand years ago but was reintroduced in 1999 to restore its ecosystem effects (tree felling, dam building); here are felled birches at one of two reintroduction sites. Photograph by Jens-Christian Svenning.

Yellowstone National Park exemplifies this approach. In recent years, numerous similar rewilding projects have been initiated in Europe, including with species such as beaver *(Castor fiber)* (Figure G4.4), European bison *(Bison bonasus)* (Figure G4.5), feral horse *(Equus ferus),* and feral cattle *(Bos primigenius).* These projects are implemented not only in wildlands but also in natural areas in densely populated regions, for example, western Europe.[52] Many projects involve some element of management, for example, to overcome area constraints and minimize human–wildlife conflicts.

Looking to other regions, where defaunation is still the dominant trend, there are also cases of successful coexistence of wild megafauna,

Figure G4.5. Bison were common and widespread in Eurasia for hundreds of thousands of years, but the European bison *(B. bonasus)* went extinct in the wild in the early twentieth century. Through reintroduction programs, the species now numbers a few thousand individuals living in more or less wild conditions, and it is increasingly reintroduced not just close to its last areas of occurrence in eastern Europe but much more widely across its former range, including in densely populated western Europe. Here is one of two recently established Danish small herds, living in a fenced natural area close to the city of Randers. Photograph by Joanna B. Olsen.

such as large carnivores, with dense human populations. As a noteworthy example, leopards *(Panthera pardus)* and striped hyenas *(Hyaena hyaena)* exist at densities of five adults each per one hundred square kilometers in a densely populated agricultural landscape in western India with little natural habitat left and no wild large prey species.[53] The carnivore populations are sustained by preying on feral dogs and livestock, the latter causing some human-wildlife conflict; in contrast, there are no reports of attacks on humans in the area.[54] In another case, a dense population (fifty-two per one hundred square kilometers) of spotted hyenas *(Crocuta crocuta)* lives in a densely populated agricultural district in Ethiopia, subsisting mainly on waste and, to a smaller extent, livestock.[55]

While megafaunas continue to decline in number in much of the world, these examples of refaunation and human-megafauna coexistence offer hope that megafaunas and their many important functions could be broadly restored in the Anthropocene. It is clear that even strongly anthropogenic landscapes have potential to host wild megafauna populations. Furthermore, trophic rewilding has strong potential to remedy defaunation and restore trophic cascades that promote self-regulating biodiverse ecosystems and is highly relevant also in populated areas close to urban centers.[56] At the same time, empirical rewilding research is still rare and fragmented, and there is a strong need to develop a broad empirical research agenda, for example, to provide a better, predictive understanding of how megafauna-rich ecosystems with high trophic complexity function.[57] Other important issues that need further study are interplays of restored megafaunas and their ecological effects with landscape settings, society, and climate change.

Overall, it is obvious that human-wildlife conflicts could restrict the extent to which megafaunas will recover in the Anthropocene. In a world with large and increasing numbers of people and increasing demands for resources, there will easily be conflicts over agricultural losses and public safety. These conflicts may occur over practical problems but may also involve cultural values and perceived rather than real risks. The strong and divisive public debate in relation to reexpanding wolf populations in the United States exemplifies this. While it is clear that wild large animals can coexist with people and live in anthropogenic landscapes, the fate of the earth's megafauna in the Anthropocene will depend on the intentional and unintentional actions of people. It is up to people across the world whether the Anthropocene will be characterized by rich megafaunas or only much simpler, defaunated ecosystems. Awareness of ecological history will be one key factor here in making people notice the absences, the ghosts—a crucial step in realizing what has been lost and what could come back.

Biologist **JENS-CHRISTIAN SVENNING** is a research leader on the ecological effects of megafaunas and the potential for bringing megafaunas back to our landscapes—to replace human-caused extinctions and restore ecosystem functions. He has published widely in macro- and community ecology, including on the effects of climate change on species diversity and distributions, with a special interest in integrating paleoecological and contemporary perspectives and data in basic and applied ecology. He is professor in ecology in the Department of Bioscience at Aarhus University and a member of the Aarhus University Research on the Anthropocene (AURA) group.

Notes

I gratefully acknowledge economic support by the European Research Council (ERC-2012-StG-310886-HISTFUNC). I also consider this chapter a contribution to the Danish National Research Foundation Niels Bohr professorship project, the Aarhus University Research on the Anthropocene (AURA) project, and the Carlsberg Foundation "Semper Ardens" project on megafauna ecosystem ecology from deep prehistory to a human-dominated future (MegaPast2Future).

1. P. J. Crutzen, "Geology of Mankind," *Nature* 415, no. 6867 (2002): 23; W. Steffen, P. J. Crutzen, and J. R. McNeill, "The Anthropocene: Are Humans Now Overwhelming the Great Forces of Nature?," *Ambio* 36, no. 8 (2007): 614–21.
2. R. Dirzo, H. S. Young, M. Galetti, G. Ceballos, N. J. B. Isaac, and B. Collen, "Defaunation in the Anthropocene," *Science* 345, no. 6195 (2014): 401–6.
3. C. Sandom, C. J. Donlan, J.-C. Svenning, and D. Hansen, "Rewilding," in *Key Topics in Conservation Biology 2,* ed. D. W. Macdonald and K. J. Willis, 430–51 (Chichester, U.K.: Wiley-Blackwell, 2013); D. Jørgensen, "Rethinking Rewilding," *Geoforum* 65 (2015): 482–88; J.-C. Svenning, P. B. M. Pedersen, C. J. Donlan, R. Ejrnæs, S. Faurby, M. Galetti, D. M. Hansen, et al., "Science for a Wilder Anthropocene—Synthesis and

Future Directions for Trophic Rewilding Research," *Proceedings of the National Academy of Sciences of the United States of America* 113, no. 4 (2016): 898–906.

4. N. J. B. Kraft, B. G. Baldwin, and D. D. Ackerly, "Range Size, Taxon Age and Hotspots of Neoendemism in the California Flora," *Diversity and Distributions* 16, no. 3 (2010): 403–13.
5. J. Zalasiewicz, C. N. Waters, A. D. Barnosky, A. Cearreta, M. Edgeworth, E. C. Ellis, A. Gałuszka, et al., "Colonization of the Americas, 'Little Ice Age' Climate, and Bomb-Produced Carbon: Their Role in Defining the Anthropocene," *Anthropocene Review* 2, no. 2 (2015): 117–27.
6. See, e.g., A. C. Roosevelt, M. L. da Costa, C. L. Machado, M. Michab, N. Mercier, H. Valladas, J. Feathers, et al., "Paleoindian Cave Dwellers in the Amazon: The Peopling of the Americas," *Science* 272, no. 5260 (1996): 373–84.
7. S. K. Papworth, J. Rist, L. Coad, and E. J. Milner-Gulland, "Evidence for Shifting Baseline Syndrome in Conservation," *Conservation Letters* 2, no. 2 (2009): 93–100.
8. E.g., J. Agustí and M. Antón, *Mammoths, Sabertooths, and Hominids: 65 Million Years of Mammalian Evolution in Europe* (New York: Columbia University Press, 2002).
9. F. A. Smith, A. G. Boyer, J. H. Brown, D. P. Costa, T. Dayan, S. K. M. Ernest, A. R. Evans, et al., "The Evolution of Maximum Body Size of Terrestrial Mammals," *Science* 330, no. 6008 (2010): 1216–19.
10. T. van Kolfschoten, "The Eemian Mammal Fauna of Central Europe," *Netherlands Journal of Geosciences* 79 (2000): 269–81.
11. Ibid.
12. R. N. Owen-Smith, *Megaherbivores: The Influence of Very Large Body Size on Ecology* (Cambridge: Cambridge University Press, 1988).
13. Smith et al., "Evolution of Maximum Body Size of Terrestrial Mammals"; S. Faurby and J.-C. Svenning, "Historic and Prehistoric Human-Driven Extinctions Have Reshaped Global Mammal Diversity Patterns," *Diversity and Distributions* 21, no. 10 (2015): 1155–66.
14. C. Sandom, S. Faurby, B. Sandel, and J.-C. Svenning, "Global Late Quaternary Megafauna Extinctions Linked to Humans, Not Climate Change," *Proceedings of the Royal Society, Series B: Biological Sciences* 281, no. 1787 (2014): 20133254.
15. P. S. Martin, "Africa and Pleistocene Overkill," *Nature* 212 (1966): 339–42.
16. Smith et al., "Evolution of Maximum Body Size."
17. See, e.g., W. L. Eiserhardt, F. Borchsenius, C. M. Plum, A. Ordonez, and J.-C. Svenning, "Climate-Driven Extinctions Shape the Phylogenetic Structure of Temperate Tree Floras," *Ecology Letters* 18, no. 3 (2015): 263–72.

18. W. F. Ruddiman, *Earth's Climate: Past and Future* (New York: W. H. Freeman, 2014).

19. P. S. Martin, "Pleistocene Overkill," *Natural History* (December 1967): 32–38.

20. A. D. Barnosky, P. L. Koch, R. S. Feranec, S. L. Wing, and A. B. Shabel, "Assessing the Causes of Late Pleistocene Extinctions on the Continents," *Science* 306, no. 5693 (2004): 70–75; P. L. Koch and A. D. Barnosky, "Late Quaternary Extinctions: State of the Debate," *Annual Review of Ecology, Evolution, and Systematics* 37, no. 1 (2006): 215–50.

21. Sandom et al., "Global Late Quaternary Megafauna Extinctions."

22. Ibid.

23. J. Tollefson, "Early Humans Linked to Large-Carnivore Extinctions," *Nature* (2012), doi:10.1038/nature.2012.10508; L. Werdelin and M. E. Lewis, "Temporal Change in Functional Richness and Evenness in the Eastern African Plio-Pleistocene Carnivoran Guild," *PLOS One* 8, no. 3 (2013): e57944.

24. Koch and Barnosky, "Late Quaternary Extinctions."

25. J. L. Gill, J. W. Williams, S. T. Jackson, K. B. Lininger, and G. S. Robinson, "Pleistocene Megafaunal Collapse, Novel Plant Communities, and Enhanced Fire Regimes in North America," *Science* 326, no. 5956 (2009): 1100–1103; S. Rule, B. W. Brook, S. G. Haberle, C. S. M. Turney, A. P. Kershaw, and C. N. Johnson, "The Aftermath of Megafaunal Extinction: Ecosystem Transformation in Pleistocene Australia," *Science* 335, no. 6075 (2012): 1483–86.

26. A. Hubbe, M. Hubbe, and W. Neves, "Early Holocene Survival of Megafauna in South America," *Journal of Biogeography* 34, no. 9 (2007): 1642–46, doi:10.1111/j.1365-2699.2007.01744.x; S. T. Turvey and S. A. Fritz, "The Ghosts of Mammals Past: Biological and Geographical Patterns of Global Mammalian Extinction across the Holocene," *Philosophical Transactions of the Royal Society, Series B: Biological Sciences* 366, no. 1577 (2011): 2264–76.

27. W. J. Ripple, T. M. Newsome, C. Wolf, R. Dirzo, K. T. Everatt, M. Galetti, M. W. Hayward, et al., "Collapse of the World's Largest Herbivores," *Science Advances* 1, no. 4 (2015): e1400103; W. J. Ripple, J. A. Estes, R. L. Beschta, C. C. Wilmers, E. G. Ritchie, M. Hebblewhite, J. Berger, et al., "Status and Ecological Effects of the World's Largest Carnivores," *Science* 343, no. 6167 (2014): 1241484; G. Bar-Oz, M. Zeder, and F. Hole, "Role of Mass-Kill Hunting Strategies in the Extirpation of Persian Gazelle *(Gazella subgutturosa)* in the Northern Levant," *Proceedings of the National Academy of Sciences of the United States of America* 108, no. 18 (2011): 7345–50.

28. E.g., J. Terborgh, J. A. Estes, P. Paquet, K. Ralls, D. Boyd-Heger, B. J. Miller, and R. F. Noss, "The Role of Top Carnivores in Regulating

Terrestrial Ecosystems," in *Continental Conservation—Scientific Foundations of Regional Reserve Networks,* ed. M. E. Soulé and J. Terborgh, 39–64 (Washington, D.C.: Island Press, 1999); J. L. Gill, "Ecological Impacts of the Late Quaternary Megaherbivore Extinctions," *New Phytologist* 201, no. 4 (2014): 1163–69, doi:10.1111/nph.12576; C. J. Sandom, R. Ejrnæs, M. D. D. Hansen, and J.-C. Svenning, "High Herbivore Density Associated with Vegetation Diversity in Interglacial Ecosystems," *Proceedings of the National Academy of Sciences of the United States of America* 111, no. 11 (2014): 4162–67; E. S. Bakker, J. L. Gill, C. N. Johnson, F. W. M. Vera, C. J. Sandom, G. P. Asner, and J.-C. Svenning, "Combining Paleo-Data and Modern Exclosure Experiments to Assess the Impact of Megafauna Extinctions on Woody Vegetation," *Proceedings of the National Academy of Sciences of the United States of America* 113, no. 4 (2016): 847–55; C. E. Doughty, J. Roman, S. Faurby, A. Wolf, A. Haque, E. S. Bakker, Y. Malhi, J. Dunning, and J.-C. Svenning, "Global Nutrient Transport in a World of Giants," *Proceedings of the National Academy of Sciences of the United States of America* 113, no. 4 (2016): 868–73.

29. D. H. Janzen and P. S. Martin, "Neotropical Anachronisms: The Fruits the Gomphotheres Ate," *Science* 215, no. 4528 (1982): 19–27.

30. Terborgh et al., "Role of Top Carnivores in Regulating Terrestrial Ecosystems"; J. A. Estes, J. Terborgh, J. S. Brashares, M. E. Power, J. Berger, W. J. Bond, S. R. Carpenter, et al., "Trophic Downgrading of Planet Earth," *Science* 333, no. 6040 (2011): 301–6; Doughty et al., "Global Nutrient Transport in a World of Giants"; F. A. Smith, J. I. Hammond, M. A. Balk, S. M. Elliott, S. K. Lyons, M. I. Pardi, C. P. Tomé, P. J. Wagner, and M. L. Westover, "Exploring the Influence of Ancient and Historic Megaherbivore Extirpations on the Global Methane Budget," *Proceedings of the National Academy of Sciences of the United States of America* 113, no. 4 (2016): 874–79.

31. B. J. Miller, H. J. Harlow, T. S. Harlow, D. Biggins, and W. J. Ripple, "Trophic Cascades Linking Wolves *(Canis lupus),* Coyotes *(Canis latrans),* and Small Mammals," *Canadian Journal of Zoology* 90, no. 1 (2012): 70–78; W. J. Ripple and R. L. Beschta, "Trophic Cascades in Yellowstone: The First 15 Years after Wolf Reintroduction," *Biological Conservation* 145, no. 1 (2012): 205–13; W. J. Ripple, R. L. Beschta, J. K. Fortin, and C. T. Robbins, "Trophic Cascades from Wolves to Grizzly Bears in Yellowstone," *Journal of Animal Ecology* 83, no. 1 (2014): 223–33.

32. A. P. Dobson, "Yellowstone Wolves and the Forces That Structure Natural Systems," *PLOS Biology* 12, no. 12 (2014): e1002025.

33. H. Okarma, "The Trophic Ecology of Wolves and Their Predatory Role in Ungulate Communities of Forest Ecosystems in Europe," *Acta Theriologica* 40, no. 4 (1995): 335–86; M. Hebblewhite and D. W. Smith, "Wolf Community Ecology: Ecosystem Effects of Recovering Wolves in Banff and Yellowstone National Parks," in *The World of Wolves: New Perspectives on Ecology, Behavior, and Management,* ed. M. Musiani, L. Boitani, and P. Paquet, 69–120 (Calgary: University of Calgary Press, 2010); R. Callan, N. P. Nibbelink, T. P. Rooney, J. E. Wiedenhoeft, and A. P. Wydeven, "Recolonizing Wolves Trigger a Trophic Cascade in Wisconsin (USA)," *Journal of Ecology* 101, no. 4 (2013): 837–45.

34. W. J. Ripple, A. J. Wirsing, C. C. Wilmers, and M. Letnic, "Widespread Mesopredator Effects after Wolf Extirpation," *Biological Conservation* 160 (2013): 70–79; Ripple et al., "Status and Ecological Effects of the World's Largest Carnivores."
35. Gill, "Ecological Impacts of the Late Quaternary Megaherbivore Extinctions"; Sandom et al., "High Herbivore Density"; Bakker et al., "Combining Paleo-Data and Modern Exclosure Experiments"; Doughty et al., "Global Nutrient Transport in a World of Giants."
36. Gill et al., "Pleistocene Megafaunal Collapse."
37. Rule et al., "Aftermath of Megafaunal Extinction."
38. C. E. Doughty, S. Faurby, and J.-C. Svenning, "The Impact of the Megafauna Extinctions on Savanna Woody Cover in South America," *Ecography* 39, no. 2 (2016): 213–22.
39. D. N. Zaya and H. F. Howe, "The Anomalous Kentucky Coffeetree: Megafaunal Fruit Sinking to Extinction?," *Oecologia* 161, no. 2 (2009): 221–26; C. E. Doughty, A. Wolf, N. Morueta-Holme, P. M. Jørgensen, B. Sandel, C. Violle, B. Boyle, et al., "Megafauna Extinction, Tree Species Range Reduction, and Carbon Storage in Amazonian Forests," *Ecography* 39, no. 2 (2016): 194–203.
40. R. Bradshaw, G. E. Hannon, and A. M. Lister, "A Long-term Perspective on Ungulate-Vegetation Interactions," *Forest Ecology and Management* 181, no. 1–2 (2003): 267–80.
41. F. W. M. Vera, *Grazing Ecology and Forest History* (Oxon, U.K.: CABI, 2000).
42. J.-C. Svenning, "A Review of Natural Vegetation Openness in North-Western Europe," *Biological Conservation* 104, no. 2 (2002): 133–48; Sandom et al., "High Herbivore Density."
43. Sandom et al., "High Herbivore Density."
44. Bakker et al., "Combining Paleo-Data and Modern Exclosure Experiments."
45. Ripple et al., "Status and Ecological Effects of the World's Largest Carnivores"; Ripple et al., "Collapse of the World's Largest Herbivores."
46. Ripple et al., "Collapse of the World's Largest Herbivores."
47. S. Deinet, C. Ieronymidou, L. McRae, I. J. Burfield, R. P. Foppen, B. Collen, and M. Böhm, *Wildlife Comeback in Europe: The Recovery of Selected Mammal and Bird Species. Final Report to Rewilding Europe by ZSL, Birdlife International and the European Bird Census Council* (Lon-

don: Zoological Society of London, 2013); G. Chapron, P. Kaczensky, J. D. C. Linnell, M. von Arx, D. Huber, H. Andrén, J. V. López-Bao, et al., "Recovery of Large Carnivores in Europe's Modern Human-Dominated Landscapes," *Science* 346, no. 6216 (2014): 1517–19.

48. L. M. Navarro and H. M Pereira, "Rewilding Abandoned Landscapes in Europe," *Ecosystems* 15, no. 6 (2012): 900–912.

49. Svenning et al., "Science for a Wilder Anthropocene."

50. E.-G. Ritchie, B. Elmhagen, A. S. Glen, M. Letnic, G. Ludwig, and R. A. McDonald, "Ecosystem Restoration with Teeth: What Role for Predators?," *Trends in Ecology and Evolution* 27, no. 5 (2012): 265–71; L. H. Fraser, W. L. Harrower, H. W. Garris, S. Davidson, P. D. N. Hebert, R. Howie, A. Moody, et al., "A Call for Applying Trophic Structure in Ecological Restoration," *Restoration Ecology* 23, no. 5 (2015): 503–7; Svenning et al., "Science for a Wilder Anthropocene."

51. See C. J. Donlan, J. Berger, C. E. Bock, J. H. Bock, D. A. Burney, J. A. Estes, D. Foreman, et al., "Pleistocene Rewilding: An Optimistic Agenda for Twenty-First Century Conservation," *American Naturalist* 168, no. 5 (2006): 660–81; C. J. Donlan, H. W. Green, J. Berger, C. E. Bock, J. H. Bock, D. A. Burney, J. A. Estes, et al., "Re-wilding North America," *Nature* 436, no. 7053 (2005): 913–14. Janzen and Martin also point to this idea in "Neotropical Anachronisms."

52. P. Jepson, "A Rewilding Agenda for Europe: Creating a Network of Experimental Reserves," *Ecography* 39, no. 2 (2016): 117–24.

53. V. Athreya, M. Odden, J. D. C. Linnell, J. Krishnaswamy, and U. Karanth, "Big Cats in Our Backyards: Persistence of Large Carnivores in a Human Dominated Landscape in India," *PLOS One* 8, no. 3 (2013): e57872.

54. Ibid.

55. G. Yirga, W. Ersino, H. H. de Iongh, H. Leirs, K. Gebrehiwot, J. Deckers, and H. Bauer, "Spotted Hyena *(Crocuta crocuta)* Coexisting at High Density with People in Wukro District, Northern Ethiopia," *Mammalian Biology* 78, no. 3 (2013): 193–97.

56. Svenning et al., "Science for a Wilder Anthropocene"; Jepson, "A Rewilding Agenda for Europe."

57. See Svenning et al., "Science for a Wilder Anthropocene."

5

LADDERS, TREES, COMPLEXITY, AND OTHER METAPHORS IN EVOLUTIONARY THINKING

Andreas Hejnol

METAPHORS ARE ALWAYS A DOUBLE BIND: they at once allow us to see and stop up our abilities to notice.

For centuries, biology has relied on a particular set of metaphors—including ladders and trees—to classify and order living beings. Such metaphors have depicted life as a slow but inexorable march upward—up a stairway of creatures with humans at the top, positioned as the most advanced beings. This hierarchical understanding of life, which defines "progress" as a linear movement from the so-called simple to the complex, has long haunted biological inquiry. At the same time that such metaphors have been essential to the development of evolutionary thinking, they have also limited biologists' queries about lateral relations, movements toward simplicity, and the lives of seemingly lesser organisms. In this chapter, I first trace the history of hierarchical notions of life and evolution and the suites of metaphors about ladders and trees that they have produced. Then, I turn to describing how new biologies are forcing us to tell very different stories with dramatically different metaphors—ones that challenge long-standing notions of hierarchy and complexity and that ultimately reconfigure our own place in the world of living creatures, past and present.

Ordering Nature: Aristotle's Ladders and Haeckaelian Trees

While human attempts to describe and order nature around them can be traced back to prehistoric cave drawings, two historical milestones of the systematization of animals were Aristotle's *Historia Animalium* and, in the eighteenth century, Carl Linnaeus's *Systema Naturae*.[1] These early, preevolutionary systematizations were mainly driven by "similarity" as a criterion of placing organisms into the same group. The groups were then ordered according to notions of "complexity."

Aristotle was the first to coin this understanding of life the "Scala Naturae"—or Great Chain of Being—but many others sketched similar images. The scale or chain typically moved from a substrate of rocks upward to plants, then animals, then people, then gods. Within a given category, such as that of animals, there were still more hierarchical orderings: lions were placed above sheep, which were placed above fish, which were placed above insects. Such rankings were based on specific notions of complexity but not on genealogy. They were ranked, but not related, without a sense of evolution. All organisms were seen to be unchangeable and a result of the gods' flawless creation—a process that allegedly culminated with humans.

Up to the eighteenth century, the classification of organisms remained largely ahistorical. In the eighteenth and nineteenth centuries, however, the scientific community began to debate historical ideas and consider that organisms might change over time. The integration of time and history into the discourse of ordering nature was a radical new view on the order of nature that became one of the foundations of modern science.[2] Darwin's nineteenth-century publications about evolutionary theory are the most well-known example of classification based on historical principles.[3] Yet Darwin's theory was only one of many. In the eighteenth century, for example, several other naturalists crafted their own versions of historically ordered nature, including Jean-Baptiste Lamarck in his *Philosophie Zoologique* (Figure G5.1).[4] Despite the integration of history, the depictions and descriptions of organisms still continued to rely conceptually on Aristotle's *Scala Naturae* and its notions of complexity. Lamarck, for instance, insisted that physical laws dictated that organisms naturally evolve toward increasing complexity.[5] The metaphor of a ladderlike, hierarchical arrangement of organisms ranging from simple, unicellular

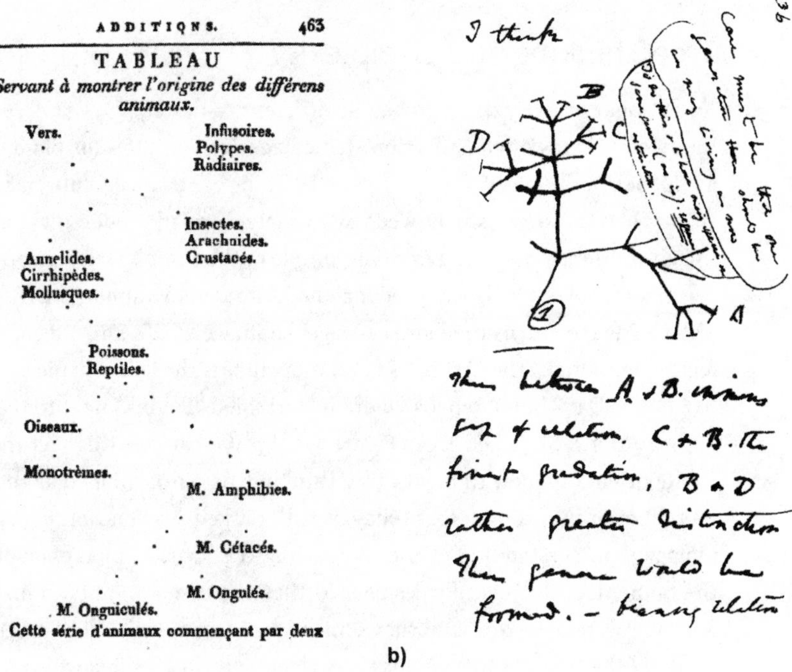

Figure G5.1. Ladders and trees. *a,* Lamarcks's depiction of connections between different animal groups. *b,* Darwin's notebook excerpt with the iconic tree of evolutionary relationships.

organisms to more complex animals remains common in scientific literature today.[6]

Although its origins lie in pre-Christian Greece, the narrative of a nature organized along a gradient from the simple to the complex strongly resonated with widespread European religious ideas of creation that positioned humans as the creatures closest to God. Even the discovery of the common genealogy of all organisms with the rise of evolutionary theory did little to change such assumptions of human superiority. Chambers has proposed that this anthropocentrism is based on our perception that the human brain is the most complex thing that has ever evolved.[7] Thus, our "perceived value of the outputs of its deliberations" leads humans to more highly value animals, such as dolphins, whales, and primates, that display similar forms of "intelligence."[8]

Tree Thinking and Complexity

Evolutionary theory itself proved insufficient to destabilize the hegemony of systems of classification that categorized animals (or plants) as "higher" and "lower" (or "primitive"). On the contrary, evolutionary theory became increasingly wedded to metaphors of ladders, which imagined life as proceeding from the simple to the complex. Evolutionary theory, however, also allowed for a novel kind of metaphor, the tree of life. The tree turns the ladder's stagelike image of life into a genealogical one. One of the first trees of relationships is the iconic depiction in Charles Darwin's notebook where he scribbled "I think" next to the depiction of a branched tree (Figure G5.1b). We can see this picture as the origin of a new thought—tree thinking. As old-fashioned as the metaphor of the tree appears today, it still marked a revolution in the biological understanding of life. Borrowing the kinship metaphor of the family tree, biological trees posited that organisms were both historical and related to each other. But in their shape, biological trees retained the sense of movement through time from a simple world to a more complex one—from a single trunk toward countless branches.

The image of the branches of a tree became the basis for another considerable contribution, namely, the so-called cladistics or phylogenetic systematics developed in the twentieth century by German entomologist Willi Hennig, who made it the foundation for a theoretico-historical systematization of organisms.[9] Although cladistics is also used in fields such as linguistics, to determine the origin of languages, they have been particularly important in biological classification.[10] Hennig proposed to order animals in a system according to their genealogy—and provided the theoretical framework for how morphological characteristics can be used as arguments for proposing particular "clades" that group together species that all share a common ancestor. Comparative anatomical methods allow researchers not only to detect shared morphological characters for a clade but also to infer evolutionary relationships. This theory of cladistics developed by Hennig provided the foundation of modern studies of the relatedness of organisms.

In the latter part of the twentieth century, however, new technologies allowed for observations not only of organisms' morphologies but also of their genes. With the development of polymerase chain reaction (PCR) and related technologies, biologists were able to see

DNA and RNA sequences, the building parts of proteins and nucleic acids in organismal genomes. Such technologies introduced a new kind of information from which scientists can reconstruct the relationships among organisms. No longer restricted to morphological traits, cladistics is now enacted through the comparison of the molecular data that become available through analysis of the nucleotide sequences of proteins that are shared between organisms.

Rethinking Trees

Although "similarity" is still the unit that is used to group and classify organisms, the theoretical frameworks for thinking about these relationships are much more elaborate than they were five decades ago. Notions of relationality are becoming more nuanced. For example, biologists no longer think in terms of "transitional" and hierarchical relationships among living species (i.e., "humans evolved from apes") but instead think through genealogical cladistics that stress the ongoing changes of all organisms (i.e., "humans and apes evolved from a common ancestor").

Perhaps most importantly, the application of cladistic methods to the vast number of molecular sequence data has led to profound changes in our understanding of animal relationships.[11] Put simply, molecular data uproot the phylogenetic tree. Not only do they strengthen the deconstruction of the teleological elements in the understanding of the processes in evolution, including hierarchical orderings of beings, they also demonstrate that evolution itself is nondirectional and unpredictable.

Here I want to focus on two animal groups—tunicates and comb jellyfish—both of which have recently been repositioned in biological orderings of life in ways that have important consequences for how we understand life more broadly. The tunicates and comb jellies have the power to teach us about evolution in a different key. They illustrate that evolution does not necessarily proceed from simple to complex; they rupture the foundational ideas that underpin the Great Chain of Being. They also force us to question the forms of anthropocentrism that still haunt evolutionary theory, even within the scientific community. Tunicates and comb jellyfish demonstrate that nature does not have an apical structure; humans have to find new ways of representing their place in nature.

Such insights are important in many ways, because hierarchical metaphors for ordering beings have shaped not only human perceptions of nature but also human strategies for managing natural worlds. Rethinking relations among organisms and the metaphors we use to describe them can shift how we value other beings—and thus change how we aim to protect our natural environment. For example, many current conservation projects focus on the protection of charismatic megafauna, an approach that, at least to some degree, has its foundation in assumptions about "higher" versus "lower" animals and, thus, their relative importance. Moving away from hierarchical metaphors for ordering organisms might therefore open up more holistic ways of protecting our environment that do not necessarily begin with animals assumed to be at the top of the ladder.

Tunicates

The so-called urochordates, or tunicates, are marine animals that are mostly sessile filter feeders (Figure G5.2). Some species also drift in the water column as colonies (salps). Urochordates are sack shaped with two large openings: water flows inward through one opening, passes a structure that filters food particles, and then eventually exits the body through the other opening. In addition to this feeding system, the animal has male and female gonads that are responsible for reproduction. For more than a century, this relatively simple morphology has meant that the tunicates were seen as evolutionary precursors to more complex groups of animals, including the so-called cephalochordates *(Branchiostoma)* and vertebrates (including humans). This is because the tunicates share, with these two groups of animals, a stabilizing structure, called chorda, but lack the other major characteristics of those groups, such as segmented packages of musculature and a closed blood vascular system. As a result, the metaphorical ladder of evolution that has organized life from simple to complex has consistently placed the urochordates as our most distant and primitive relatives.

About a decade ago, however, the first large-scale molecular analyses of animal relationships began to hint at a different arrangement, an arrangement that has been confirmed by follow-up studies.[12] These studies revealed that the relatively simple urochordates are indeed the closest relatives of the vertebrates. The more complex

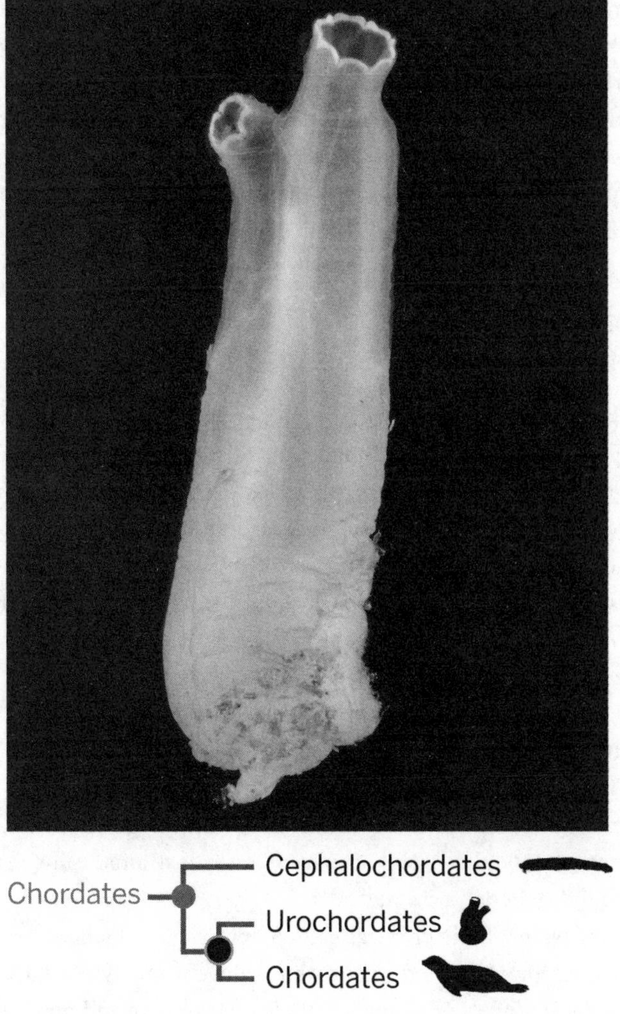

Figure G5.2. Representatives of urochordates: the urochordate sea squirt *Ciona intestinalis* and, below, its relationships to other animals. Photograph by Andreas Hejnol, Sars Centre.

cephalochordates, which share many more morphological characteristics with vertebrates than the tunicates, are actually far more distantly related to vertebrates.[13] The most parsimonious explanation for this apparently illogical arrangement is that the urochordates have lost the shared characteristics of the vertebrate group, such as the closed blood vascular system, without leaving any visible remnants.

Examples of such "regressive" evolution that includes the loss of whole organ systems, such as brains and hearts, had been previously noted in parasitic animals, for example, tapeworms or flukes, but it was also assumed to be limited to them. Tapeworms and flukes lost their digestive systems, including their mouth openings, as they evolved to take up nutrients from their hosts via their skin. By similarly evolving to have less so-called complexity, tunicates force biologists to reconsider long-standing assumptions about directionality in evolution.

Comb Jellyfish and Sponges

Comb jellyfish, or *Ctenophora,* is another animal group whose placement has recently provoked taxonomic discussions. Comb jellies are pelagic predators that show a fair variety of cell types, such as nerve cells, individual muscle cells, and elaborate sensory organs. One species—*Mnemiopsis leidyi*—has become particularly famous as an invasive species, originally from the U.S. East Coast, that moves around the world (Figure G5.3). Traveling in the ballast water tanks of trading ships, this highly reproductive species and devastating predator is likely introduced annually into the Baltic Sea. In the 1980s, it was introduced into the Black Sea, where it destroyed the region's anchovy industry.

Thirty years later, these gelatinous ctenophores became famous for another reason: with the advent of the first large-scale molecular reconstructions, this group of marine animals dramatically changed their phylogenetic position.[14] Although gelatinous and mostly composed of water, their physical cell composition—which was seen as relatively "complex"—led scientists to conventionally place them as close relatives of other animals that also show nerve and muscle cells. Indeed, they were assumed to represent a sister group to all remaining animals. Yet, massive comparisons of gene sequences revealed that, despite their morphologic features, these animals were actually among the most distant relatives of so-called complex animals, such as vertebrates.

Sessile sponges, in comparison, had been seen as much less "complex." Sponges lack nerve cells and musculature and thus have been treated as more ancient animals than comb jellies. As a result, zoologists had labeled the sponges as the animal group most distantly related to vertebrates.

Recently, however, the positions of comb jellies and sponges have

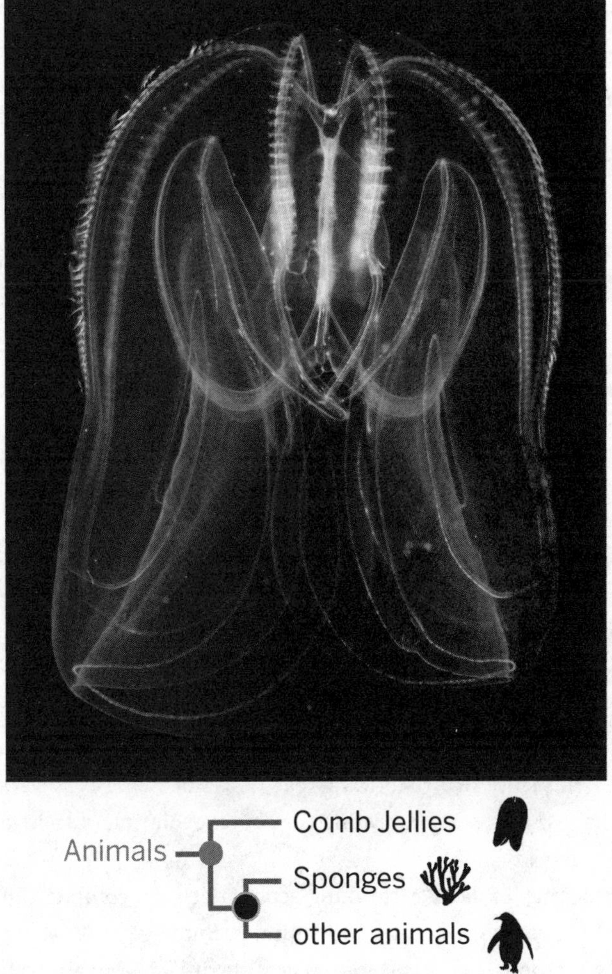

Figure G5.3. The comb jelly *M. leidyi* as representative of the ctenophores and its relationship to the other groups. Photograph by Andreas Hejnol, Sars Centre.

been reversed. The comb jellyfish may have a variety of cell types similar to those of other animals, but it is genetically quite distant. Sponges, meanwhile, may look primitive and strange, but they are genetically closer to the remaining animals, excluding ctenophores. The way that sponges have evolved implies that something unexpected happened in evolutionary processes. One hypothesis for understanding the genetic similarity of sponges with more complex animals is that sponges once possessed, but have since lost, a diversity of neuronal and muscular cell types. In this case, nerve cells and musculature may be much older than previously thought, with cellular diversity preceding the early split of sponges and creatures like comb jellies. The second possibility, even more heretical, is that these rather complex cell types may have an independent origin: cellular tissue diversity evolved, independently, twice in evolutionary history.[15] Despite evidence, this latter possibility is still met with considerable resistance from the scientific community.

When Metaphors Fall Apart

How could the relative simplicity of sponges and urochordates possibly be the result of a long evolutionary process by which complexity was lost? How could it conceivably be possible for relatively similar, and fairly complex, cell types, such as muscles and neurons, to evolve twice independently? Genetic technologies have raised a suite of new questions about the ordering of organisms, and they have shifted the ways through which we understand evolutionary relations. They force us to rethink the anthropocentric notions of complexity, with their origins in ancient Greece, that continue to haunt biological thinking today. The figure of Aristotle's Great Chain of Being persists in the ongoing resistance to illustrating the actual evolution of sponges and urochordates.

According to ladder thinking, comparatively complex animals should not come prior to simpler animals. But what, after all, is complexity? If there is such a thing as "complexity," it is an adaptation to specific ecological conditions, not the outcome of a teleological process. Furthermore, in any use of the term, complexity should not be defined as morphological or behavioral similarity to humans.

It is important to emphasize that all animals alive today have had similar time to evolve from the last common ancestor. Evolution is an ongoing process; no group of animals got stuck in their ancestral state, even those, such as the horseshoe crab (see Funch, in *Monsters*), that appear to be morphologically similar over a long period of time. The notion that some animal got stuck is intrinsic to the idea of a stepped evolutionary process leading inexorably in one direction, namely, complexity. From an evolutionary perspective, today's comb jellies are as distant from the last common ancestor of all animals as are today's humans. In other words, comb jellies and other contemporary animals have spent the same amount of time developing their individual and specific traits and relative complexity. Evolution is an ongoing process and has not placed a hold on the adaptation of any lineage.

Yet, too often, assumptions about hierarchy and complexity continue to limit biological thinking. For example, in Haeckel's depiction of the evolutionary tree in Figure G5.4a, animal groups that have evolved more recently are assigned to the stem of the tree (e.g., "Vermes," "Chorda-Animals," "Amphibia"), which suggests that these represent intermediate forms from which other recent animals have evolved.

This would correspond to the narrative that humans evolved from apes—which is indeed how the relationship is depicted in Haeckel's treetop. In Haeckel's illustration, recent animal groups have a one-to-one correspondence to human ancestors and would thus be resistant to evolutionary change. A similar motif can be found in the newest edition of a leading textbook about animal evolution.[16] Figure G5.4b is an illustration from this textbook that is evocative of a ladder, with the names of animal groups forming its rungs. The diagram put "Bilateria," that is, animals with bipolar bodies such as humans, on top. From this placement, students are easily led to believe that humans are the most evolved or most complex beings. Animals listed on the left appear as if they wandered off the path of Progress, forming evolutionary dead ends.

Such misconceptions about animal evolution, rooted in the ideas that evolution proceeds from simple to complex and that less complex animal groups are eventually superseded, remain not only in textbooks but also among zoologists. They deserve our attention because they matter: they shape our thinking about the biology of life and have consequences for what we find important to investigate in animals, plants, fungi, and bacteria. In a recent review of the impacts of new insights into the evolution of sponges, Casey Dunn and collaborators highlight how the assumption that sponges are less complex animals prevents us from investigating the unique complexity and diversity of these animals.[17] Evolution has likely produced—in lineages distant from humans—complex adaptations that remain undiscovered, because we view these animals from an anthropocentric perspective as being at the lowest and most primitive rungs of an imagined evolutionary ladder and thus not as a place to look for evolutionary insights.

PEDIGREE OF MAN.

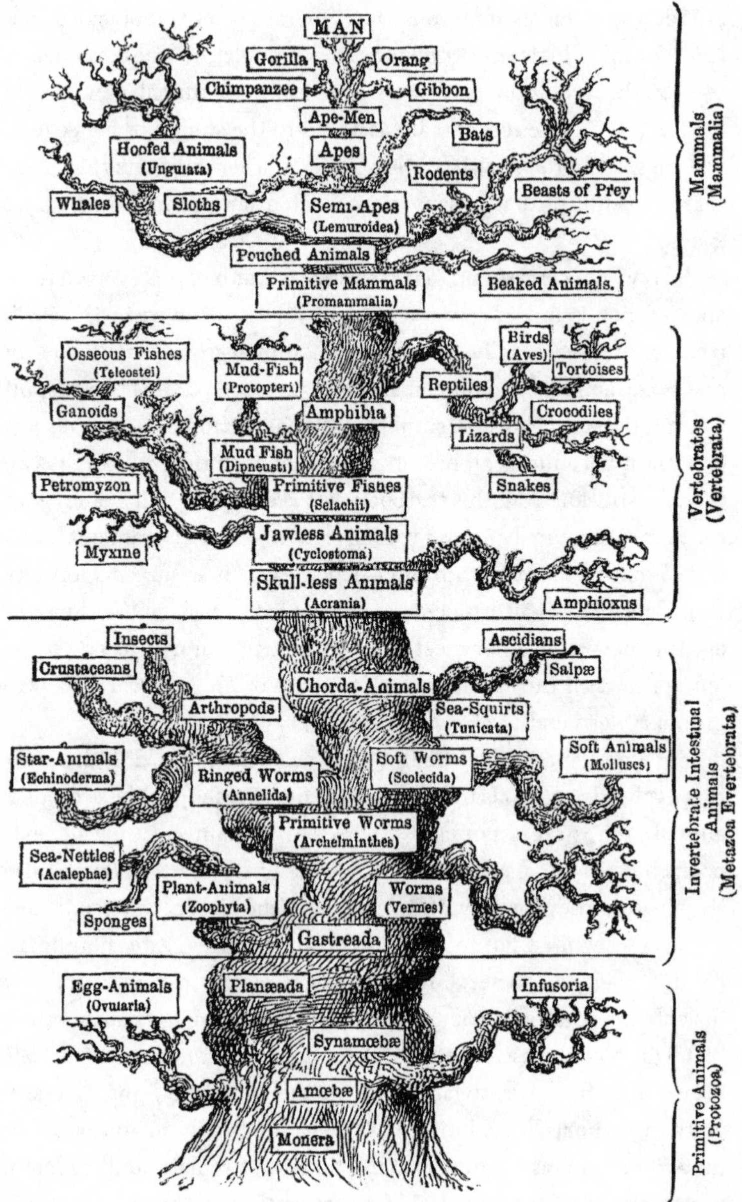

a

Figure G5.4. Trees containing elements of a chain of being. *(a)* Haeckel's tree in *Descent of Man,* with humans on top. *(b)* Evolutionary tree of a leading zoology textbook of Nielsen with growing complexity during evolutionary time. Reprinted from Nielsen, *Animal Evolution,* with permission from John Wiley & Sons and Oxford University Press.

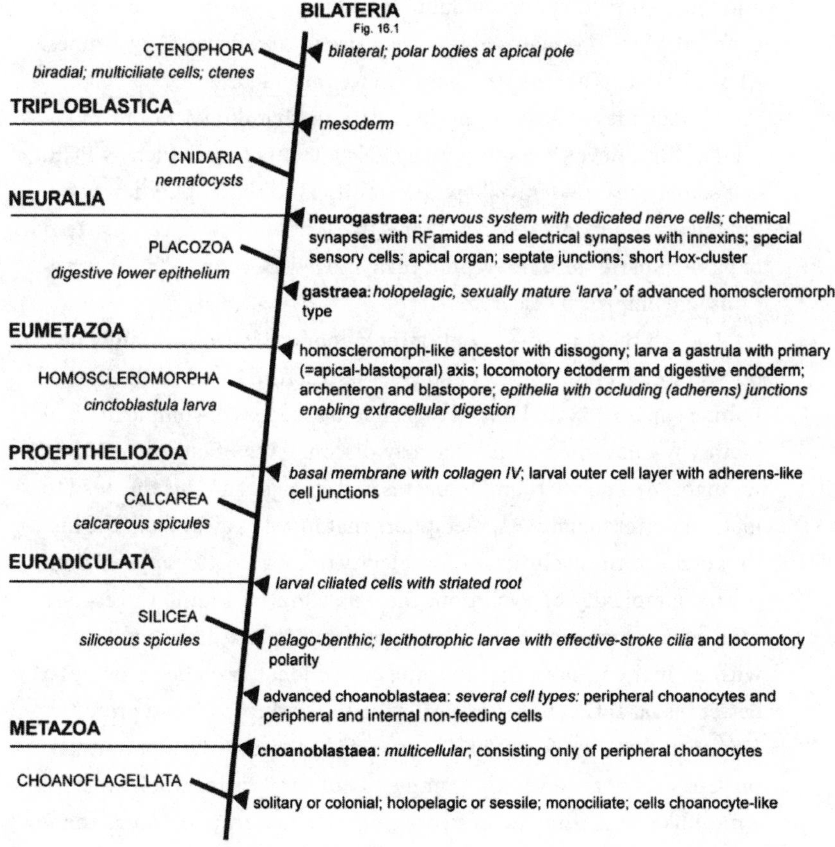

b

From the Evolution of Complexity to the Complexity of Evolution

This chapter has tried to sketch how old metaphors continue to shape views of animal evolution. It is not, however, an attack on metaphors as such. Metaphors are an important tool for conceptualizing our worlds, yet at the same time, they always betray us because they cannot capture the complexity of the world. Metaphors are not static; they change because the world always exceeds them. While our vision and forms of inquiry are undoubtedly shaped by the metaphors of our times, the world remains able to surprise us and disrupt our frames. The empirical is always stranger than we imagined.

Although it is harder to notice relations that do not fit our metaphors, it has never been impossible. New technologies, such as PCR, can sometimes suddenly allow the empirical to flash up in new ways; but curiosity and long-term noticing, like that of Darwin, can also lead to paradigmatic shifts. Metaphors can, indeed, be changed by observations and empirical research.

Though there is no way out of metaphors, there are certainly better and worse ones. Ladders and trees—structured around the idea of human superiority and linked to problematic ideas of complexity and hierarchy—have proved particularly discouraging of curiosity. When we manage to notice them, tunicates and sponges show us the need to look for better metaphors, metaphors that move us away from the history of life as the evolution of complexity toward a better appreciation of the complexity of evolution. Biologists are beginning to respond with new metaphors. Some have suggested that a branched "coral" with its many intersecting and multidirectional branches provides a better visualization of the evolution of animal life, past and present.[18] However, for some forms of life, such as bacteria with their horizontal gene exchanges, even the complexity of corals is insufficient. Here a meshlike or rhizomatic network might be better.[19] So, too, is there likely still a place for tree metaphors, read nonteleologically, as a way to visualize particular aspects of evolutionary thinking. Indeed, any one image may always be insufficient. We need many metaphors—rhizomes and corals as well as bushy, gnarly trees—to capture the complexity of evolution.

ANDREAS HEJNOL explores the evolution of biodiversity through a focus on unloved and often unnoticed creatures, including bryozoans, brachiopods, xenacoelomorphs, priapulids, and ribbon worms. Internationally known for his work in taxonomy, he is the leader of the Comparative Developmental Biology of Animals group at the Sars International Center for Marine Molecular Biology at the University of Bergen, Norway.

Notes

1. Carl Linnaeus, *Systema Naturæ* (Stockholm, 1759).
2. Michel Foucault, *Les mots et les choses: Une archéologie des sciences humaines* (Paris: Gallimard, 1966).
3. Charles Darwin, *The Descent of Man and Selection in Relation to Sex* (London: Murray, 1871).
4. Jean-Baptiste Lamarck, *Philosophie Zoologique* (Paris: Museum d'Histoire Naturelle [Jardin des Plantes], 1809).
5. Ibid.
6. Emanuele Rigato and Alessandro Minelli, "The Great Chain of Being Is Still Here," *Evolution: Education and Outreach* 6 (2013): 18, doi:10.1186/1936-6434-6-18.
7. Geoffrey K. Chambers, "Understanding Complexity: Are We Making Progress?," *Biology and Philosophy* 30 (2015): 747–56, doi:10.1007/s10539-014-9468-5.
8. Ibid.
9. Willi Hennig, *Phylogenetic Systematics* (Urbana: University of Illinois Press, 1966).
10. Olivier C. Rieppel, "Fundamentals of Comparative Biology" (Basel: Birkhäuser, 1988).
11. Casey W. Dunn, Gonzalo Giribet, Gregory D. Edgecombe, and Andreas Hejnol, "Animal Phylogeny and Its Evolutionary Implications," *Annual Reviews in Ecology, Evolution, and Systematics* 45 (2014): 371–95, doi:10.1146/annurev-ecolsys-120213-091627.
12. Frédéric Delsuc, H. Brinkmann, D. Chourrout, and H. Philippe, "Tunicates and Not Cephalochordates Are the Closest Living Relatives of Vertebrates," *Nature* 439, no. 7079 (2006): 965–68, doi:10.1038/nature04336.
13. Thomas Stach, "Chordate Phylogeny and Evolution: A Not So Simple Three-Taxon Problem," *Journal of Zoology* 276 (2008): 117–41, doi:10.1111/j.1469-7998.2008.00497.x.
14. Casey W. Dunn, Andreas Hejnol, Dave Q. Matus, Kevin Pang, William E.

Browne, Stephen. A. Smith, Elaine Seaver, et al., "Broad Phylogenomic Sampling Improves Resolution of the Animal Tree of Life," *Nature* 452, no. 7188 (2008): 745–49, doi:10.1038/nature06614.

15. Andreas Hejnol, "Evolutionary Biology: Excitation over Jelly Nerves," *Nature* 510, no. 7503 (2014): 38–39, doi:10.1038/nature13340.
16. Claus Nielsen, *Animal Evolution* (Oxford: Oxford University Press, 2012).
17. Casey W. Dunn, Sally P. Leys, and Steve H. Haddock, "The Hidden Biology of Sponges and Ctenophores," *Trends in Ecology and Evolution* 30, no. 5 (2015): 282–91, doi:10.1016/j.tree.2015.03.003.
18. Horst Bredekamp, *Darwins Korallen: Frühe Evolutionsmodelle und die Tradition der Naturgeschichte* (Berlin: Klaus Wagenbach, 2005).
19. Florian Maderspacher, "The Tree View of Life," *Current Biology* 25, no. 19 (2015): R845–47, doi:10.1016/j.cub.2015.09.031.

6

NO SMALL MATTER

MUSHROOM CLOUDS, ECOLOGIES OF NOTHINGNESS, AND STRANGE TOPOLOGIES OF SPACETIMEMATTERING

Karen Barad

MATTER FELL FROM GRACE DURING THE TWENTIETH CENTURY. What was once labeled "inanimate" became mortal. Very soon after that, it was murdered, exploded at its core, torn to shreds, blown to smithereens. The smallest of smallest bits, the heart of the atom, was broken apart with a violence that made the earth and the heavens quake. In an instant, in a flash of light brighter than a thousand suns, the distance between heaven and earth was obliterated—not merely imaginatively crossed by Newton's natural theophilosophy but physically crossed out by a mushroom cloud reaching into the stratosphere. "I am become death, the destroyer of worlds."[1]

"Space Is Never Empty and Time Is Never Even": Haunted Landscapes and Spacetimemattering

The clocks were arrested in Hiroshima on August 6, 1945, at 8:15 A.M.[2] Time stopped. The internal mechanisms melted. Time was frozen with a heat as intense as the sun. Time died in a flash. Its demise captured

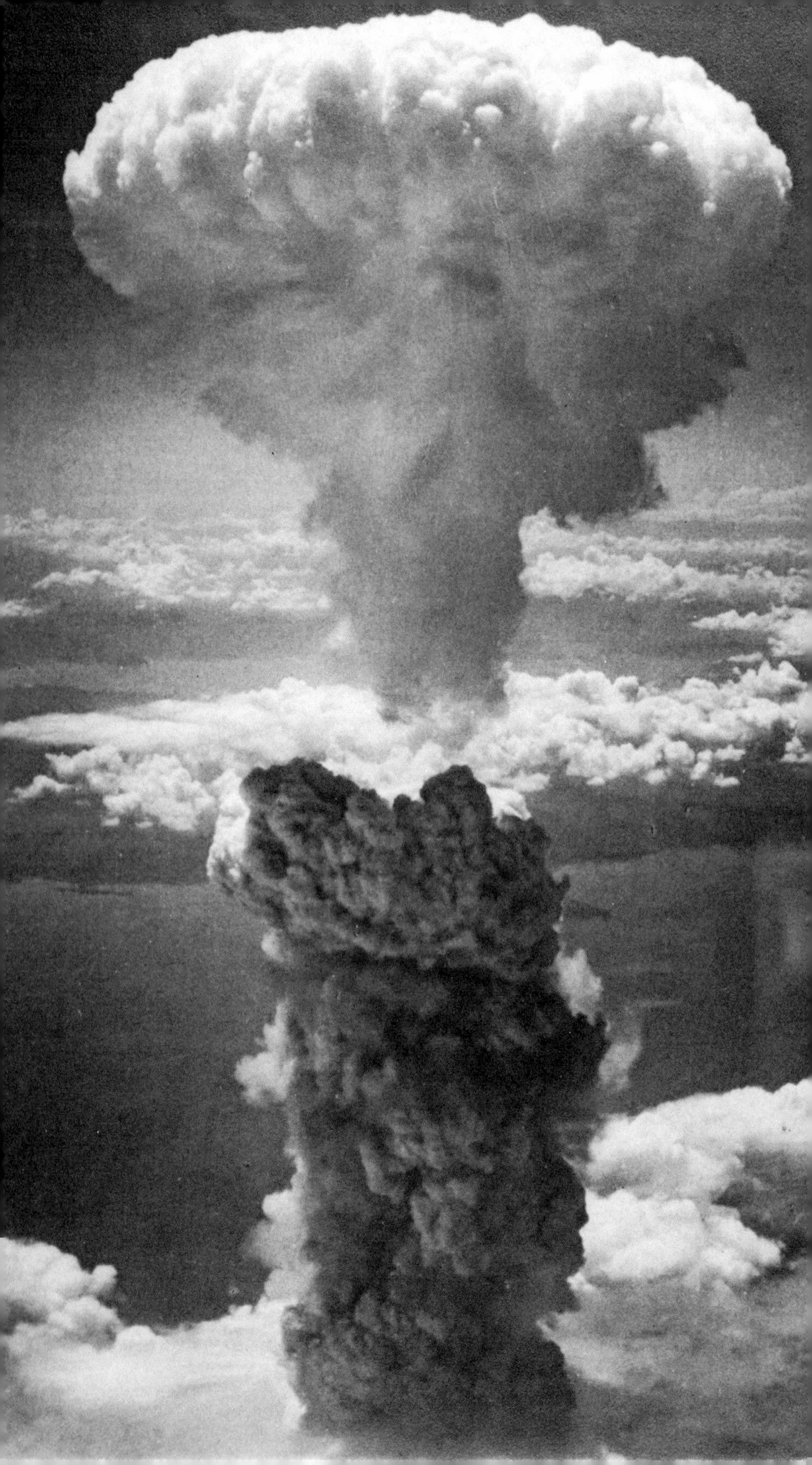

Figure G6.1. Mushroom cloud, Nagasaki, August 9, 1945. U.S. Army Air Force, http://www.getty.edu/education/teachers/classroom_resources/curricula/headlines/ib_nagasaki.html.

in shadows: silhouettes of people, animals, plants, and objects, its last moment of existence emblazoned on walls. Never before was it possible to kill time, not like this. Atomic clocks. Doomsday clocks. The hands of time indeterminately positioned as creeping toward the midnight of human and more-than-human existence, moving and no longer moving.[3]

Frozen clock faces have become emblematic of nuclear destruction. In the Hiroshima Peace Memorial Museum, there are watches in showcases and larger-than-life pictures of clocks, all forever fixed at 8:15. And in front of the Hiroshima National Peace Memorial Hall is a sculpture of a clock: the face of time forever set at 8:15. Hiroshima–8:15 makes up one space-time point. But there is also Trinity Test site–5:30, Nagasaki–11:02, and Fukushima–2:46 . . . reverberations of time being stopped, coming in waves.

What happens to time when nuclear forces are harnessed and unleashed? Is (space)time(mattering) not shattered, torn, broken into dis/connected pieces? Vaporized, dispersed, made particulate, whisked away on the breeze? Condensed into raindrops that fall to the ground making puddles on streets and quenching the soil's thirst? Sent up in smoke as the water

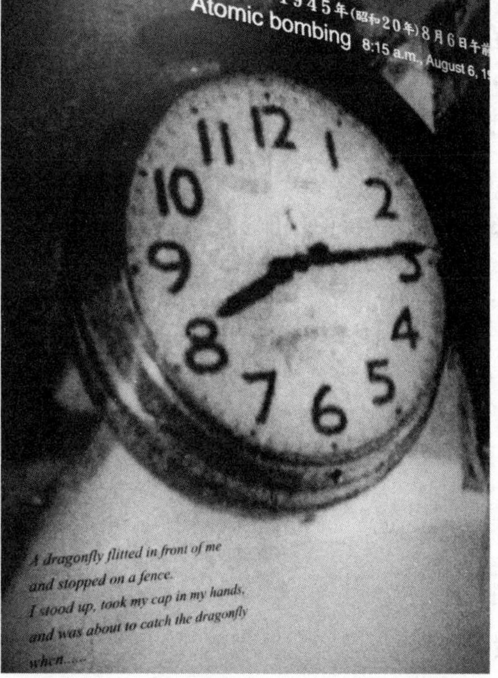

Figure G6.2. Photograph of a clock frozen at 8:15 A.M., the time the atom bomb was dropped, as displayed at the Hiroshima Peace Memorial Museum. Photograph by Amanda Patterson.

invades the electrical systems of the nuclear plant? Leaked into the groundwater as the nuclear core melts?

Time ~~is/had been~~ crossed out. Time drawn out like taffy, twisted like hot metal, cooled, hardened, and splintered. In the twentieth century, time is given a finite lifetime, a decay time. Moments live and die. Time, like space, is subject to diffraction, splitting, dispersal, entanglement.[4] Each moment is a multiplicity within a given singularity. Time will never be the same—at least for the time-being.[5]

The body clocks of *hibakusha* have been synchronized to the bomb; their cells tick with the rhythms of radioactivity.[6] *Hibakusha* have been robbed of their pasts, their homes, their city, their health, and their future. Used by the postwar Atomic Bomb Casualty Commission to set standards for radiation exposure, *hibakusha* bodies have been reduced by U.S. officials to a yardstick for measuring bodily tolerance limits, although they have proven inadequate; in actuality, these figures are more a measure of entangled capitalist-imperialist-racist forms of violence and exploitation.[7] Radioactivity worldwide is now synchronized to the bombings in Japan. The entire world is entangled with the explosion, a global dispersal of the bombing. The bomb continues to go off everywhere (but not everywhere equally). The whole world is downwind.

When it comes to nuclear landscapes, loss may not be visibly discernible, but it is not intangible. There are losses emblazoned on walls: shadows of what once was become eternal . . . the flash so bright, the heat so hot, nearly every surface becomes a photographic plate. Loss is not absence but a marked presence, or rather a marking that troubles the divide between absence and presence.

A speaking silence permeates the spacetimemattering-scape, like the forgotten movements of the wind that trouble any static notion of landscape.[8] "In Hiroshima and Nagasaki, it is . . . 'deathly silence' or 'ghastly stillness' that many survivors speak of as one of the most chilling parts of the atrocity. . . . The silence atomic bomb writers refer to might also represent the massive, instant or rapid deaths that overwhelmed the survivors and descended over the time and space of the burned Hiroshima and Nagasaki. These deaths are in every *hibakusha*'s body and mind and the awareness of this negative space is a part of their living."[9] Histories, geopolitics, nothingness, written inside each cell.

These devastated landtimescapes are surely haunted, but not

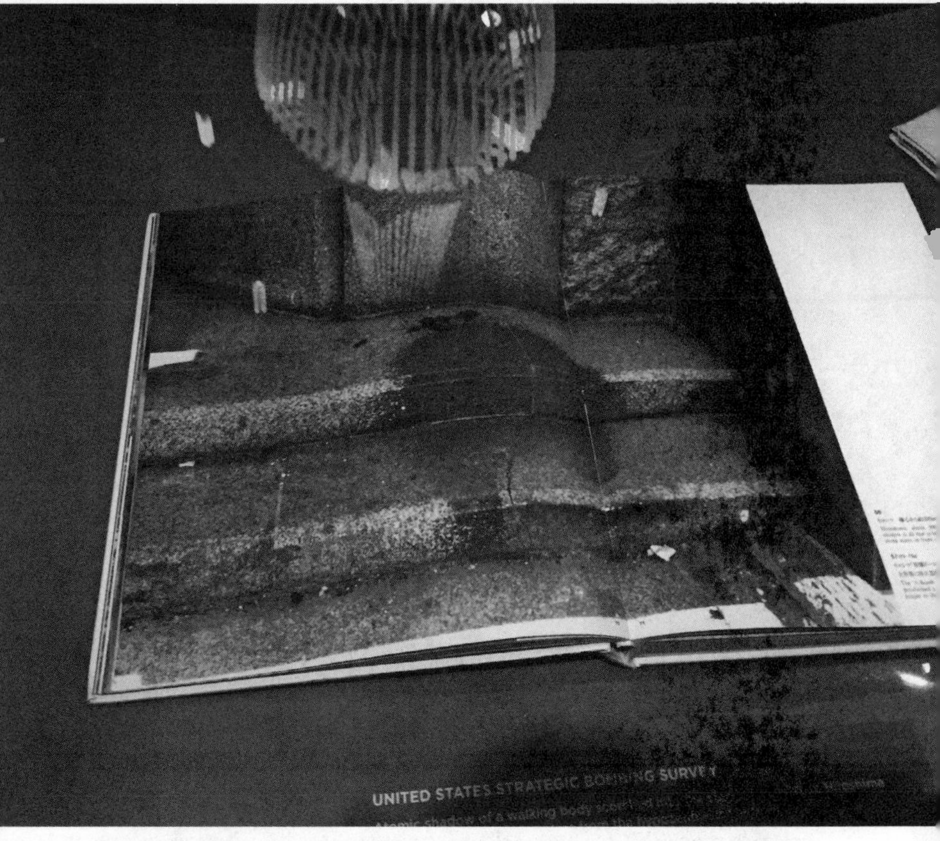

Figure G6.3. Superposition of a green-glowing "radioactive mushroom cloud" and a photograph of a photograph of a photograph; the latter is an "atomic shadow" imprinted on a staircase, the trace of a victim photographed by the atomic bomb. The green mushroom cloud is the inverted reflection of an antique chandelier frame refitted with uranium glass beads and UV bulbs, created by Japanese–Australian artists Ken and Julia Yonetani in the wake of the Fukushima disaster. *Camera Atomica,* Toronto, 2015, curated by John O'Brien, in association with Sophie Hackett. Photograph by Karen Barad.

merely in the sense that memories of the dead, of past events, particularly violent ones, linger there. Hauntings are not immaterial. They are an ineliminable feature of existing material conditions. In the aftermath of Fukushima, for example, nuclear time, decay time, dead time, atomic clock time, doomsday clock time—a superposition of dispersed times cut together-apart—are swirling around with the radioactivity in the Pacific Ocean. Time itself is nationalized, racialized, out of joint. The entanglements of nuclear energy and nuclear weapons, nationalism, racism, global exchange and lack of exchange of information and energy resources, water systems, earthquakes, plate tectonics, geopolitics, criticality (in atomic and political senses), and more are part of this ongoing material history, which is embedded in the question of Japan's future reliance on nuclear energy, where time itself is left open to decay.

No Small Matter

> The trace elements of Los Alamos weapons science now saturate the biosphere creating an atomic signature found in people, plants, animals, soils, and waterways. The Manhattan Project not only unlocked the power of the atom, creating new industries and military machines, it also inaugurated a subtle but total transformation of the biosphere . . . we need to examine the effects of the bomb not only at the level of the nation-state but also at the level of the local ecosystem, the organism, and ultimately, the cell. . . . America's nuclear project has witnessed the transformation of human "nature" at the level of both biology and culture . . . turning the earth into a vast laboratory of nuclear effects that maintain an unpredictable claim on a deep future.
>
> —*Joseph Masco*

What is the scale of nuclear forces? When the splitting of an atom, or more precisely, its tiny nucleus (a mere 10^{-15} meters in size, or one hundred thousand times smaller than the atom), destroys cities and remakes the geopolitical field on a global scale, how can anything like an ontological commitment to a line in the sand between "micro" and "macro" continue to hold sway on our political imaginaries? When incalculable devastation entailing uncountable deaths is unleashed in the harnessing of a force that is so fantastically limited in extension its job is merely to hold together the nucleus of an atom—a tiny fraction of a speck, a mere wisp of existence, a near-nothingness—then surely anything like some preordained geometrical notion of scale must have long ago been blown to smithereens, and the tracing of entanglements might well be a better analytical choice than a nested notion of scale (neighborhood $\subset$ city $\subset$ state $\subset$ nation) with each larger region presuming to encompass the other, like Russian dolls. That is, when a force extending a mere millionth of a billionth of a meter in length reaches global proportions, destroys cities in a flash, and reconfigures geopolitical alliances, energy resources, security regimes, and other large-scale features of the planet, this should explode the geometrical notion of nested scales that remains operative when the question arises as to what quantum physics has to do with the "macro-world."[10]

What is time's measure?[11] In a flash of an eye (a blinding flash, a flash that has been known to melt eyes), the explosion is over but forever lives on. Bodies near ground zero "become molecular"—nay, particulate, vaporized—while *hibakusha,* in the immediate vicinity and downwind, ingest radioactive isotopes that indefinitely rework body molecules all the while manufacturing future cancers, little time bombs waiting to go off.[12] The bomb that went off, the cascading energy of the nuclei that were split, lives on and continues its explosion in the interior and exterior of bodies. The temporality of radiation exposure is not one of immediacy; rather, it reworks this notion, which must then include generations before and to come. Radioactivity inhabits time-beings and resynchronizes and reconfigures temporalities/spacetimematterings. Radioactive decay elongates, disperses, and exponentially frays time's coherence. Time is unstable, continually leaking away from itself.

What is the scale of matter, of spacetimemattering? We are stardust—made of atoms cooked inside of stars through a process of nuclear fusion—all the while, a brilliance "brighter than a thousand suns" resides inside the nucleus of an atom.[13] The largest of space-time-matter measures, the smallest of space-time-matter measures: *each contained inside the other, each threaded through the other. A strange topology.*

If, as anthropologist Joseph Masco urges, "we need to examine the effects of the bomb, not only at the level of the nation-state, but also at the level of the local ecosystem, the organism, and ultimately, the cell"—indeed, at every scale, from the grandest cosmological and astrophysical scales to the scale of subatomic particles—what analytical tools might we use to understand not merely the entanglements of phenomena across scales but *the very iterative (re)constituting and sedimenting of specific configurations of space, time, and matter, or rather, spacetimematter(ing), and the (iterative re)making of scale itself*?[14]

In contrast to the universality and homogeneity of space, time, and matter in Newtonian physics, quantum physics—in particular, quantum field theory—holds that every bit of matter, every moment in time, every location (if we for a moment forget that we cannot speak of these conceptions separately), is diffractively/differentially constituted; or more precisely, *every "morsel" of spacetimemattering is diffractively/differentially constituted, each "bit" specifically entangled inside all others. Spacetimemattering is not a set of static points, coordinates of a void, but a dynamism of* differancing.

If quantum physics provides useful conceptual tools for understanding the politics of matter and the matter of politics in the "nuclear age," it is not because quantum physics gets right what Newtonian physics got wrong (apropos some modernist notion of progress), nor is it a question of providing a politically neutral frame, nor even one with inherently better, necessarily more radical politics (as if this could be determined in advance); rather, it is because it is fully implicated in, and arguably marked by, the making of the atomic bomb. Quantum physics and the atom bomb are directly and deeply entangled.[15] Indeed, the point is that *the theory and the bomb materially inhabit and help constitute each other.* Indeed, just like the ontology (hauntology) it suggests, quantum physics, too, and any measure or analytical tool it might provide, is shot through with the political (i.e., by virtue of the very nature of its ontology). This is *why* it might be helpful.

Quantum Physics and Inseparability

The Newtonian nature of space, time, matter, and the void are undone by quantum physics. In particular, it undoes the Newtonian assumptions of separability and metaphysical individualism. There are no self-contained individual entities running in the void. Matter is not some givenness that preexists its interactions. Matter is always already caught up with nothingness. Bodies, space, time, and the void are not ontologically separate matters.

Contra Newton's conception that there are external forces acting on inert matter, according to quantum physics, matter is understood to be agential, and forces, in their multiplicity, are "immanent in the sphere in which they operate."[16] Interpretations of quantum physics differ vastly, but at least on my *agential realist* reading—the result of a diffractive reading of quantum physics through contemporary theories of social justice—ontology is not a matter of givenness. On the contrary, agential realism understands the very nature of matter and the very matter of nature as (iteratively re-)constituted through a(n iteratively reconfigured) multiplicity of force relations. This by no means invalidates notions such as entity, force, time, scale, boundary, resistance, or resilience. Rather, the point is to get underneath as it were, to have an analytical frame for asking a set of *prior questions* about how to understand such notions in their materiality and to ask how such things come into existence, rather than starting the analysis after they've arrived on the scene. Entities, space, and time exist only within and through their specific *intra-actions*; this is not to say that they are mere transient and fleeting effects but rather that they are specifically materially constituted.[17] On this account, *quantum entanglements* are not mere contrivances, nor simply the outcome of highly technical laboratory practices, but rather the core of this relational agential ontology.[18] Entanglements are not the mere intertwinings of, or linkages between, separate events or entities or simply forms of interdependence that point to the interconnectedness of all being as one. *Entanglements are the ontological inseparability of intra-acting agencies. Naturalcultural phenomena are entanglements, the sedimented effects of a dynamics of iterative intra-activity,* where *intra-actions* (contra interactions do not assume separability, but rather) *cut together-apart, differentiate-entangle. Phenomena are specific material relations of the ongoing differentiating of the world, where "material" needs to be understood as iteratively constituted through force relations. Phenomena are not located in space and time; rather, phenomena are material entanglements enfolded and threaded through the spacetimemattering of the universe. Entanglements are the iterative intra-active (re)configurings and enfoldings of spacetimemattering.*

Not only does *spacetimemattering* mark the inseparability of space, time, and matter in a radical troubling of Newtonian metaphysics and epistemology; also, the verb form is intended to signal the dynamic (re) making of spacetimemattering through the iterative sedimentation of intra-actions in their specificity. *Spacetimemattering* is a dynamic ongoing reconfiguring of a field of relationalities among "moments," "places," and "things" (in their inseparability), where *scale is iteratively (re)made in intra-action.*

Quantum field theory, the attempt to make a coherent theory that combines the insights of quantum mechanics with those of the theory of relativity and field theory, takes this Newtonian undoing even further, producing a radical rethinking of the nature of being (time-being) and nothingness.

Quantum Field Theory: How Big Is an Infinitesimal?

In the mid-twentieth century, the nature of change changed. The design of new physics, a *quantum field theory* (QFT), from 1927 to 1947 and beyond, indeed, to the present day, had a profound impact on the nature of temporality and change, to say nothing of the technoscientific dimensions of World War II, and vice versa. In fact, there was a striking overlap between the physicists who worked on the Manhattan Project and those who worked on the development of QFT. During this time, being and time were together remade. No longer an independent parameter relentlessly marching forward in the future, time is no longer continuous or one. *Time is diffracted, imploded/exploded in on itself: each moment made up of a superposition, a combination, of all moments* (differently weighted and combined in their specific material entanglements).[19] And directly linked to this indeterminacy of time is a shift in the nature of being and nothingness.

According to QFT, matter is not eternal. Birth and death are not merely the inevitable fate of the animate world; so-called inanimate beings are also mortal. Particles have finite lifetimes, decay times. "Particles can be born and particles can die," explains one physicist. In fact, "it is a matter of birth, life, and death that requires the development of a new subject in physics, that of quantum field theory. . . . Quantum field theory (QFT) is a response to the ephemeral nature of life."[20]

Particles are born out of the void, go through transformations, die, return to the void, and are reborn, all the while being inseparable from the wild material imaginings of the void. At the core is the *indeterminacy of time-being* (i.e., the reciprocally related indeterminacy of time and energy/matter/being), and this gives rise to the fact that nothingness is not empty but flush with *virtuality—the indeterminate play of the non/presence of non/existence.* As a result of a primary ontological indeterminacy, the void is not nothing but a desiring orientation toward being/becoming, flush with yearning and innumerable imaginings of what could be/might yet have been. Nothingness is a material presence, belying any insinuation of emptiness—an indeterminate movement, an intra-active self-touching of no-thingness. It is a matter of time-being itself that is at stake in the play of indeterminacy, where an event is not one and living and dying are inseparable (though not the same): the dying is within the living within the dying.[21]

The fact that the void is not empty, mere lack or absence, matters. The question of absence is as political as that of presence. When has absence ever been an absolute givenness? Is it not always a question of what is seen, acknowledged, and counted as present, and for whom? The void—a much-valued colonialist apparatus, a crafty and insidious imaginary, a way of offering justification for claims of ownership in the "discovery" of "virgin" territory—the notion that "untended," "uncultivated," "uncivilized" spaces are *empty* rather than plentiful, has been a well-worn tool used in the service of colonialism, racism, capitalism, militarism, imperialism, nationalism, and scientism.

The void is not the background against which something appears but an active, constitutive part of every "thing." As such, *even the smallest bits of matter—for example, electrons, infinitesimal point particles with no dimensions, no structure—are haunted by, indeed, constituted by, the indeterminate wanderings of an infinity of possible configurings of spacetimemattering in their specificity. Entire worlds inside each point, each specifically configured. Infinitesimals are infinite. Matter is spectral, haunted by all im/possible wanderings, an infinite multiplicity of histories present/absent in the indeterminacy of time-being.*

Much has been made lately of hauntings. Some understand hauntings as one or another form of subjective human experience—the epistemological revivification of the past, a recollection through which the past makes itself subjectively present. But according to QFT, *hauntings are lively indeterminacies of time-being, materially constitutive of matter itself*—indeed, of everything and nothing. Hauntings, then, are not mere rememberings of a past (assumed to be) left behind (in actuality) but rather the dynamism of ontological indeterminacy of time-being/being-time in its materiality. *And injustices need not await some future remedy, because "now" is always already thick with possibilities disruptive of mere presence. Each moment is thickly threaded through with all other moments, each a holographic condensation of specific diffraction patterns created by a plethora of virtual wanderings, alternative histories of what is/might yet be/have been. Re-membering,* then, is not merely subjective, a fleeting flash of a past event in the inner workings of an individual human brain; rather, it is a constitutive part of the field of spacetimemattering.[22]

Mushroom Clouds

Almost immediately after the bombs hit Hiroshima and Nagasaki, images of mushroom clouds spread across the front pages of newspapers and magazines more quickly than any fungal spores were ever carried by the wind. In 1960, *Time* magazine described the cloud as cauliflower-shaped. But perhaps it's not accidental that the toadstool association won out. Mushrooms are the ultimate *pharmakon*—traditionally associated with life and death, food and poison—matter with occult virtues. Historian of physics Spencer Weart traces the association of nuclear energy with cosmic creation back to medieval alchemy's contention that it is in fact possible to unlock the secrets of a cosmic life force.[23] At the crux of the matter is a fascination with and anxiety over the alchemical notion of transmutation, hitched to fantasies of human control over life and death. Harkening back to the birth of modern science in the crucible of these desires, alchemy itself was transformed into a mature, rational, and mechanistic philosophy. Transformation is arguably a particularly charged kind of change, and an entire history of modern sciences, persecution of witches, and more is packed into it.

But when the atom bomb exploded, the mushroom cloud that connected heaven and earth was a condensation of matters that were more than merely symbolic. "When Hiroshima was destroyed by an atomic bomb in 1945, it is said, the first living thing to emerge from the blasted landscape was a matsutake mushroom."[24] Whether or not this story is historically accurate, it has been verified that mushrooms were found not only in the immediate area surrounding the Chernobyl nuclear reactor after the accident in 1986 but also growing *inside* the reactor, on its walls.[25] What is it about mushrooms and radioactivity?

Radioactivity, like the mushroom, is a *pharmakon,* and they are entangled with each other in specific ways. It turns out that fungi that contain melanin actually *thrive* on radioactive emissions, using beta and gamma rays (ionizing radiation) as a digestive aid of sorts. "Radiotrophic" fungi use melanin ionization rather than photosynthesis for "food." The highly concentrated radioactive materials in mushrooms make their way up the food chain and thereby hitch a ride outside of contaminated areas (for example, via wild boar gobbling up mushrooms around Chernobyl). At the same time, some

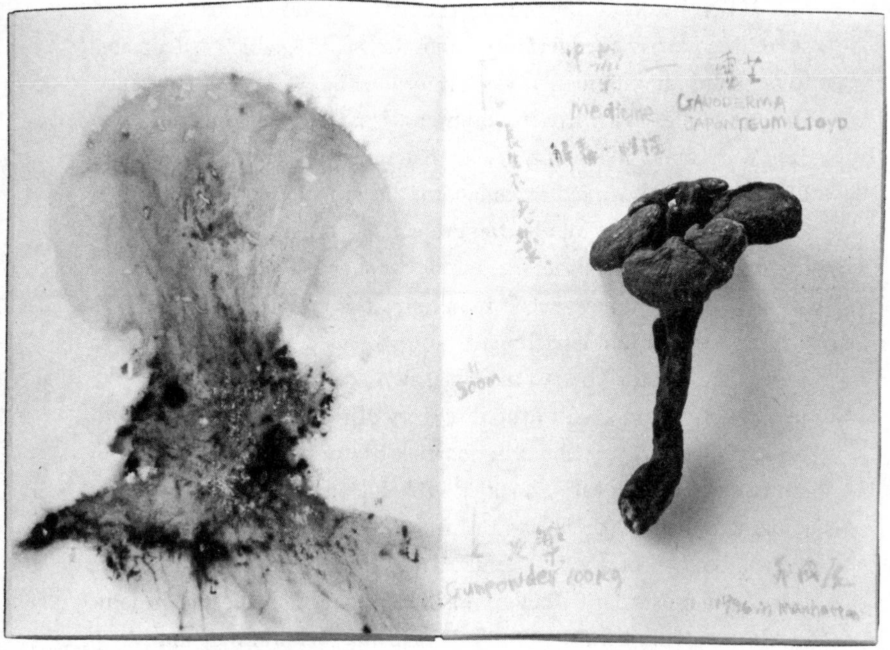

Figure G6.4. Cai Guo-Qiang, *Mushroom Cloud and Mushroom*, 1995–96. Gunpowder, ink, and dried lingzhi mushroom on paper, 20-page folding album. Drawing for "The Century with Mushroom Clouds: Project for the 20th Century." Photograph by Hiro Ihara. Courtesy Cai Studio and Solomon R. Guggenheim Museum, New York. Partial gift of the artist and purchased with funds contributed by the International Director's Council and Executive Committee Members.

scientists see the possibility of using mushrooms for purposes of nuclear remediation (a project being considered to help remediate areas of Fukushima).[26]

There are also specific material connections between mushrooms and clouds. "Clear-cut the land and you . . . clear-cut the sky!" warn scientists who have been following a hunch that mushrooms may be responsible for the cloud cover over the Amazon.[27] A sample of rain forest air was brought to the Lawrence Berkeley National Laboratory in California and placed in the facility's synchrotron, where X-rays of varying energies were fired at the collected specks.[28] The analysis supports a linkage between the potassium released from fungi living in the rain forest and cloud formation. The Lawrence Berkeley National

Lab (LBNL) (owned by the University of California, as is its offshoot, the Lawrence Livermore Laboratory [LLL], which is dedicated to weapons development) was the brainchild of Ernest Lawrence. (Edward Teller of hydrogen bomb and Strategic Defense Initiative/Star Wars fame collaborated with Lawrence in helping to establish LLL.) Lawrence worked in collaboration with Robert Oppenheimer (lead scientist on the Manhattan Project) and other Berkeley theoretical physicists to strengthen the bonds between particle physicists and military research. Lawrence's radiation lab is credited with finding a process for uranium enrichment. After the war, Lawrence sought to strengthen military ties to his lab. Lawrence and his "Rad Lab" are also directly connected to the discovery of the process of photosynthesis.[29] The Rad Lab, and other particle physics labs, also made miniature clouds in the lab—so-called Wilson cloud chambers—to detect ionizing radiation (the kind mushrooms feed on). Wilson clouds—so named because of a visual similarity to the detector—are a specific feature of mushroom clouds. Mushroom clouds inside mushrooms inside clouds . . . infinities of infinities inside each infinitesimal.

These are just some small bits of a very entangled story. The earth and the heavens are connected in oh so many ways.

It is not simply that there is a homology between terrestrial and atmospheric mushrooms; rather, there is *an uncanny material topology: each inhabiting the other.* When a nuclear bomb explodes, each radioactive bit of matter is an imploded diffraction pattern of spacetimemattering, a mushrooming of specific entangled possible histories. Tiny radioactive particles raining down from the sky, radiotrophic mushrooms thriving in nuclear contaminated areas, wildlife thriving around the reactors in Chernobyl, mushrooms living inside reactors, Fukushima multiple reactor meltdowns, remediation by mushrooms, radioactive particles traveling ocean currents to North America, *Bulletin of Atomic Scientists* Doomsday Clock resynchronized to include the climate crisis, cloud formation over the Amazon rain forest sustaining millions of species tagged to microscopic bits of potassium emitted from mushrooms, tested at LBNL, connected to LLL, the University of California, cyclotrons, particle accelerators, uranium enrichment, particle physics, nuclear physics, quantum field theory, the Manhattan Project, bomb testing and uranium mining on native lands, racism, internment camps, war, militarism, imperialism, fascism, capitalism, industry expansion, GI Bill, housing boom, boom in racial disparity in

U.S. housing markets, nuclear annihilation of cities with single bombs, security state, nuclear power plants, power plant failure and uninhabitable areas, mushrooms growing. And much more.

All these material-discursive phenomena are constituted through each other, each in specifically entangled ways. This is not a mere matter of things being connected across scales. Rather, matter itself in its very materiality is differentially constituted as an implosion/explosion: a superposition of all possible histories constituting each bit. The very stuff of the world is a matter of politics. Matter is not only political all the way up and all the way down; it has all matters of matter inside it. Planetary geopolitics inside a morsel—a strange topology, an implosion/explosion of no small matter.

How big is infinitesimal? What is the measure of nothingness?

KAREN BARAD is an expert in theoretical particle physics and quantum field theory who has opened up transdisciplinary theories and practices. She is professor of feminist studies, philosophy, and history of consciousness at the University of California, Santa Cruz, and author of *Meeting the Universe Halfway: Quantum Physics and the Entanglement of Matter and Meaning*, a groundbreaking book that rethinks central questions in philosophy and social theory through rigorous attention to the physical sciences.

Notes

1. The remarkable discovery that matter is not eternal is an insight that comes from quantum field theory, discussed later. This line from the Bhagavad Gita was famously quoted by physicist J. Robert Oppenheimer (his translation from Sanskrit) in the wake of the first atomic bomb explosion.
2. Section subtitle inspired by the artists Eiko and Koma; *Time Is Not Even: Space Is Not Empty* is a retrospective exhibition, first shown at Wesleyan University's Zilkha Gallery in 2009.
3. The Doomsday Clock of the *Bulletin of Atomic Scientists,* introduced in 1947, represents scientists' estimation of the proximity to global catastrophe.
4. See Karen Barad, "Troubling Time/s, Ecologies of Nothingness: On the Im/Possibilities of Living and Dying in the Void," in *Eco-Deconstruction: Derrida and Environmental Philosophy,* ed. Matthias Fritsch, Phil

Lyons, and David C. Wood (Bronx, N.Y.: Fordham University Press, 2017). Variations on this theme were explored for keynotes for the Society for Literature Science and the Arts and the Society for Phenomenology and Existential Philosophy and invited lectures at Columbia, Johns Hopkins, Rice, and Emory universities during 2014–2015.

5. With a wink to Ruth Ozeki regarding her novel *A Tale for the Time Being* (New York: Penguin, 2013).
6. *Hibakusha* is Japanese for "explosion-affected people."
7. See, e.g., Gayle Green, "Science with a Skew: The Nuclear Power Industry after Chernobyl and Fukushima," *The Asia-Pacific Journal* 10(1), no. 3 (2011), http://apjjf.org/2012/10/1/Gayle-Greene/3672/article.html.
8. "The word landscape is a bit strange, it gives the impression of static. But for the Japanese, landscape is not static image. Landscape is *hu-kai*, wind-scape. So, landscape is always moving, trembling by wind." Vdrome video *The Radiant* (The Otolith Group, 2012).
9. Eiko Otake, "Artistic Representation of Human Experiences of the Atomic Bombings" (MA thesis, New York University, 2007), 29–30.
10. There is no scale at which the laws of physics change from quantum physics to Newtonian physics, from "microworld" to "macroworld"; to the best of our knowledge, quantum physics holds at all scales (Newtonian physics is simply a good approximation in some cases). Furthermore, entanglements call into question the geometrical notions of scale and proximity; topology with its focus on issues of connectivity and boundary becomes a more apt analytical tool. It's not that scale doesn't matter; the point is that it isn't simply given and that what appears far apart might actually be as close as the object in question, indeed, it may be an inseparable part of it. Scale—indeed, spacetime(mattering) itself—is iteratively intra-actively materialized and (re)configured. See the concept of *spacetimemattering* in Karen Barad, *Meeting the Universe Halfway: Quantum Physics and the Entanglement of Matter and Meeting* (Durham, N.C.: Duke University Press, 2007).
11. For a more detailed discussion of how the notion of time is reworked according to agential realism, see especially Barad, "Troubling Time/s."
12. "In this flash the body becomes molecules." Bill Johnston and Eiko Otake, video of class at Wesleyan University, 4:58, https://vimeo.com/10387574.
13. "Brighter than a thousand suns"—this phrase is from a verse from the Bhagavad Gita said to have been quoted by J. Robert Oppenheimer at the Trinity nuclear test.
14. Joseph Masco, "Mutant Ecologies: Radioactive Life in Post-Cold War New Mexico," *Cultural Anthropology* 19, no. 4 (2004): 520–22.
15. Karen Barad, "Infinity, Nothingness, and Justice-to-Come" (unpublished manuscript).

16. Michel Foucault, *History of Sexuality, Volume I: An Introduction*, trans. Robert Hurley (New York: Vintage Books, 1980), 92. The force relations I have in mind here are not merely social, and their effects (in a reworking of cause-effect) are not limited to the formation of human subjects. On the contrary, it is only through the workings of apparatuses of bodily production that forces come to be distinguished as social, biological, geological, political, etc., as the case may be. For more details on agential realism, see Barad, *Meeting the Universe Halfway.*
17. *Intra-action* does not simply refer to a change from presumed separability to nonseparability (a relational ontology) but entails a radically different understanding of causality and an ontoepistemological framework with implications for thinking about questions of justice. See Barad, *Meeting the Universe Halfway.*
18. The technical notions of "wavefunction collapse" and "decoherence" do not constitute fundamental objections here. See Barad, *Meeting the Universe Halfway,* chapter 7.
19. For more on time diffraction, see Barad, "Troubling Time/s, Ecologies of Nothingness," and Barad, "Infinity, Nothingness, and Justice-to-Come."
20. Anthony Zee, *Quantum Field Theory in a Nutshell,* 2nd ed. (Princeton, N.J.: Princeton University Press, 2010), 3–4.
21. See Karen Barad, *What Is the Measure of Nothingness? Infinity, Virtuality, Justice* (dOCUMENTA [13]: 100 Notes—100 Thoughts, book 099, 2012), and Barad, "On Touching—The Inhuman That Therefore I Am," *differences: A Journal of Feminist Cultural Studies* 23, no. 3 (2012): 206–23, for a detailed presentation of my agential realist understanding of quantum field theory (which also entails a further elaboration of agential realism).
22. This brings to mind the work of *(hibak) denshosha*—memory keepers in Japan (and also elsewhere). The question of how to keep memories alive as survivors die is poignant and urgent.
23. Spencer Weart, *The Rise of Nuclear Fear* (Cambridge, Mass.: Harvard University Press, 2012).
24. Anna Lowenhaupt Tsing, *The Mushroom at the End of the World: On the Possibility of Life in Capitalist Ruins* (Princeton, N.J.: Princeton University Press, 2015), 3.
25. Albert Einstein College of Medicine, "'Radiation-Eating' Fungi Finding Could Trigger Recalculation of Earth's Energy Balance and Help Feed Astronauts," *ScienceDaily,* May 27, 2007, http://www.sciencedaily.com/releases/2007/05/070522210932.htm; "Radiation-Eating Fungi. They Kill Trees and They Kill People," *Bobby1's Blog,* November 14, 2012, http://optimalprediction.com/wp/radiation-eating-fungi-they-kill-trees-and-they-kill-people/.

26. Paul Stamets, "Using Fungi to Remediate Radiation at Fukushima," *Permaculture* (blog), November 26, 2015, http://www.permaculture.co.uk/articles/using-fungi-remediate-radiation-fukushima.
27. Veronique Greenwood, "How Fungi May Create the Amazon's Clouds," *Discover Magazine*, September 5, 2012, http://discovermagazine.com/2012/sep/04-mushroom-cloud.
28. Ibid.
29. The mechanism of photosynthesis was decoded by chemist Melvin Calvin in the Rad Lab using radioactive carbon on the encouragement of Lawrence.

7

HAUNTED GEOLOGIES

SPIRITS, STONES, AND THE NECROPOLITICS OF THE ANTHROPOCENE

Nils Bubandt

IF YOU TRAVEL SOUTH BY CAR FROM SURABAYA, Indonesia's second-largest city located on the sweltering north coast of Java, toward the cool mountain town of Malang, you will, after about twenty-five kilometers, come upon a vast elevated landscape of mud. From the road, your view will be blocked by the massive dikes that have been erected to stem the mud. But if you climb to the top of the twenty-meter containment walls, you will see a barren and flat landscape, stretching eastward toward the horizon and the shallow coastline of the Madura Strait. The smell of petrol, emanating from the petroliferous components in the mud, is mixed with a faint but distinctive smell of rotten eggs.[1] If you scan the horizon, you will see, off in the distance to the right, the source of the smell: a plume of steam, pulsating at irregular intervals, at the center of the mudflat. The plume, consisting of methane mixed with hydrogen sulfide and sulfur dioxide, comes from the main vent, one of five initial eruption sites of the mud volcano that since May 2006 has spewed out enormous amounts of gas, water, and mud. Eleven meters of sludge over an area of seven square kilometers now bury what used to be twelve villages. The mud has displaced 39,700 people and caused damage estimated to be 30 trillion rupiah (US$2.2 billion).[2] As mud has built up within the containment walls, underground cave-ins have occurred. In one such event, in November 2006, the natural gas pipeline to Surabaya ruptured and exploded, killing thirteen.[3] Initially projected to continue for centuries, recent estimates suggest that the mud volcano may self-plug within the next two decades.[4] By this time, however, the weight of the 140 million cubic meters of mud from the volcano will likely have caused the affected area to subside at least ninety-five meters.[5] Except for some species of coliform and thermophile bacteria, nothing today lives in the sulfuric and heavy metal–rich mud.

The mud volcano is not only the largest of its kind in the world. It is also by far the most controversial, and it has experts, residents, politicians, activists, and industrialists split into two camps. Some people claim the mudflows were triggered by an earthquake, whereas others maintain that it was caused by oil drilling. As such, the mud volcano is a tragic and dystopic, but also illuminating, illustration of the Anthropocene, conventionally described as the geological period in which human activity exceeds the forces of nature.[6] What better example of such excess than if humans caused a disastrous volcanic eruption? The Indonesian mud volcano, however, also highlights another, equally important and unsettling feature of the Anthropocene, namely, the increasing impossibility of distinguishing human from nonhuman forces, the *anthropos* from the *geos.* For the volcano is simultaneously a national disaster at the center of a continuing political scandal and

Figure G7.1. The mudflats in East Java. The plume of the main vent is visible on the right. Photograph by Nils Bubandt.

the object of an ongoing geological dispute about whether its eruption was, in fact, anthropogenic or natural. An undecidability haunts the mud volcano. Is it an effect of human industry or of tectonic forces? Is it an effect of life or of nonlife? It is the undecidability of the mud volcano, and of the Anthropocene, that is the subject of this chapter. For undecidability, I will argue, is simultaneously the signature characteristic, the curse, and the promise of our current moment.

Spirits and the Necropolitics of the Anthropocene

The different names of the mud volcano index its undecidability. Some people refer to the volcano as Lumpur Lapindo ("Lapindo Mud"), after the oil company, PT Lapindo Brantas Incorporation, that drilled for petroleum nearby and that may have caused its eruption.[7] Lumpur Lapindo names an anthropogenic and political event tainted by industrial greed, mismanagement, and corruption. A second, equally used name for the mud volcano is Lumpur Sidoarjo ("Sidoarjo Mud"), after the sprawling nearby district capital. If the first name highlights the human agency and political liabilities of the mud disaster, Lumpur Sidoarjo is a geographical name used to denote where a "natural disaster" happened to strike. But "natural" figures awkwardly here, for not only is this name as political as the previous one but the name also points directly to the world of spirits. The name Lumpur Sidoarjo is thus frequently shortened into the portmanteau "Lusi." Pronounced like the common woman's name "Lucy," it names an earth being with

a will of its own, and victims of the mud disaster speak its name with as much deference as political acerbity. Lusi is, in other words, equal parts spirit name and political critique. In a play on the name of the Malaysian capital, Kuala Lumpur (literally "Muddy Estuary"), people in East Java, for instance, joke that Lusi is their Kualat Lumpur, literally their "Cursed Mud." The cursed mud is clearly the inverse image of the shining cosmopolitan dream conveyed by the Malaysian capital: a stinking, muddy, and failed modern. But more than metaphors are at play here, for "curses" *(kualat)* belong to a very real realm of the world in Indonesia, namely, that of occult forces and spirits *(batin)*. *Kualat* is a calamity you bring upon yourself by behaving inappropriately. The curse of the mud volcano is in that sense a response to a moral transgression of some sort, an explanation that encapsulates condemnation of industrial mismanagement, critique of political corruption, and anxieties about cosmological punishment.

Like Fukushima, Bhopal, Chernobyl, and other contemporary disasters where the forces of nature and human politics act to exacerbate each other, Lusi is the name for a monstrous geography haunted by the natural as well as the unnatural.[8] But more so than other recent disasters with an anthropogenic component, the ontologies of the natural and the unnatural (whether human or spiritual kinds of "unnature") coalesce in Lusi's muddy ferment. On the mudflats of East Java, the realms of geology, politics, industry, divination, lawsuits, spiritual revenge, and corruption are inextricably entangled in each other. Indeed, the inability to separate one from the other—nature from politics, geothermal activity from industrial activity, human corruption from spiritual revenge—is a constituent part of the volcano's necropolitics.

Achille Mbembe, in his founding article on the term, defined necropolitics as the subjugation of human life to the powers of death in the context of war, terror, and weapons of mass destruction.[9] But in a time of global warming, ocean acidification, and mass extinction, I suggest necropolitics has come to cover a much broader and much more stochastic politics of life and death. Humans, animals, plants, fungi, and bacteria now live and die under conditions that may have been critically shaped by human activity but that are also increasingly outside of human control. I use the notion of a necropolitics of the Anthropocene to indicate the life-and-death effects—intended as well as unintended—of this kind of ruination and extinction. Nature may increasingly be human-made, but humans have not only lost control of this nature making and unmaking; we have increasingly lost the ability to tell the difference between our own world and the natural worlds we make and destroy. As each new scientific discovery reveals more details of the complex interplay between human worlds and natural worlds, we are also increasingly faced with our inability to tell these worlds apart. In the Anthropocene, necropolitics operates under the sign of metaphysical indeterminacy rather than certainty, unintended consequences rather than control.

As it so happens, spirits exist under the same conditions of uncertainty and possibility. Spirits are never just "there." They are both manifest and disembodied, present and absent. Spirits thrive, as a result, in conditions of doubt rather than belief.[10] "I do not believe in ghosts, but . . ." is, after all, the conventional start to accounts of experiences with ghosts and spirits. How striking, in light of this, that the Anthropocene is so clearly associated with spirits. Take the figure of Gaia, the self-regulating, sympoetic superorganism of earth's biosphere named after a Greek goddess by climate scientist James Lovelock and biologist Lyn Margulis.[11] Or take Donna Haraway's *chthulus,* those earthly "myriad intra-active entities-in-assemblages" that inhabit the Anthropocene.[12] These tentacular beings of the earth are so named by Haraway to point to the overlap between indigenous spirits—from Pachamama, the Incan goddess of fertility, to A'akuluujjusi, the mother creator of all animals in Inuit thought—and new biological insights into the evolutionary co-becoming of life (see the chapters by Haraway and Gilbert in *Monsters*). In the Anthropocene, both climate science and biology seem to bring spirits, once thought to have been killed by secular thought, back to life. This chapter argues that geology in similar ways brings spirits into being. By paying attention to the spirits that abound in and around the Lusi mud volcano, we may yet learn to see, and live with, the ghosts that abound in the necropolitical landscapes of the Anthropocene.

The Story of a Mud Volcano—in Two Parts

The Lusi mud volcano is a geological event with two histories. The volcano is essentially a two-part story. Part one, the "unnatural history" of the volcano, as it were, begins in 2006. In the early hours of the morning on May 29, the mud volcano erupted, shortly after the oil company PT Lapindo Brantas Incorporation had begun exploratory drilling for gas in a late Miocene stratum twenty-eight hundred meters below the surface of the earth. Studies later showed that the drilling operation fractured a high-pressure aquifer, allowing the rapid influx of formation fluids and gases into the open drill hole, which, contrary to standard practice, lacked a protective steel casing over a one kilometer stretch.[13] The pressurized gas, liquids, and mud, mainly from the Pleistocene period, that filled the drill hole eventually caused a series of blowouts 150 meters away from the drilling rig Banjar Panji-1. It is from these blowout vents that an unstoppable flow of mud has since been burying the surrounding landscape.

This first part of the story is a very recognizable Anthropocene. It is an Anthropocene in which human activity (in this case, an oil company) exacerbates the forces of nature, causing what has been called "the first humanly-made volcanic eruption in planetary history."[14] The eruption is in this account an anthropogenic perversion of the historical relationship between mud and oil. For oil and gas exploration has always been intimately tied to mud volcanoes. In the nineteenth century, early prospectors discovered that mud volcanism was related to active underground petroleum systems, and they began to use mud volcanoes as indicators for potential oil fields.[15] Now, it seemed, this historical relationship had been turned on its head. Instead of mud volcanoes being the sign of a petroleum system ready for extraction, fossil carbon extraction was itself perversely creating mud volcanoes.

Until recently, the notion that humans could have an impact on the tectonics of the earth itself was laughable. Not so anymore. Industrially produced tectonics have become an increasingly recognized anthropogenic risk, since fracking and high-pressure injection wells have been shown to generate an increase in earthquake activity in the United States.[16] But Lusi was the first case in which conventional drilling was established as the cause of geothermal activity. As a result, the East Javanese mud volcano quickly became the global icon for a carbon-craving world gone awry, testimony to an oil industry that characterized by mismanagement, greed, and corruption was inadvertently tampering with the very makeup of the earth itself. Indeed, the link between cooperate greed and tectonic disaster seemed embarrassingly obvious. Lapindo Brantas, the oil company linked to the blowout, was controlled by the Bakrie Group, a consortium in which Aburizal Bakrie, then Indonesia's richest man, was a key stakeholder. The fact that Aburizal Bakrie was also minister for people's welfare (Menkosra) in the coalition government of President Susilo Bambang Yudhoyono, but refused to visit the site or assume any cooperate responsibility for the damages, made the disaster a striking example of the hypocrisy of capitalist carbon extraction. An "unnatural disaster," the magazine *National Geographic* called it.[17]

But there is also a second part to the story of Lusi. This part—its "natural history"—paradoxically only adds to Lusi's uncanny nature. This second part of the story begins in the early morning of May 27, 2006, roughly forty-eight hours before the eruption of Lusi, when a massive earthquake measuring 6.3 on the Richter scale shook the ground near Yogyakarta, killing 5,749 people and injuring more than thirty-eight thousand. Mud volcanoes, a global phenomenon, are often caused by seismic activity, and some studies therefore argued that the near-synchronicity of the earthquake and the volcanic eruption indicated that the two were causally linked.[18] The island of Java is traversed by a geological depression along its east-west axis.[19] The depression, which has been filled with sediments over the last 23 million years, closely follows a subduction zone between the Indian Oceanic and the Eurasian continental plates. This has created one of the world's most seismically active areas but also the conditions for the presence of rich underground petroleum resources that have been exploited for a hundred years. The same region is home to numerous naturally occurring mud volcanoes associated with the presence of petroleum. The Sidoarjo mud volcano, in this scenario, was a "natural" event in an unstable geothermal region: the earthquake near Yogyakarta caused a so-called strike-slip movement of the Watukosek fault, one of many tectonic fault lines in this area, triggering the eruption of the mud volcano some 250 kilometers away.[20]

This second account of the eruption was favored by a number of Indonesian experts, including the senior drilling advisors of the oil company, who published their findings in the same prestigious journals as their opponents.[21] It was also supported by a number of the Indonesian government's own geological experts, allegedly under the influence of the investors behind the oil company, who were eager to establish the mud volcano as a "natural disaster" in a bid to evade legal responsibility.[22] Opponents of this explanation countered that synchronicity in itself failed to establish a causal link between the earthquake and the mud volcano and that the geographical distance between the two events exceeded other known cases in which mud volcanism had been triggered by seismic activity.[23] The pedigree of those who sought to establish that the mud volcano was a "natural fact" suggested that they were "merchants of doubt," scientists paid by industry to deny the truth of global warming, the harmful effects of smoking, or, in this case, the anthropogenic origins of volcanism.[24] Indeed, the theory that Lusi was caused by tectonic activity was haunted by accusations of poor science and corrupt politics.

As a result, the truth of the anthropogenic origin of Lusi seemed secure. Until recently, that is, when independent, computer-based studies showed that the curved underground rock formation in the area could have focused the seismic waves of the Yogyakarta earthquake to produce enough seismic stress on the fault line to trigger the eruption, even if it was more than two hundred kilometers away.[25] This analysis seriously challenges those who maintain that the volcano was triggered by drilling and lent credibility from an unexpected and unbiased source to the industrial merchants of doubt. In its wake, uncertainty rules more than ever.[26] As one geologist concludes, "we may never know what the final trigger was, whether it would have happened anyway, nor even if an early trigger averted a greater disaster, had pressures continued to build up."[27] When it comes to Lusi, geology, the science behind the concept of the Anthropocene, is haunted by undecidability. This epistemological undecidability is coupled with high political stakes: the oil company wants the eruption to be a natural disaster to escape liability, while victims want it to be an industrial disaster to enforce payment of compensation. The question essentially is whether Lusi is a political event with a geothermal afterlife or a geothermal event with a political afterlife. At the moment, it is both.[28] I suggest calling this a "spectral moment," a time of undecidability but also a time of spirits and ghosts.

The Hope of Stones

On quiet afternoons, you are likely to see people scour the Lusi mudflats. Once in a while, they will stoop to pick up a pebble and inspect it closely before either dropping it again or putting it in a fanny pack around their waist. People say the stones are just trinkets, children's marbles. And yet, they keep collecting them, carefully polishing them smooth with sandpaper in an evident labor of love and dedication to bring out the proper contours, the shades of meaning that hide within. Some stones come to assume the shape of a dolphin, others a human face. Yet others have organic filaments or veins of quartz that take the shape of a dragon or a lion or the eye of a dead king. Mas Hadi is one of the people collecting stones. He is also a descendant of royalty from the mythical Majapahit empire and a diviner *(waskitó)* with "spirit eyes" that see into the otherworld *(mata batin)*. Having spirit eyes also enables Mas Hadi to distinguish ordinary stones from unique treasures, a skill in high demand on the mudflats.

One day I sat with Mas Hadi when a *tukang ojek,* a driver of a motorbike taxi, dropped by with an object he had found on the mudflat. It looked like a fossilized shark tooth. The concavity of the labial face, the lack of serration along the edges, and the robustness of the root suggested it was from a mako shark (L. *Isurus oxyrinchus*), probably one who lived and died around 2 million years ago to became part of the Pleistocene stratum from where most of the volcanic mud originates.[29] To Mas Hadi, however, it was something else. For along the center of the crown of the tooth was the outline of something, a pointed object. "This," he declared after some pause, "is special. Do you see the *kris* inside? It comes from the Majapahit empire." What the *ojek* driver had inadvertently stumbled upon was a double *kris,* a

Figure G7.2. Shark tooth containing a magical double *kris.* Photograph by Nils Bubandt.

dagger associated with royalty and a powerful magical object. "Take it, and keep it safe," Mas Hadi instructed the man, closing the man's palm with his own around the object.

Objects such as this tooth-dagger become personal treasures, part of one's arsenal of heirlooms and amulets. Such objects are kept hidden or are fitted and worn in rings for protection. In particular, they are seen to have a magical capacity *(khasiat)* to confer upon the finder good fortune *(rezeki)*. The objects are precious because they are full of life, fossilized proof of a spirit life that thrives in an otherwise toxic landscape. The stones are said to come from Lusi's main vent. A giant spirit snake, it is said, dwells within it. Or more accurately, the vent itself is a snake, the guardian spirit *(penunggunya)* of the volcano, from whose belly deep underground the stones and objects emerge. The treasures are essentially bezoars from a spirit snake. Traded from Asia to Europe for medicinal purposes since the Renaissance, snake stones *(mustika ular)* and other bezoars are regarded as powerful magical antidotes throughout Indonesia.[30] The petrified objects that are spewed from the giant snake spirit at the center of the mud volcano are like such bezoars, objects that hold potentially great spiritual power *(kesaktian)*.

Searching for spirit shapes in the stones on the mudflats is one among a panoply of means through which you may acquire good fortune through magical means in Java. Good fortune or *rezeki* can take many forms, not all of which are financial. *Rezeki* may be to acquire a spouse, a child, a job, recognition, success, or money. It is about leading the good life, about being fulfilled, calm, and happy. *Rezeki* is about destiny. It is existential and social rather than merely financial. The pursuit of *rezeki* by magical means is called *pesugihan* and can be acquired from a veritable multispecies salon of spirits. On the sacred mountain of Kawi, you may, for instance, acquire good fortune if you observe a leaf of the *dewandaru* tree (L. *Eugenia uniflora*) fall to the ground. Or you may take up relations with the black boar spirit called *babi ngepet*. The spirit will enable you to turn into a black boar that inconspicuously can steal from other people. Trees, boars, and snakes may all provide good fortune, but they also require compensation, a reciprocal payment *(tumbalan)*, to be pacified. The black boar is said to ask for a human baby in return for its riches. Mas Hadi claimed that the children's graves vandalized in a Sidoarjo cemetery in 2012 had been emptied of human remains by people in search of such compensation gifts.

Spiritual anxiety has been the constant companion of dreams of good fortune at Lusi since its eruption in 2006. While engineers from global mining consultancies have dropped hundreds of cement balls and iron chains into the vent in an unsuccessful attempt to plug it, people throughout Indonesia worry that human heads—procured by government headhunters—have also been surreptitiously thrown into the vent as reciprocal payment *(tumbalan)* to its spirit guardian.[31] For like most volcanoes in Indonesia, the Lusi mud volcano is a spiritual as well as a geothermal entity—a vengeful and angry geospirit.[32] Calming the spirit of such a massive disaster requires magic of a special kind. A hundred mystics from all over Java thus participated in a locally organized event in 2006 that attempted to use "paranormal" powers, reciprocal payments, and soothing ritual offerings *(sesajen)* in an effort to stop the mudflow.

The Politics of Mud

The search of good fortune through magical means is one of many strategies that people pursue to offset the disastrous effects of the mudflow on their lives. Mas Hadi is fifty-one years old and makes a meager living as a self-appointed parking guard at a local school. He spends his afternoons on the mudflats, and when he does not divine stones, he is one of a few dozen men, all displaced by the mud, who sell pirated DVDs about Lusi's eruption and offer paid motorbike rides to the mainly Indonesian disaster tourists who come to see the mudflats. Mas Hadi is married for the second time. His first wife died, "of stress" as he puts it, when social obligations forced the family to share with distant relatives the money they had received as the first installment of a compensation payment from the oil company. The money gone, the family had been unable to build a new house, and Mas Hadi's wife had died of grief.

Mas Hadi's story is a common one. The victims' struggle to receive compensation for their lost livelihoods has been long and frustrated. In response to the mudflow, a presidential decree from 2007 (Perpres 14/2007) divided the disaster area in two. The decree required the Lapindo oil company to pay 3.8 trillion rupiah (US$338 million) in compensation to people who used to live inside the so-called affected area map. Meanwhile, the state agreed to pay almost twice as much (6 trillion rupiah, or US$534 million) from the state budget to villagers living outside of the "affected area." The decision was widely considered part of a politically brokered deal between the government of Susilo Bambang Yudhoyono (SBY) and its coalition partner, Golkar. Aburizal Bakrie was thus not only co-owner of Lapindo but also a key figure of Golkar.[33] The suspicion was that SBY protected the Bakrie conglomerate from full liability, asking the Bakrie Group to pay only a tenth of the overall estimated cost of the disaster in exchange for Golkar's support for SBY's shaky government.[34] Deals such as these are standard in Indonesian politics and the basis for widespread accusations of corruption.[35]

Despite the generous political deal, Lapindo sought through a variety of political, legal, and strong-arm tactics to defer payment of the government-ordered compensation to the victims. The company set up a subsidiary, PT Minarak Lapindo Jaya, to handle the compensation, but locals feel that the company's main purpose has been to infiltrate the victims' protest groups and divide them internally by paying full compensation to the most vocal victims in return for political loyalty. For the people looking for stones on the mudflats, their informal motorcycle taxi association, which takes tourists around the site, doubles as a political organization. It is the only remaining victims' group, so they say, that has resisted company payoffs.

Other stakeholders, including the police and courts, have been less stalwart. In 2009, the regional police in East Java gave up its criminal investigation against Lapindo, a decision that was widely suspected of being made under pressure and influenced by oil company bribes.[36] The Constitutional Court in 2014 upheld the 2007 decree allowing the new parliament, led by President Joko Widodo, to put pressure on Lapindo to pay the remaining 781 billion rupiah (US$65 million) that the company still owes to the victims.[37] A victory for democracy, one might claim, but the court's decision maintains the injustice of the initial decree in which the government essentially exonerated the oil company in exchange for political support—Indonesian "politics-as-usual" *(politik seperti biasa),* as one of the victims told me indignantly in a text message.

A Multiplicity of Ghosts

Deprived of adequate compensation, the victims now make a living and seek good fortune on top of the toxic mud that covers what used to be their villages. In their struggle for compensation, mud has become

Figure G7.3. Indonesian disaster tourists pose in front of a papier-mâché replica of Aburizal Bakrie. Dressed in the yellow jacket of the Golkar Party, of which he has been a longtime member and, since 2009, chairman, Aburizal Bakrie is popularly held responsible for the mudflow. Photograph by Nils Bubandt.

a frequent symbol of political protest, and demonstrators regularly smear their bodies in mud as a sign of protest against a cynical oil company and a corrupt government. But mud is not just a symbol of political corruption; it is also an index of it. The mud at the vent will boil more violently, it is said, when government bureaucrats come to visit. The higher the position and moral liability of the official, the more violently the mud will boil.[38]

Mud is cosmopolitical: at once a political symbol and a cosmological agent. The political agency of mud is deeply entangled with the world of spirits. The popular narrative that the eruption of Lusi was the result of spiritual revenge from a murdered labor activist highlights this cosmopolitical agency.

The district of Sidoarjo is a densely populated area of East Java, and the abundance of cheap labor has for decades attracted numerous companies, foreign and domestic. East Java has also always been a political hot spot, and it has a long history of labor disputes as well. One of the twenty-five factories that now lie buried under the mud is PT Catur Putra Surya (CPS), a manufacturer of wristwatches made infamous for being the employer of labor activist Marsinah, who was kidnapped, raped, and killed by unknown assailants in 1993. Although the murder was never solved, it was likely ordered by a New Order network of military, government, and employer representatives to silence labor protesters.[39] However, Marsinah's murder galvanized the Indonesian labor movement during the 1990s, and Marsinah herself posthumously became a national celebrity.[40] Mas Agus, one of the *ojek* drivers and stone prospectors on the mudflats, told me that the mudflow was Marsinah's curse against her murderers. Indeed, Mas Agus claimed that the Chinese owner of the watch company went insane after the mud drowned his factory. In the Lusi mud, environmental

Figure G7.4. A protester smears mud on the logo of the Bakrie Group at its Jakartan head office. Photograph by Sapariah Saturi Harsono.

disaster, political protest, and the curses of spirits are remolded. The power of geothermal mud to speak through spirits to an unjust political world is legendary; its power is, as the victims put it, "strange but true" *(aneh tapi nyata).* The 2012 movie *Hantu Lumpur Lapindo* (The ghost of the Lapindo mud) exploits this idea. An example of *film mistik,* a popular movie genre that combines soft eroticism with horror stories featuring the many varieties of spirits and ghosts in the Indonesian mystical universe, *Hantu Lumpur Lapindo* is the story of a striptease dancer who is murdered by a gang of organ thieves after they have removed her heart. The gang dumps her body in the Lapindo mud, but the ghost rises, smeared in mud, to haunt the gang and kill its members one by one. In the movie, mud is the spiritual index of vengeance against capitalist murk, personal greed, and social betrayal.

From Necropolitics to Symbiopolitics

Lusi's muddy landscape is haunted. Her "cursed mud" *(kualat lumpur)* is the mark of a necropolis, and people see in it an explicit contrast to the metropolis of Kuala Lumpur, a betrayal of people's dream of modernity. In this ruined landscape, destroyed by a heady mix of greedy industry, corrupt politics, tectonic forces, and chthonic spirits, body politics fuse with geopolitics: protesters smear their bodies in mud, while a murdered labor unionist turns into a muddy avenging ghost; an employer goes mad when his factory drowns in mud; the government employs headhunters whose prize heads are used to plug what the cement balls of international engineers were unable to stop; a snake guardian in a geothermal vent offers gifts of good fortune, while the mud itself is strangely alive and seems to be able to tell corrupt politicians from those who are honest.

The strange life of stones and mud speaks to a spectral moment in Indonesia in which geology is political, politics is corrupt, and corruption is haunted by spirits. But the life of mud and stone is also the sign of a spectrality that characterizes the Anthropocene more generally. The Anthropocene, after all, invites us to imagine a world in which an alien geologist from the future detects in the strata of the ground evidence of the presence of humans long after we have gone extinct.[41] This science fiction–like character of the concept of Anthropocene opens up to a retrospective reading of the current moment, a "paleontology of the present" in which humans themselves have become geological sediments or ghosts.[42] In the Anthropocene, life is already geologic. In this geological ghost vision, the present proceeds from the future, because the possibility of co-species survival depends crucially on what we humans are going to do now, in the midst of an increasingly given fate of ruination and extinction.

Mas Hadi and the other people looking for fossil spirits in a haunted landscape are in that sense not unlike contemporary geologists. Take Jan Zalasiewics, the geologist who, in his book *The Planet in a Pebble,* discovers in a single pebble the ingredients for all life on earth.[43] Zalasiewics is not any geologist; he is chair of the Anthropocene Working Group of the International Commission on Stratigraphy, the organization in charge of deciding whether to accept "Anthropocene" as the scientific name for our time. When he is not busy with this work, Zalasiewics looks at stones. And for him, too, every pebble is full of ghosts.[44] Like fossil fuel, the building blocks of every pebble are constituted—in addition to minerals—by a complex of amorphous organic matter, traces of the ancient and strange biology trapped within: acritarchs, chitinozoans, graptolites. Zalasiewics, like Mas Hadi, is interested in the ghostly contours of life in stones not merely because they are telltale remnants of a past but because stones allow him to dream of a different future at the brink of disaster, a future in which livelihood and good fortune do not come at the expense of devastation and death. Geology here performs the job of *pesugihan,* the magical pursuit of good fortune, in a ruined landscape. In the necropolitics of the Anthropocene, geology is as entangled with politics as it is with ghosts. In the same movement that the Anthropocene is being established as a geological fact, geology itself is becoming political. As geologists have to choose which of the many radioactive, industrial, and chemical signals in the ground, in the sea, and in the air define our time, it is also becoming increasingly apparent that geology can no longer perform what Donna Haraway has famously called the "god trick" of remaining outside of what it studies. Like the other sciences of the Anthropocene, geology's diagnosis of our time mires it in contemporary politics.

The question is what kind of politics to choose: the ghostly necropolitics of the current moment or a politics informed by other kinds of spirits. It seems to me that the spectrality of the Anthropocene is full of ghosts of many kinds. There are the old ghosts of carbon-based industry, the specters of corrupt politics, and the God-tricks of conventional science, to be sure. But there are also the spirits of a different, emergent kind of politics, a symbiopolitics. The Anthropocene presents us with the geological possibility that humans are the graptolites of the future, fossil colonial animals that are engineering our own demise. This shift in perspective is important. If modernity dreamed of the future, the Anthropocene dreams of the present as seen from the future, a perspectival shift that makes our necropolitics apparent to ourselves in the starkest of lights. As the deep time of geology becomes the political history of the present, this also changes what geology, along with other sciences, can and should be.[45] We are all inhabitants of the same mudscape, the same geological sludge, as it were. Anthropocene landscapes of death and extinction are, however, also inhabited by emergent and unexpected constellations of life, nonlife, and afterlife. Before mud becomes our only future, we need to learn from stones to notice all the forms of life and possibility that exist in the midst of death: that, as I see it, is the message and the magic of the geology of the present. It is also the message of East Javanese people's engagement with spirits, as I read it.

The spirits that reside in the stones and mud of Lusi remind us that the scientific, political, and legal inability to differentiate the *anthropos* from the *geos* has its own metaphysics. This metaphysics may be the brainchild of our current troubles and thus the product of a long history of exploitation, colonialism, and extermination. But a metaphysics that has lost the ability to distinguish the *bios* from the *geos*, the human from the nonhuman, also holds a promise. For the kind of symbiopolitics that this metaphysics makes visible offers the chance for a novel kind of collaboration between science and the politics of the otherwise, a politics that we might learn from spirits. The indigenous spirits of the Indonesian mud volcano and the secular spirits of the Anthropocene seem to me to form an awkward alliance here. For both indigenous spirits and the spirits of the new geological idea of the Anthropocene ask us to notice the magic of the forces, human and nonhuman, that shape the atmosphere, biosphere, and lithosphere. The spirits highlight how the inexorable logic of carbon-based business-as-usual that brought us into our current predicament is inherently spectral. But they offer a dissenting voice to this conjuring as well, and here is the basis for a common front between indigenous spirits and the emergent sciences of the Anthropocene, one that grows from a shared recognition of the magic of being-with, the magic of symbiopolitics.

As an anthropologist, **NILS BUBANDT** has learned to be equally at home with witches, protesters, and mud volcanoes. Co-convener of Aarhus University Research on the Anthropocene (AURA), with Anna Tsing, he is professor at Aarhus University and editor in chief of the journal *Ethnos* (with Mark Graham). His books include *The Empty Seashell: Witchcraft and Doubt on an Indonesian Island* and *Democracy, Corruption, and the Politics of Spirits in Contemporary Indonesia.*

Notes

1. Geoffrey S. Plumlee, Thomas J. Casadevall, Handoko T. Wibowo, Robert J. Rosenbauer, Craig A. Johnson, George N. Breit, Heather A. Lowers, et al., *Preliminary Analytical Results for a Mud Sample Collected from the LUSI Mud Volcano, Sidoarjo, East Java, Indonesia,* USGS Open-File Report 2008-1019 (Reston, Va.: U.S. Geological Survey, 2008).
2. Hans David Tampubolon, "Mudflow Erupting after 7 Years," *Jakarta Post,* March 5, 2013, http://www.thejakartapost.com/news/2013/03/05/mudflow-erupting-after-7-years.html.
3. Jim Schiller, Anton Lucas, and Priyambudi Sulistiyanto, "Learning from the East Java Mudflow: Disaster Politics in Indonesia," *Indonesia* 85 (April 2008): 53.
4. Jonathan Amos, "Mud Volcano to Stop 'by Decade's End,'" BBC News, December 20, 2013, http://www.bbc.com/news/science-environment-25188259.
5. Richard Davies, Simon Mathias, Richard Swarbrivk, and Mark Tingay, "Probabilistic Longevity Estimate for the LUSI Mud Volcano, East Java," *Journal of the Geological Society* 168 (2011): 517–23.
6. Will Steffen, Paul Crutzen, and John McNeill, "The Anthropocene: Are Humans Now Overwhelming the Great Forces of Nature?," *Ambio* 36, no. 8 (2007): 614–21.
7. Italicized words are in Indonesian or Javanese. Latin names are prefaced with "L."
8. For analyses of these disasters, see Theodore C. Bestor, "Disasters, Natural and Unnatural: Reflections on March 11, 2011, and Its Aftermath," *The Journal of Asian Studies* 72, no. 4 (2013): 763–82; Kate Brown, *Plutopia: Nuclear Families, Atomic Cities, and the Great Soviet and American Plutonium Disasters* (Oxford: Oxford University Press, 2013); Kim Fortun, *Advocacy after Bhopal: Environmentalism, Disaster, New Global Orders* (Chicago: University of Chicago Press, 2001); Adriana Petryna, *Life Exposed: Biological Citizens after Chernobyl* (Princeton, N.J.: Princeton University Press, 2013).
9. Achille Mbembe, "Necropolitics," *Public Culture* 15, no. 1 (2003): 11–40.

10. For an extended case study of this from Indonesia, see Nils Bubandt, *The Empty Seashell: Witchcraft and Doubt on an Indonesian Island* (Ithaca, N.Y.: Cornell University Press, 2014).
11. James Lovelock and Lynn Margulis, "Atmospheric Homeostasis by and for the Biosphere: The Gaia Hypothesis," *Tellus, Series A* 26, no. 1–2 (1974): 2–10.
12. Donna Haraway, "Anthropocene, Capitalocene, Plantatiocene, Chthulucene: Making Kin," *Environmental Humanities* 6 (2015): 160.
13. Richard Davies, Maria Brumm, Michael Manga, Rudi Rubiandini, Richard Swarbrick, and Mark Tingray, "The East Java Mud Volcano (2006 to Present): An Earthquake or Drilling Trigger?," *Earth and Planetary Science Letters* 272 (2008): 627–38.
14. Michael Northcott, "Anthropogenic Climate Change and the Truthfullness of Trees," in *Religion and Dangerous Environmental Change: Transdisciplinary Perspectives,* ed. S. Bergmann and D. Gerten (Münster: LIT, 2010), 103.
15. Guiseppe Etiope and Alexei Milkov, "A New Estimate of Global Methane Flux from Onshore and Shallow Submarine Mud Volcanoes to the Atmosphere," *Environmental Geology* 46 (2004): 1692.
16. Eric Hand, "Injection Wells Blamed in Oklahoma Earthquakes," *Science* 345, no. 6192 (2014): 13–14.
17. Andrew Marshall, "Drowning in Mud: An Unnatural Disaster Erupts with No End in Sight," *National Geographic* 213, no. 1 (2008): 58–63.
18. More than one thousand terrestrial and shallow-water mud volcanoes have been identified, and they occur in virtually every part and climatic zone of the world. Even more mud volcanoes occur in the world's oceans, and as many as one hundred thousand mud volcanoes may exist in deepwater environments. Recent estimates suggest that the annual methane release from terrestrial and shallow-water mud volcanoes is between six and nine megatons, between 3 and 4.5 percent of the total release of an estimated two hundred megatons of methane to the atmosphere from natural sources. Etiope and Milkov, "A New Estimate of Global Methane Flux." Some four hundred megatons of methane are released annually from anthropogenic sources.
19. Awang Harun Satyana and Asnidar, "Mud Diapirs and Mud Volcanoes in Depressions of Java to Madura: Origins, Natures, and Implications to Petroleum System," paper presented at the annual meeting of the Indonesian Petroleum Association, IPA08-G-139, Jakarta, May 2008.
20. A. Mazzini, A. Nermoen, M. Krotkiewski, Y. Podladchikov, S. Planke, and H. Svensen, "Strike-Slip Faulting as a Trigger Mechanism for Overpressure Release through Piercement Structure: Implications for the Lusi Mud Volcano, Indonesia," *Marine and Petroleum Geology* 26 (2009): 1751–65.
21. Nurrochmat Sawolo, Edi Sutriono, Bambang P. Istadi, and Agung B.

Darmoyo, "The LUSI Mud Volcano Triggering Controversy: Was It Caused by Drilling?," *Marine and Petroleum Geology* 26 (2009): 1766–84.

22. Jim Schiller, Anton Lucas, and Priyambudi Sulistiyanto, "Learning from the East Java Mudflow: Disaster Politics in Indonesia," *Indonesia* 85 (April 2008): 62.
23. Michael Manga, Maria Brumm, and Maxwell Rudolph, "Earthquake Triggering of Mud Volcanoes," *Marine and Petroleum Geology* 26 (2009): 1785–98.
24. Naomi Oreskes and Erik Conway, *Merchants of Doubt: How a Handful of Scientists Obscured the Truth on Issues from Tobacco Smoke to Global Warming* (New York: Bloomsbury Press, 2010).
25. M. Lupi, E. H. Saenger, F. Fuchs, and S. A. Miller, "Lusi Mud Eruption Triggered by Geometric Focussing of Seismic Waves," *Nature Geoscience* 6 (August 2013): 642–46.
26. Indeed, the assertion that the eruption was earthquake induced has recently been challenged by M. R. P. Tingay, M. L. Rudolph, M. Manga, R. J. Davies, and Chi-Yuen Wang, "Initiation of the Lusi Mudflow Disaster," *Nature Geoscience* 8, no. 7 (2015): 493–94.
27. Paul Davis, "Natural Hazards: Triggered Mud Eruption?," *Nature Geoscience* 6, no. 8 (2013): 593.
28. In 2008, the American Association of Petroleum Geologists took the unusual step of voting about the cause of the mudflow. Chaired by a Scottish geologist who was a soccer umpire in his spare time, a majority of forty-two at the Cape Town meeting agreed that Lusi was an anthropogenic phenomenon caused by the oil company. Three geologists found that the mud volcano was natural and caused by the Yogyakarta earthquake. But a significant minority of twenty-nine scientists found that the evidence was either "inconclusive" or that a combination of the two causes was to blame. See James Morgan, "Mud Eruption 'Caused by Drilling,'" BBC News, November 1, 2008, http://news.bbc.co.uk/2/hi/science/nature/7699672.stm. This inability to distinguish a natural disaster from an anthropogenic one points to a key feature of the Anthropocene: nature is losing its epistemological position as "natural fact" and increasingly becoming a contested reality—much like spirits.
29. I would like to thank Professor Gilles Cuny, expert on fossil sharks from Université Claude Bernard Lyon 1, for his help with the paleontological identification of this tooth.
30. Peter Borschberg, "The Euro-Asian Trade in Bezoar Stones (Approx. 1500–1700)," in *Artistic and Cultural Exchanges between Europe and Asia, 1400–1900: Rethinking Markets, Workshops, and Collections*, ed. M. North, 29–43 (Surrey, U.K.: Ashgate, 2010).
31. Gregory Forth, "Heads under Bridges or in Mud," *Anthropology Today* 25, no. 6 (2009): 3–6.

32. Judith Schlehe, "Cultural Politics of Natural Disasters: Discourses on Volcanic Eruptions in Indonesia," in *Culture and Changing Environment: Uncertainty, Cognition, and Risk Management in Cross-Cultural Perspective,* ed. M. Casimir, 275–99 (New York: Berghahn Books, 2008).
33. In 2009, Aburizal Bakrie was elected chairman of Golkar, which he used as a platform for his campaign to become president of Indonesia in 2014. This campaign in large part failed because of the stigma of the Lapindo disaster, which continued to make Bakrie a figure of political power, greed, and corruption.
34. Hasyim Widhiarto, "Aburizal Could Be Forced to Settle Lapindo Mudflow," *Jakarta Post,* September 30, 2014, http://www.thejakartapost.com/news/2014/09/30/aburizal-could-be-forced-settle-lapindo-mudflow.html.
35. See Nils Bubandt, *Democracy, Corruption, and the Politics of Spirits in Contemporary Indonesia* (London: Routledge, 2014).
36. Bosman Batubara, "Resistance through Memory," *Inside Indonesia* 101 (July–September 2010), http://www.insideindonesia.org/resistance-through-memory-2.
37. Widhiarto, "Aburizal Could Be Forced to Settle Lapindo Mudflow."
38. Maksum H. M. Zuber, *Titanic Made by Lapindo* (Jakarta: Lafadl Pustaka, 2009).
39. Tim Lindsey, "The Criminal State: Premanisme and the New Indonesia," in *Indonesia Today: Challenges of History,* ed. G. Lloyd and S. Smith (Singapore: Institute of Southeast Asian Studies, 2001), 287.
40. Leena Avonius, "From Marsinah to Munir: Grounding Human Rights in Indonesia," in *Human Rights in Asia: A Reassessment of the Asian Values Debate,* ed. D. Kingsbury and L. Avonius, 99–119 (Basingstoke, U.K.: Palgrave MacMillan, 2008).
41. See Oreskes and Conway, *Merchants of Doubt*; Heather Swanson, Nils Bubandt, and Anna Tsing, "Less Than One but More Than Many: Anthropocene as Science Fiction and Scholarship-in-the-Making," *Environment and Society: Advances in Research* 6 (2015): 149–66; Jan Zalasiewics, *The Earth after Us: What Legacy Will Humans Leave in the Rocks?* (Oxford: Oxford University Press, 2009).
42. W. J. T. Mitchell, *What Do Pictures Want? The Lives and Loves of Images* (Chicago: Unversity of Chicago Press, 2005), 124.
43. Jan Zalasiewics, *The Planet in a Pebble: A Journey into Earth's Deep History* (Oxford: Oxford University Press, 2012).
44. Ibid., 86.
45. Dipesh Chakrabarty, "The Climate of History: Four Theses," *Critical Inquiry* 35, no. 2 (2009): 197–222.

WHAT REMAINS

HOW SHALL WE SHELTER THE REMAINS of Holocene ecologies, that is, multispecies ecologies in the midst of human settlement and cultivation? We can't shelter anything we don't notice. This section explores how much we can learn by refining our practices of fieldwork and historical observation, human and not human.

Mathews, both a forester and an anthropologist, describes the "practices of walking, looking, and wondering" he uses to trace histories of encounter in the chestnut forests of Monti Pisani, Italy. Once carefully tended by local communities that harvested their nuts, the forests now lie in ruins. Residents moved out, and a series of plant diseases moved in. Yet, within the stumps and crumbling stone walls, stories of past and present relations remain vibrant. "If you know how to see," Mathews tells us, "you can see fascinating stories of human/animal/plant communication embedded in the forms of living and dead trees." To read such histories, Mathews shows us how to take up "an alert practice of natural history," combining arboreal sketches and ethnographic observations in a single notebook. As he reminds us, noticing the ghosts of agro-sylvo-pastoral systems is itself a political act: uncovering how human-forest relations have differed in the past allows us to imagine how they might be otherwise in the present.

Like Mathews, Anne Pringle demonstrates how humble tools—here including markers and plastic sheets—can lead us into the dynamics of anthropogenic landscapes. Pringle studies lichen growth on tombstones, a sign of death for at least some humans but a convenient substrate for at least some lichens. For more than a decade, Pringle has tracked the births, deaths, and growth of about eight hundred lichens by tracing their outlines on transparent film. What this method of long-term observation allows her to notice is remarkable: lichens may potentially be immortal, that is, capable of dying but not designed to die at a particular age. Upending our expectations of what it means to "age," lichens do not seem to be any more likely to die when they are old than when they are young. Patient, detailed fieldwork, she shows, is what brings us into such stunningly different ways of being in time. If we want to nurture landscapes in which humans and other species augment each other's survival, we will need appreciation of such discrepant multispecies rhythms. ●

8

GHOSTLY FORMS AND FOREST HISTORIES

Andrew S. Mathews

THE PINE AND CHESTNUT FORESTS OF MONTI PISANI, only five kilometers south of Lucca, in central Italy, feel very far from the tourist sights of the city center and from the industrial sprawl of paper, furniture, and shoe factories that spreads across the plain. As in many Mediterranean places, mountains and valleys are near each other, but they are in many ways different worlds.[1] These are certainly not the landscapes that most people think of when I tell them I am working in northern Tuscany, nor do many people come here. The few human visitors are mushroom pickers, hunters, and the occasional mountain biker. Although these forests are often empty of people, they are empty in a particular way; evidence of former human use is omnipresent. This is a place where people, trees, and other nonhumans have been entangled for a very long time. Traces of these past relationships are visible in the forms of trees, of areas of forest, of drystone terrace walls and of drainage systems. Through my practices of walking, looking, and wondering, I have been tracing the ghostly forms that have emerged from past encounters between people, plants, animals, and soils. These ghostly forms are traces of past cultivation, but they also provide ways of imagining and perhaps bringing into being positive environmental futures.

I find walking through these abandoned chestnut *(Castanea sativa)* and pine *(Pinus pinaster)* forests a little sad. It is the feeling of lack of care that makes these places somewhat melancholic. The forms of remaining large chestnut trees are fragments of cultivated chestnut orchards *(selve)*.[2] Large ancient stumps (stools/*ceppi*) from which multiple pole-sized chestnut stems grow (coppices/*cedui*) tell a story of peasant firewood cutters or perhaps of more recent industrial biomass cutting.[3] Ancient terracing and drainage systems are covered in a thick scrub dominated by *Ulex europaeus* or *Erica arborea* regrown after forest fires. On lower slopes, pine trees loom over the sprouts of ancient chestnut trees, testifying to the abandonment of chestnut orchards and to the stubborn resilience of chestnut trees in the face of neglect. In other places, chestnut smoking sheds recount the complex agro-sylvo-pastoral systems that formerly linked peasant chestnut growers, premodern Italian states, upland sheep and goat grazing, and lowland farmers in need of fertilizer.[4]

The forms of individual trees, the age and size structure of forests, the physical structures of terraces and buildings, are evidence of *longue durée* encounters between humans, plants, animals, fungi, bacteria, and soils. This is a story of the relations between capitalism, state formation, and plant colonization; of the lively capacity of nonhumans to escape human imaginations; and of the ways that different forms of human politics have emerged from encounters between particular humans and particular nonhumans. Over the last fifteen hundred years, peasant agriculturalists across Italian mountains have tried to sculpt chestnut, oak, and pine trees into the particular forms that produce nuts, timber, fodder, and fuelwood, while also providing sufficient pasture for sheep and goats, linking mountain wood pastures with lowland or valley-bottom agriculture through the fertilizing dung provided by the animals.[5] Over the last one hundred and fifty years, industrialization, rural outmigration, the arrival of alternative forms of fertilizer, and the arrival of successive epidemic diseases have undermined chestnut cultivation and transhumance, leaving a ruined landscape that is haunted by material and linguistic ghosts.[6]

As we come to study complex Anthropocene landscapes, where the ruins of past landscapes of cultivation remain ghostly presences, I suggest that the fieldwork practices of natural history, and historical ecology, are helpful in showing how we can pay attention to the partial and historical relations between plants, animals, soils, and politics.[7] This is a historical ecology where the forms of trees are emergent from partial relations between multiple actors and where the actors themselves are constantly changing as a result of relations with others. These relations are partial, because they emerge from the texture of particular interactions and could be a different kind of relation at another moment or in relation to another kind of being. A farmer who grafts and prunes a chestnut tree is not as interested in the form of this tree for firewood. A subterranean fungal mycorrhizal associate is not interested in the form of the tree above the ground but in its relations with the tree roots. Each relation has a particular texture and gives rise to particular reactions from both parties. The tree may take a certain shape; the farmer may claim a certain tax exemption from the state. Such relations are partial because they are not exhaustive of the forms that each actor might take; an alert practice of natural history is my way of attending to such relations.

How to Read Ghost Forests

Walking through the forest with my botanist assistant, Francesco Roma-Marzio, I do what natural historians have done for hundreds of years and historical ecologists more recently. I deliberately walk across an ecological transect, from the valley bottom near Lucca, across the top of the Monti Pisani at nine hundred meters elevation, and then down the other side to the plain of Pisa. I note down what tree, shrub, and understory plant species I see and what forms they have, and I wonder what past histories might have produced these particular shapes. I look at the texture of tree bark for the evidence of disease, of grafting, or of fire. I note down what plants are flowering; I look closely at the textures and forms of walls and ditches, of houses and ruins. This is hard work; it requires constant attention to form, texture, and color, constant speculation as to pattern. I walk with a dozen speculative possibilities in mind, some of which strengthen into impressions, many more of which I soon dismiss or remain speculations. Is this tree like that one, this house like that one, this wall like that one? Do all the trees on this valley have fire scars on their base? Does that color of lichen grow only on oaks or on other trees also? This is mentally exhausting work that requires close attention, and yet, paradoxically, it also contains an element of speculation. It reminds of nothing so much as participant observation or ethnographic interviewing, with its constant tension between here and elsewhere, accompanied by close attention to the indeterminacy of what is going on in a particular encounter. What kind of thing is this person telling me? What kind of thing is this tree? Curiously, ethnographic field notes and natural history field notes are quite similar. I keep mine in the same black notebook. As a poor artist, I take numerous photographs, but I cross-index these with sketches, where I summarize what the key features of the photographs are and why they matter. A few pencil strokes can summarize the patterns that I noticed when I took a picture, whereas photographs are notoriously less helpful for highlighting key features of a complex plant-landscape assemblage. If you know how to see, you can see fascinating stories of human-animal-plant communication embedded in the forms of living and dead trees. Let me show you a stump that tells a story (Figure G8.1).

This image may not be very imposing. Perhaps it looks like a slightly blurry field of shades of gray and black (in the image you are looking at), certainly nothing to remark upon if you walk by something similar on your next walk in the countryside. For me, primed as

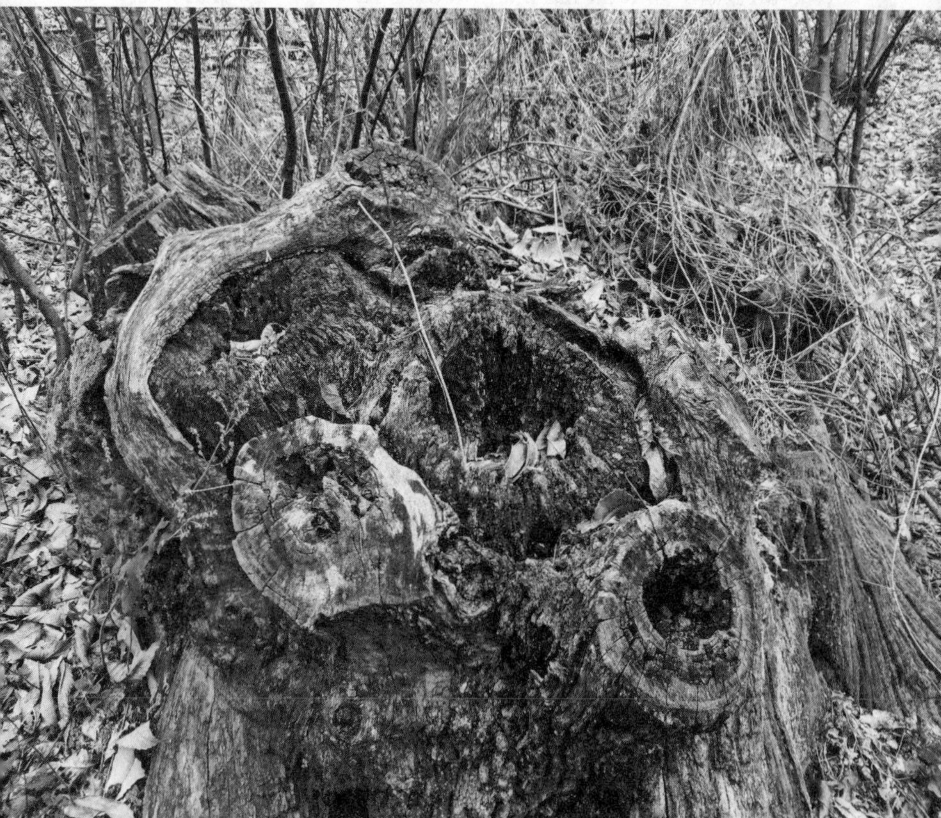

Figure G8.1. An ancient stool *ceppo,* Pizzorna, Lucca, 2014. Photograph by Andrew Mathews.

I was by talking to small farmers, by my training as a forester, and by many walks across similar landscapes, this particular stump told a fascinating story. Let me tell you a bit more about what you are looking at, as you may not have walked through this forest or other forests like it. Perhaps this brief account may help you see something different the next time you go for a walk.

This picture is of an ancient chestnut stool/*ceppo,* and it tells me a story of centuries of careful grafting, pruning, and cultivation of an ancient tree that was more than ninety centimeters in diameter and was probably at least two hundred years old when it was cut. In the 1950s, perhaps because chestnut cultivation began to be uneconomical or perhaps because of the arrival of the fungal chestnut cancer *Cryphonectria parasitica,* the peasants who cared for this tree decided to cut it down and sell the massive trunk to the tannin factory at Bagni di Lucca, about fifteen kilometers away. Cutting the tree did not kill it. On the contrary, cutting caused a sudden wild flourishing of dormant buds that had lurked as potential growing shoots in the cambium layer beneath the bark. Cutting the main trunk removed hormonal signals from the dominant meristems in the upper crown and shed a flood of sunlight on the bark, triggering the emergence and rapid growth of shoots near the stump. Gradually, the number of shoots/*polloni* diminished due to competition that left perhaps half a dozen stems of about ten to fifteen centimeters in diameter, allowing the tree to change form to become a coppice/*ceduo.* The circular structures on the stump are the remains of pole-sized stems from the successive cuts that have taken place about every twenty years. Older cuts leave more textured and eroded circles. If you look closely, on the top left of the image, you will see a new round of shoots emerging: this tree is not yet dead. Or perhaps it is dying—there are no shoots on the near side of the stump. Certainly such chestnut stools/*ceppi* can persist for centuries, in places where they are not outcompeted by other tree species. This could be a new landscape where beech *(Fagus sylvatica),* oak *(Quercus cerris),* or black locust *(Robinia pseudoacia)* gradually displaces chestnut but where bikers and hunters are quite happy.

Figure G8.2. Healthy ancient cultivated chestnut orchard/ *selva,* Fosciandora, Lucca, 2013. Photograph by Andrew Mathews.

Figure G8.3. Mature chestnut coppice/ *ceduo,* Monti Pisani, Lucca, 2013. Photograph by Andrew Mathews.

What Shape Is a Chestnut and How Long Will It Live?

The forms of trees, as of other beings, emerge from relations with others. As Oliver Rackham pointed out long ago, individuals of the same tree species take a particular shape depending upon where they are growing and upon their history of encounters with animals, fires, and diseases.[8] This is certainly true of sweet chestnut trees in Italy, where cultivated trees *(selve)* look dramatically different from coppice/*ceduo* (see also Figures G8.2 and G8.3).[9]

During what we might call the golden age of chestnut cultivation in central Italy, very loosely between about 1000 and 1800 C.E., sweet chestnut was tightly integrated into an agro-sylvo-pastoral system of cultivation. Figure G8.2 shows what such a cultivated chestnut *selva* looks like. In such a *selva,* desired chestnut varieties are grafted onto the wild root stock.[10] Grafted trees require continual care: peasant cultivators describe how shoots emerge from the root stock below the graft scar and have to be continually pruned if they are not to displace the grafted variety. When you see an imposing ancient tree with a graft scar, you should also imagine centuries of patient peasant cultivators cutting away at shoots every year or two, cutting off dead branches and grafting new ones every twenty years or so, perhaps even cutting down an entire tree and regrafting the stump if the crown lost vigor. Chestnut orchards are also shaped by pastoralism: goats and sheep are used to browse grasses and fertilize soils; leaves are raked up and burned or (formerly) used for stable litter. Where chestnut husks have been burned for centuries there are areas of blackened soil, perhaps with different plant species growing on them. Additional actors in this shifting multispecies assemblage are mycorrhizal fungi that allow the tree to absorb mineral nutrients; the tree reciprocates with gifts of carbohydrates and other chemicals. The edible *Boletus* mushrooms that are gathered in chestnut groves are also a sign of collaboration between trees and fungi. Finally, it is worth lingering to consider the nonliving actors that are part of this story. The forms of the stone walls have emerged from relations between trees, soils, and water; chestnut trees require moisture and root protection, so in some areas, peasant cultivators built different kinds of stone retaining walls to trap soil and moisture. Centuries of cultivation also transform soils, although modern soil science is rather reluctant to acknowledge this. It is also worth remarking that although chestnut trees can live for a very long time (perhaps two thousand years), they are ecologically rather particular, and that extensive chestnut forests are usually replaced by other tree species when human cultivators are no longer willing to work with chestnuts.[11]

Capitalism and international trade have left a mark on these forests through the arrival of exotic fungal diseases, first with the arrival of the "ink disease" *(Phytophthora cambivora)* in the 1850s, then with the arrival of chestnut cancer *(C. parasitica)* in the 1950s. These diseases were complex actors; in some areas they caused devastating mortality, making cultivation for food no longer feasible and pushing peasants and foresters to see chestnut trees as a source of firewood for coppice or of wood for tannin production. In other areas, trees were more disease resistant (perhaps as a result of local soil, water, and cultivation conditions). In the case of chestnut cancer, the disease agent *Cryphonectria* has acquired its own viral disease that causes the fungus to become hypovirulent (less virulent), allowing affected trees to scar over and enclose the diseased area.[12]

Linguistic Forms as Ghost Forests

Italian, like English, is rich in technical terms for tree and landscape forms, but these terms are now known mainly to old people, to some foresters and biologists, and to those who study landscape history. These terms are a sign of the capacity of particular plant and landscape forms to elicit language; the loss of common knowledge of the full richness of this linguistic field is a sign of the changing relationship between people and plants. Words are an index of the degree to which people and plants are entangled. The fact that for this essay I have to carefully define a few words from this rich linguistic array is a sign of the distance between historic rural peasant practitioners and present-day urban audiences.[13] In rural Italy, a few important terms are *ceduo* (coppice), *ceduo con matricine* (coppice with standards), *alto fusto* (high forest), *selva* (cultivated chestnut forest), and *bosco* (wild forest).[14] These classifications of forest forms are embedded in state forestry regulations as to when and how trees may be cut.[15] These terms are not merely a curiosity; they tell us something about what happens when humans become intensely involved in relationships with particular multispecies assemblages and about what might happen to us if we engage in a practice of looking at ruined landscapes. A practice of reading landscapes helps us see plants and landscapes differently; knowledge of changing landscapes gives us words for describing how forests have been used in the past and how they might be used in the future. These words are resources for contemporary environmental politics and for producing different visions of livable futures.

What Ghost Forests and Natural History Have to Say to the Anthropocene

In this chapter, I have described my empirical practice of reading forest landscapes around Lucca, showing how I look for and record evidence of past land use, of particular cultivation practices, and of histories of partial encounters between humans and nonhumans. This practice of reading ghost forests is also a practice of paying close attention to the ways that attending to landscape requires a close and speculative attention not only to the emergence of patterns in what I see in the landscape but also to the categories through which I see and describe these things to other people. I can perhaps offer a few provocations that emerge from this practice.

First of all, historical ecology and natural history should be understood as a practice not of describing the relations between pregiven entities but rather of attending to the multiple forms that emerge from partial relations between different plants, animals, and people. As we have seen, chestnut trees take different forms depending on where they have grown and what other beings they have encountered along the way, and I could tell a similar story for pines, oaks, and other plants. A chestnut is not one thing: it can be a gnarled ancient tree that is in a set of partial relations with goats, people, sheep, and terraces; a chestnut can also be a dense forest of pole-sized stems of "wild" coppice/*ceduo,* cut repeatedly to produce firewood for local household consumption or perhaps to produce woodchips for biomass energy plants that produce electricity. New diseases may change social relations, but these diseases may themselves change, as in the transformation of chestnut cancer into its hypovirulent form.

My outline of how we might cultivate an attention to the coemergence of plant forms and of the words we have for talking about these forms has clues for how the social sciences and humanities might engage with the Anthropocene. I suggest that one way of looking at the Anthropocene is to pay attention to the coemergence of material forms and linguistic terms, of causal accounts, and of histories that can multiply our ways of thinking and acting in the face of overwhelming environmental change. Social scientists have rightly pointed out that the Anthropocene, as it has emerged from earth systems modeling, has produced a strikingly singular and impoverished language of politics.[16] It is clear from this chapter that reading ghost landscapes in Lucca does not lead to this singular vision of Anthropocene politics and that the stories of disease and economic change might give modelers pause. I prefer to think of "Anthropocenes" as irreducibly multiple rather than of a singular "Anthropocene." We can remember the sustainable Anthropocene of peasant chestnut cultivation over the last fifteen hundred years or the less hopeful Anthropocene of chestnut abandonment in the wake of industrialization and globalization over the last two hundred. Both are Anthropocenes; each gives rise to a diverse set of political imaginations and causal accounts of how the environment might change in the future. In Italy, popular responses to climate change policies draw on imaginations of both of these Anthropocenes and might come up with still another form of landscape cultivation. Paying close attention to the ghostly forms of past histories in present-day forests allows us to consider the many forms of political and economic life that these forests are or might be connected to, including imagining multiple possible Anthropocene futures. The texture and form of our material surroundings are full of speculative politics and causal accounts, not only in forests but in other places, if we can only attend to them.

Anthropologist **ANDREW S. MATHEWS**'s training in forest ecology has allowed him to pioneer a distinctive mix of natural history observation and social analysis. His book *Instituting Nature: Authority, Expertise, and Power in Mexican Forests* offers a subtle analysis of forestry and forests in action and won the Harold and Margaret Sprout Award of the International Studies Association. He is associate professor of anthropology at the University of California, Santa Cruz.

Notes

1. Fernand Braudel, *The Mediterranean and the Mediterranean World in the Age of Philip II* (New York: Harper and Row, 1972).
2. In what follows, I will pair the closest English word with its Italian complement. This rich vocabulary emerged from a field of agri-sylvo-pastoral practices over the last several thousand years. A. T. Grove and A. T. Rackham, *The Nature of Mediterranean Europe: An Ecological History* (New Haven, Conn.: Yale University Press, 2001); O. Rackham, *Woodlands* (London: HarperCollins, 2006).
3. The capacity of numerous tree species to *coppice,* that is, to resprout from the stump, has allowed people around the world to produce a vast variety of products from many tree species, depending on the length and diameter of the stems desired. Baskets, barrels, construction timber, fence posts, and vine poles—all of these could be produced from chestnut coppice. Firewood, however, was perhaps one of the most important products.
4. Italy comprised a patchwork of states until reunification between 1859 and 1870. For most of the last thousand years, Lucca was an aristocratic republic; Pisa was first an independent republic and then part of Florentine territory from 1406. Antonio Mazzarosa, *Le Pratiche della Campagna Lucchese* (Lucca: Tipografia di Giuseppe Giusti, 1846).
5. Paolo Squatriti, "Water, Nature, and Culture in Early Medieval Lucca," *Early Medieval Europe* 4, no. 1 (1995): 21–40; Giuliana Puccinelli, "All'origine di Una Monocultura: L'espansione del Castagneto nella Valle del Serchio in Eta Moderna," *Rivista di Storia Dell'Agricoltura* 50, no. 1 (2010): 3–65; Raffaello Giannini and Antonio Gabbrielli, "Evolution of Multifunctional Land-Use Systems in Mountain Areas in Italy," *Italian Journal of Forest and Mountain Environments* 68, no. 5 (2013): 259–68.
6. Gregory T. Cushman, *Guano and the Opening of the Pacific World: A Global Ecological History* (Cambridge: Cambridge University Press, 2014).
7. Michael R. Canfield, *Fieldnotes on Science and Nature* (Cambridge, Mass.: Harvard University Press, 2011). Historical ecology is a large field with multiple traditions. W. Balée, "The Research Program of Historical Ecology," *Annual Review of Anthropology* 35, no. 1 (2006): 75–98. For present purposes, I find it most helpful to adopt the deliberately eclectic approaches suggested by Oliver Rackham in *Woodlands* (London: HarperCollins, 2006) and by Roberta Cevasco and Diego Moreno, "Rural Landscapes: The Historical Roots of Biodiversity," in *Italian Historical Rural Landscapes,* ed. Mauro Agnoletti, 141–52 (Dordrecht, Netherlands: Springer, 2013). See also Diego Moreno, Pietro Piussi,

and Oliver Rackham, eds., "Boschi: Storia e Archeologia," special issue, *Quaderni Storici* 4a, no. 1 (1982).

8. A. T. Grove and Oliver Rackham, *The Nature of Mediterranean Europe: An Ecological History* (New Haven, Conn.: Yale University Press, 2001), 45–71.
9. In the Lucchese dialect, this multistemmed tree might be termed part of a *vernacchia,* a coppice/*ceduo* that has been left to grow longer than usual, producing larger construction timbers rather than the pole-size stems usually cut from coppice. The biological mechanism of resprouting from the stump remains the same.
10. Nino Breviglieri, "Indagini e Osservazioni Sulle Cultivar di Castagno," in *Studio Monografico Sul Castagno Nella Provincia di Lucca,* 65–138 (Florence: Centro di Studio Sul Castagno, 1958); Massimo Giambastiani, F. Occhipinti, M. Armanini, F. Bagnoli, F. Camangi, C. Cossu, S. Fineschi, et al., *Le Cultivar di Castagno Della Provincia di Lucca—I Parte* (Lucca: IRF, 2011).
11. Giovanna Pezzi, Giorgio Maresi, Marco Conedera, and Carlo Ferrari, "Woody Species Composition of Chestnut Stands in the Northern Apennines: The Result of 200 Years of Changes in Land Use," *Landscape Ecology* 26, no. 10 (2011): 1463–76.
12. Ibid.
13. For English, Rackham, *Woodlands,* helpfully lists key terms for tree and landscape forms: *coppice, standards, pollards, spinney, hanging, wood pasture.* Italian is similarly rich, with additional terms for types of terraces. Charles Watkins, "The Management History and Conservation of Terraces in the Val di Vara, Liguria," in *Ligurian Landscapes,* ed. R. Balzaretti, M. Pearce, and C. Watkins (London: Accordia Specialist Studies on Italy, University of London, 2004), 141–53.
14. Moreno et al., "Boschi: Storia e Archeologia."
15. Provincia di Lucca, *Tabella Riepilogativa dei Tagli nei Cedui (sia puri che misti)* (2014), 2.
16. Eva Lövbrand, Silke Beck, Jason Chilver, Tim Forsyth, Johan Hedren, Mike Hulme, Rolf Lidskog, and Eleftheria Vasileiadou, "Who Speaks for the Future of the Earth? How Critical Social Science Can Extend the Conversation on the Anthropocene," *Global Environmental Change* 32 (April 2015): 211–18.

9

ESTABLISHING NEW WORLDS

THE LICHENS OF PETERSHAM

Anne Pringle

Lichens Are Ancient Worlds

What was once thought to be a mutualism involving two species may be an entangled symbiosis of thousands of species, interacting in every conceivable fashion. A lichen is not just a fungus and its photosynthetic algae. Lichens house hundreds, thousands, or perhaps tens of thousands of other species within the thallus, including other kinds of fungi and myriad bacteria. Bacterial diversity peaks at the center of a thallus, while the various edges house relatively fewer taxa.[1] Bacterial communities at the centers of different lichens resemble each other, while edges house more random assemblages. Even something as basic as reproduction may involve bacterial communities. Lichens often reproduce asexually, and within asexual propagules, the lichen packages its bacteria.[2] Perhaps the countless additional bacteria and fungi associated with a thallus provide benefits to the lichen, or perhaps many are antagonists. While we are far from a perfect understanding of the natural history within a lichen, it is clear that lichens are worlds unto themselves.

Moreover, these worlds have existed for a very long time. The earliest known fossil lichen dates to the Early Devonian. *Chlorolichenomycites devonicus* was discovered in rock 415 million years old from the Welsh Borderland.[3] This species, along with other lichen species, species of nonlichenized fungi, and free-living algae and cyanobacteria, may have formed extensive, diverse communities across large swaths of Early Devonian landscapes. These landscapes would have looked almost unlike anything we know today. At a time before modern plants evolved, these communities would have seemed like living, textured, and tufted carpets, and perhaps in some places one would have also seen the gigantic, three-meter-tall *Prototaxites,* an enigmatic life-form sometimes interpreted as the sporocarp of an enormous fungus.

I study a complex of lichens within the genus *Xanthoparmelia.* The taxonomy of the genus is controversial, and one paper uses *Xanthoparmelia* to argue against morphology as an appropriate descriptor of species boundaries.[4] To avoid the debate, I will simply use the generic name *Xanthoparmelia.* I work with populations growing on tombstones of the North Cemetery in Petersham, Massachusetts (Figure G9.1). Because I am curious about the natural histories of groups of lichens growing on different tombstones, I go to the cemetery each year and track the status of approximately eight hundred individuals.

Figure G9.1. The North Cemetery of Petersham, Massachusetts. Tombstones are novel habitats for lichens. Photograph by Anne Pringle.

Landscapes Are Changed

New England *Xanthoparmelia* grow on rocks, or structures made from rock; in Petersham the lichens are commonly found on stone walls as well as in cemeteries. Before European settlement, the lichens would have grown on rocks or boulders within forest clearings or at forest edges. Exposed boulders are a regular feature of primary forests but are not abundant (Figure G9.2).

When Europeans settled in Petersham, coming west from Boston and east from the Connecticut River Valley, forests were cleared to make way for farms. Most land was cleared for pasture, but around North Cemetery, up to 15 percent of the land was used for agriculture.[5] Agriculture requires workable soil, and to farm a field, a farmer must first dig the rocks out of the ground. These rocks were often used to build massive stone walls (Figure G9.3; note both Figure G9.2 and Figure G9.3 offer a view from the same vantage point).

In Petersham, farmers built 436 miles of rock walls. Whereas walls were built to suit the individual farmer, New England cemeteries were built according to very specific town plans;[6] often, in addition to a central (or center) cemetery, four cemeteries were built away from the town center in each of the cardinal directions (north, south, east, and west). Thus, the North Cemetery of Petersham is approximately three miles north of Petersham's Center Cemetery. Whereas there are hundreds of tombstones within the Center Cemetery, the North Cemetery houses only a few dozen.

The lichens followed the rocks. New England tombstones are often covered with a profusion of lush lichen growth, and lichens are a nearly universal feature of old stone walls. If the habitats of *Xanthoparmelia* were relatively rare before European settlement, by 1850, they were not, and although there are no data that track the rise of lichen populations, today a sunny meter of stone wall easily houses several hundred *Xanthoparmelia*. When Petersham was a town of 436 miles of exposed stone walls, it's quite likely there were hundreds of thousands of *Xanthoparmelia* on the walls.

New England agriculture peaked in the early nineteenth century. After about 1850, the Louisiana Purchase, California Gold Rush, and industrialization of cities including Holyoke opened up new opportunities for Petersham farmers.[7] The Erie Canal and railroads made it possible for midwestern farms to move farm goods east, making local

Figure G9.2. A diorama of pre-European settlement forest, circa 1700 C.E., on display at the Fisher Museum of Harvard Forest. Note the rock outcrop at bottom left; these kinds of surfaces are the original habitat of *Xanthoparmelia* lichens. Photograph by John Green, Harvard Forest, Harvard University.

Figure G9.3. A diorama of forest clearing and agriculture, circa 1830 C.E., on display at the Fisher Museum of Harvard Forest. Note the extensive stone walls around fields and along roads; lichens would have quickly established on these new habitats. In Petersham, approximately 436 miles of stone walls remain, but often walls are found within regenerated forest. Photograph by John Green, Harvard Forest, Harvard University.

farming less profitable. Agricultural lands were abandoned, and forests grew up around the stone walls. Although some cemeteries were also abandoned, for example, Petersham's Poor Farm Burial Plot, many remain in continuous use, for example, both North and Center Cemeteries. Probably many lichens died as stone walls and abandoned cemeteries were shaded—*Xanthoparmelia* need light to grow. But once again, there are no data to suggest exactly how populations changed during and after farm abandonment.

Worlds Are Moved

As the lichens dispersed to the tombstones and walls, they would have carried their bacteria with them. One class of bacteria found in *Xanthoparmelia* is the *Alphaproteobacteria.* The class is enormously diverse, but its species often grow in symbiosis, for example, as plant mutualists capable of fixing nitrogen from air into metabolically useful compounds or as insect parasites capable of turning insect males into females. The class was the source of mitochondria and plant chloroplasts, and the origin of eukaryotes is intimately associated with *Alphaproteobacteria.* In this class, horizontal gene transfer, or the movement of genes across species boundaries, is rampant, and one species is used to facilitate the movement of foreign DNA into plant genomes. Although we have a poor understanding of what the *Alphaproteobacteria* do within *Xanthoparmelia* (and it is unlikely the *Xanthoparmelia* house insect parasites!), we do know the lichens enabled the movement of entire, complex communities across the New England landscape.

Immortality

How long does each world last? I started to work with lichens because I am fascinated by an idea that filamentous fungi are immortal. Aging, or senescence, is defined as a decreased probability of reproduction, and increased probability of death, with time. An immortal organism never ages, and the probabilities of reproduction or death may take unusual patterns. An immortal organism can still be killed, for example, it can be run over by a bus or chopped to bits with an ax, but it is no more likely to die of natural causes at age $x + 1$ than it was at age x. Human aging is an intuitive concept; an eighteen-year-old is more likely to have a baby and less likely to die than an eighty-year-old.

What it means for a fungus to age may be less intuitive. Remember that the fungus forming the foundation of any lichen thallus grows as a network. It's a filamentous fungus. In contrast, yeasts, like the yeasts you use to bake bread or brew beer, are unicellular fungi. Each cell eventually stops budding and dies. Yeasts are not immortal. But it's not clear if filamentous fungi stop reproducing, or are more likely to die, as they grow older. Moreover, because there are hundreds of thousands of species of filamentous fungi, it's not clear whether or how data from one species would translate across the kingdom.

To date, research on aging within the filamentous fungi has focused on the dung fungus *Podospora anserina.* This fungus grows only on animal poop, and these habitats are ephemeral; dung quickly decays. Because its habitats constantly appear and disappear, *P. anserina* hasn't had a chance to maintain or evolve immortality. In fact, if you grow it in a petri dish, you will find that it ages. The dynamics shaping the evolution of aging in *P. anserina* are not completely understood, but perhaps the species has lost the ability to repair DNA damaged by the normal wear and tear of daily growth. Because it doesn't last to old age in nature, there would be no reason to maintain repair mechanisms. In fact, most fungi from ephemeral habitats appear to age.[8]

But boulders and rock walls last a very long time. Have the filamentous fungi that form the basis of the *Xanthoparmelia* escaped senescence? To understand that question, I decided a long time ago that I needed to understand the dynamics of natural populations. I wanted demographic data: data describing the births, growths, and deaths of many individuals. Aggregate data would enable me to calculate the probabilities of reproduction and death over time and understand if demographic patterns are typical, for example, if they look like human demographic patterns, or if patterns are somehow different from the patterns of animal or plant species.

Methods

When I go to the cemetery, I focus on a subset of tombstones. I have permission from both the Petersham Cemetery Commission and families to trace and photograph the lichens, although nothing I do causes any damage to either the tombstones or the lichens. Many of the tombstones belong to past directors of the Harvard Forest, including Ernie Gould and Hugh Raup (Figures G9.4 and G9.5).

Figure G9.4. The Gould tombstone. *Xanthoparmelia* are green, but black and gray lichens also grow on the tombstone. Photograph by Anne Pringle.

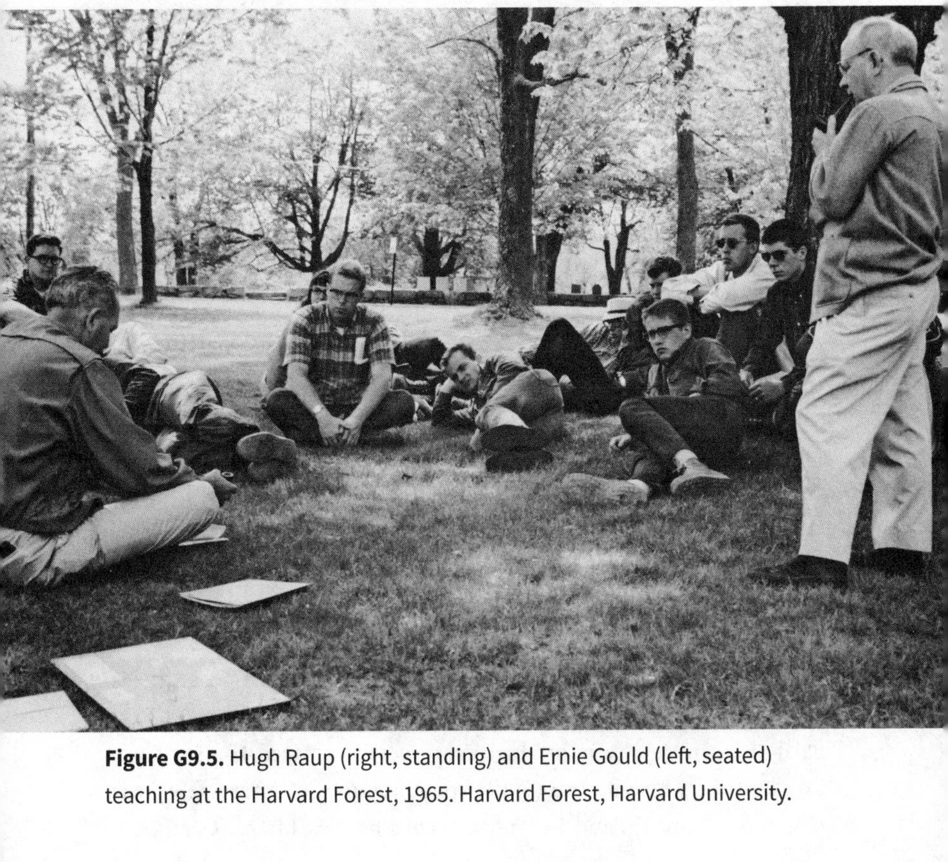

Figure G9.5. Hugh Raup (right, standing) and Ernie Gould (left, seated) teaching at the Harvard Forest, 1965. Harvard Forest, Harvard University.

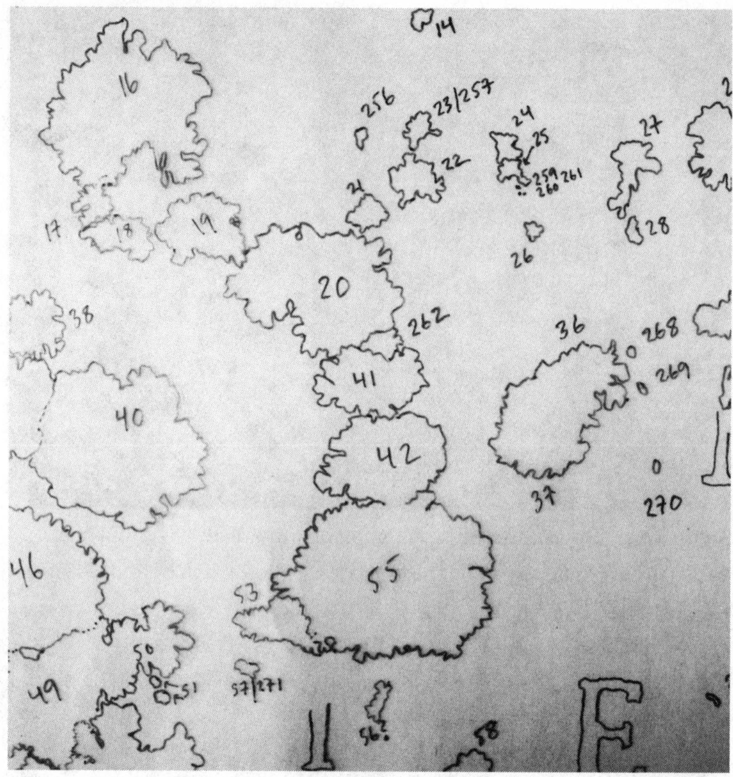

Figure G9.6. Every year, I trace maps of the *Xanthoparmelia* lichens growing on a set of tombstones. From year to year, I measure growth, reproduction, and death. Photograph by Anne Pringle.

Because Gould and Raup were foresters, concerned with ongoing changes to local landscapes, I like to think they wouldn't mind having research done on their tombstones; in fact, I like to think they'd find it hilarious.[9] I know I would. Each fall, I draw maps of the lichen populations. I place transparent, plastic sheets over the various tombstone surfaces and carefully trace the perimeter of target individuals (Figure G9.6).

The letters and numbers carved into the rock enable accurate orientation of the previous year's maps at each new census, and I can easily identify every individual every year. I started the survey in 2005. At first, I would trace entire populations, but after a few years, this was too much, and I stopped recording new births. Changes in individual perimeters from year to year provide growth rate data, but I also note every individual's reproductive status and health and occasionally collect data from various subsets of individuals to answer other questions.

The Persistence of Dispersed Worlds

Patterns of reproduction and death among the *Xanthoparmelia* are very different from patterns described for many animals, and often different from patterns described for plants. As an individual lichen grows older, it grows larger, and as the lichen grows larger, it reproduces more. The Petersham *Xanthoparmelia* are largely asexual, and as an individual grows larger, more surface area is available for the construction of asexual propagules. Lichens occasionally fragment, but in this cemetery, it doesn't happen very often, perhaps because disturbance is rare. While I've recorded the disappearance of small lichens, which may wash off the rock in bad weather or be overgrown by larger neighbors, I've yet to record the death of a large lichen. As lichens grow older, the probability of death seems to decrease.

But as the years in the cemetery passed, I began to wonder about how I was describing death, really because of the data of a few individual lichens. I recorded death when an entire thallus disappeared. Lichens grow at their edges, and if there are no barriers to growth, a thallus will expand in a more or less circular fashion. Centers are older than edges, and by now, I can date with some exactness the age of particular tissues.[10] Remember that bacterial communities at older centers are different from bacterial communities at younger edges. Perhaps succession is at work within this world, in the same way that abandoned New England farms start as grasslands but eventually become forests. Time influences landscapes, and as a lichen grows older, perhaps bacterial communities harmful to the foundation fungus form at the center. In response, the fungus may kill its center, and sometimes I do see large lichens with missing centers. In fact, this is the classic definition of death in lichenology: if the center falls out, the lichen is dying. I'm still not sure that's true—won't the edges grow on forever? But I wonder if aging is a feature of parts, or modules, of an individual thallus, even if it isn't a feature of the thallus itself.

Petersham is beautiful, and it's easy to go for long walks in the surrounding forests, but Petersham is a human habitat, and there isn't one square inch that hasn't been influenced by people. Trails often cross abandoned stone walls. Lichens on old stone walls are a quintessential feature of the landscape, signals of clean air and stability. New stone walls don't have lichens, and neither do walls in polluted environments, but the lichens and sometimes mosses of Petersham's stone walls can form an almost continuous carpet of green.

These lichens are nature. The word *nature* is complicated, I know, but if in its simplest sense nature means something alive, and not built by humans, then lichens are nature, even if they grow in altered landscapes. We built Petersham and, by doing so, built the lichens substrates on which to multiply, but they came on their own. Entire worlds established, and the worlds aren't static. While the interactions within each extant *Xanthoparmelia* may play out almost imperceptibly, they will influence the growth of the species within, creating a dynamic interior ecology. And whether it is the interactions or something else shaping the evolution of aging, it's clear these dynamic worlds will persist for a very long time.

ANNE PRINGLE works with species whose life histories and body plans seem very different from our own. She explores the fungi: their associations as lichens and with plants, their ecological roles in a changing world, and the nature of their individuality. She is associate professor of botany and bacteriology at the University of Wisconsin, Madison.

Notes

1. A. A. Mushegian, C. N. Peterson, C. C. M. Baker, and A. Pringle, "Bacterial Diversity across Individual Lichens," *Applied and Environmental Microbiology* 77 (2011): 4249–52.
2. I. A. Aschenbrenner, M. Cardinale, G. Berg, and M. Grube, "Microbial Cargo: Do Bacteria on Symbiotic Propagules Reinforce the Microbiome of Lichens?," *Environmental Microbiology* 16 (2014): 3743–52.
3. R. Honegger, D. E. Edwards, and L. Axe, "The Earliest Records of Internally Stratified Cyanobacterial and Algal Lichens from the Lower Devonian of the Welsh Borderland," *New Phytologist* 197 (2012): 264–75.
4. H. T. Lumbsch and S. D. Leavitt, "Goodbye Morphology? A Paradigm Shift in the Delimitation of Species in Lichenized Fungi," *Fungal Diversity* 50 (2011): 59–72.
5. D. R. Foster and J. D. Aber, ed., *Forests in Time: The Environmental Con-*

sequences of 1,000 Years of Change in New England (New Haven, Conn.: Yale University Press, 2004).

6. D. R. Foster, pers. comm., November 2015.
7. Ibid.
8. T. D. Geydan, A. J. M. Debets, G. J. M. Verkley, and A. D. van Diepeningen, "Correlated Evolution of Senescence and Ephemeral Substrate Use in the Sordariomycetes," *Molecular Ecology* 21 (2012): 2816–28.
9. H. M. Raup, "The View from John Sanderson's Farm: A Perspective for the Use of the Land," *Forest History* 10 (1966): 2–11.
10. Mushegian et al., "Bacterial Diversity across Individual Lichens."

CODA

CONCEPT AND CHRONOTOPE

Mary Louise Pratt

MONSTERS AND GHOSTS ARE THE RUBRICS organizing this volume. As I compose this note, a new mini-monster, the Zika virus, wends its way across the planet, leaving a generation of microcephalic babies in its wake. Its spread is born of collaboration between humans and *Aedes aegypti* mosquitos, who thrive in the wet spots of urban slums. Meanwhile, artistic awards rain down on a Hollywood film titled *The Revenant*—"the ghost," literally "the one who returns." The film recounts the gruesome endurance-and-survival story of a nineteenth-century white fur trader in the far north. The movie itself is a revenant, replaying an exhausted white man frontier fantasy in an ecology that is rapidly dissolving as the planet warms. At the same time, the relentless brutality of the film gestures forward toward our imagined apocalyptic future of environmental collapse, where civilization fails and all are pitted against all.[1] Complicating that grim but plausible story is one of the preoccupations of this book.

A kind of rough draft of the Anthropocene story, complete with revenants, appeared in 1968, in Franklin Schaffner's classic film *Planet of the Apes*. After nearly four thousand years in space, a crew of American astronauts lands on a planet where humans are despised and enslaved by a highly developed (English-speaking) society of apes. Most of us remember the climactic moment when, on a beach in the "Forbidden Zone," Charlton Heston encounters the half-buried remains of the Statue of Liberty and realizes he is on earth in the long aftermath of nuclear holocaust. The film also includes an ape archeologist, Cornelius, who is conducting a dig in the Forbidden Zone. He is puzzled because the more deeply he digs, the more highly developed the culture becomes. He has found the Anthropocene, *avant la lettre*. Once again, art precedes life.

What is at stake in these essays is not what the Anthropocene *is* but how it will be *lived*. Whether the stratigraphic authorities authorize the term will make no difference. In this respect, "Anthropocene" is what Deleuze and Guattari would call a concept. "All concepts," they say, "are connected to problems without which they would have no meaning, and which can themselves only be understood as their solution emerges."[2] As Elizabeth Grosz elaborates, this means that concepts are never true; they only enable. "Concepts are ways of adding ideality to the world," she explains, "transforming the givenness of chaos, the pressing problem, into various forms of order, into possibilities for being."[3] They "enable us to surround ourselves with possibilities of being otherwise."[4] The point of "Anthropocene" is to enable reflection in Western academic circuits on what this volume calls "arts for living on a damaged planet." The concept starts a conversation on "what we humans are going to do now, in the midst of an increasingly given fate of ruination and extinction" (Bubandt).

For most of the writers here, the question of how to live the Anthropocene is inseparable from the question of how to write it. Indeed, writing becomes the way of posing the question of how to live. The Anthropocene is also what narrative theorist Mikhail Bakhtin called a *chronotope*, a particular configuration of time and space that generates stories through which a society can examine itself. Bakhtin studied novels. "In the literary artistic chronotope," said Bakhtin, "spatial and temporal indicators are fused into one carefully thought-out, concrete whole. Time, as it were, thickens, takes on flesh, becomes artistically visible; likewise, space becomes charged and responsible to the movements of time, plot and history. The intersection of axes and fusion of indicators characterizes the artistic chronotope."[5] New chronotopes, Bakhtin said, create "previously nonexistent meanings." Old ones "continue stubbornly to exist" even after they have "lost any meaning that was productive in actuality or adequate to later historical situations."[6] *The Revenant* recycles the frontier chronotope, a time-space configuration that generates plot after plot of gendered whiteness. The Anthropocene creates a new chronotope with a multipolar time-space configuration. The human in the present imagines a subject who, long after humans are gone, reconstructs our era through what it will have left behind. Our detritus, to some hypothetical future and probably nonhuman geologist, will reveal a world that became increasingly "shaped by human activity but . . . also increasingly outside human control" (Bubandt). There may be multiple chronotopes at play here. The human mastery of nuclear fission in the 1940s marks one new time-space configuration, explored here by Brown and Barad; genetic science (Svenning, Hejnol) obliterates the vertical imaginings of evolutionary theory. In our deep future (Barad's beautiful term), we may be sponges. Throughout, though, the question becomes, how will people get from here to there? What material, ethical, political, esthetic, affective, choices are we and will they be called upon to make? What will be possible? Through questions like these, the Anthropocene calls forth "previously nonexistent meanings" for human experience, new possible futures, configurations of desire, action, value, intent. It promotes, in Mathews's words, the "coemergence of material forms and linguistic terms, of causal accounts and of histories that can multiply our ways of thinking and acting in the face of overwhelming environmental change." It is a device, and an invitation, for Western-identified subjects to resituate themselves in the space-time-matter of the planet.

Bakhtin would applaud Mathews's insistence on language and narrative. It is no accident that many of the essays here are experiments not just in thought and action but also in genre and style. In what discursive forms will this newly conceived relationship of humans to the planet and the future be expressed? Many of the writers experiment with nature writing. They look for ways of reading (i.e., writing) landscape not as detached from humans but as densely populated by ghosts and afterlives of human activity that have been absorbed and enmeshed into the landscape's own generativity. In the Anthropogenic chronotope, human and nonhuman agency are no longer distinguished, visually or analytically. What Rose calls "multispecies entanglements" are among the new plot elements. "Forest" in Mathews's Italian scenario is a space filled with old walls, rewilded chestnut trees, elders who remember the words for long-gone husbandries. Stern's Tijuana border zone is a ruin of dead tires and salvaged wetlands, where human detritus generates the unpredictable and uncanny, what in Spanish we call the *insólito.* Bubandt explores the aftermath of ecodisaster in Indonesia, coproduced (no one quite knows how) by some combination of human and nonhuman agency. Here, too, the thing to be grasped is the unpredictable generativity of the wreckage-strewn time-space for the humans who inhabit it and are inhabited by it. Javanese arts of living, peopled by ghosts and spirits, become mirrors for the emergent Western Anthropocenic subject. Pringle, meanwhile, wanders a New England graveyard and finds eternal life not on the tombstones but in the lichens. These writers are not discovering Anthropocenic landscapes; they are creating them, in language. They are realized in the aesthetic (vs. anesthetic) responses of the reader, driven by the artistic principle of estrangement, the making of wonder.

The figure of the nineteenth century naturalist reappears in repurposed form in these essays as reader not of primal nature but of the history-laden, ghost-ridden spaces of Anthropocene earth. Our "worlds of loss," says Rose, call for "radically reworked forms of attention." Stern calls for "new ways of seeing." These naturalist revenants retain key aspects of their predecessors: curiosity, the practice of reading landscape as it is *walked,* a deep love of the earth and its creatures, and, perhaps above all, the desire to find magic, to enchant or reenchant the world, to make it possible to inhabit it with love. Look at the opening of Alexander von Humboldt's famous nature essay "On Steppes and Deserts" (1805):

> At the foot of the lofty granitic range which, in the early ages of our planet, resisted the irruption of the waters on the formation of the Caribbean gulf, extends a vast and boundless plain.

Now read Bubandt's first sentence:

> If you travel south by car from Surabaya, Indonesia's second-largest city located on the sweltering north coast of Java, toward the cool mountain town of Malang, you will, after about twenty-five kilometers, come upon a vast elevated landscape of mud.

I cannot be the only one who hears the echo of Humboldt in Bubandt—and the same impulse to enchant. (Students of literature learn that the raw material for writing is other writing.) Apocalypses are the territory of the poetics of the sublime, the merging of beauty and terror. That is the shimmer Brown finds in the ghastly underworld of Chernobyl or the ghastly trace of Madame Curie's radioactive fingers on her notebook in a library in Japan. Alongside these new naturalists, it becomes easy to imagine an Anthropocenic flâneur who remakes the city as a multispecies chronotope where human, animal, plant, fungal, and viral life-forms negotiate their cohabitation.

These naturalist revenants remind us that the Anthropocene chronotope leaves humans, modern, Occidental humans, at the center of the narrative. Its story is still all about an "us." Perhaps its revelatory powers will exhaust quickly. But Rose reminds us that the quest for enchantment, poetry, "shimmer," is not a uniquely human trait. The orchestration of desire, the impulse to attract, she argues, is the way of life in all its forms.[7] Trees erupt into perfume and flower to attract the animals whose bodies are adapted to pollinating them. Life perpetuates itself through orchestrated aesthetic display. Music, color, dance, belong as much to birds as to people.

In the course of the days I spent writing these words, Honduras's most powerful environmental activist, Betty Cáceres, was assassinated in her home, the tenth such activist to be killed in Honduras this year;[8] nature lovers in Portland, Oregon, discovered that the moss on their trees is full of toxic chemicals from factories; and two Monterey marine scientists announced that Southern California's offshore oil rigs are teeming with flourishing marine life. How will we slouch toward our deep future, toward an almost certain demise whose script we are writing but cannot imagine? How to define the stakes for the life-forms that are not human but inextricably enmeshed with humans? By what scales can human action now be gauged? The writers of the Anthropocene, like the diviners on the Javanese mud pile, are seeking the meaning machines and desiring machines through which the dramatic, unknowable trajectory on which we are embarked can become a story and be lived.

MARY LOUISE PRATT's pioneering book *Imperial Eyes: Travel Writing and Transculturation* transformed the study of colonial encounters across the humanities and social sciences. She has developed important theoretical tools for postcolonial scholarship, most notably the concept of "contact zones" to describe the spaces where colonizers and indigenous people meet. She has contributed often to the dialogue between anthropology and the humanities. She is Silver Professor (emerita) of Social and Cultural Analysis, Spanish and Portuguese, and Comparative Literature at New York University.

Notes

1. Accepting the Academy Award for Best Actor for his role in the film, Leonardo DiCaprio spoke of its allusion to the climate change emergency.
2. Gilles Deleuze and Félix Guattari, *What Is Philosophy?* (New York: Columbia University Press, 1994), 16. Cited in Elizabeth Grosz, *Becoming Undone: Darwinian Reflections on Life, Politics, and Art* (Durham, N.C.: Duke University Press, 2011), 78.
3. Grosz, *Becoming Undone,* 78.
4. Ibid.
5. Mikhail Bakhtin, "Forms of Time and of the Chronotope in the Novel," in *The Dialogic Imagination: Four Essays* (Austin: University of Texas Press, 1981), 84.
6. Ibid., 85.
7. Her compatriot, feminist philosopher Elizabeth Grosz, makes the same point, also in reference to Australian Aboriginal painting and to her interactions with a group of women painters. Grosz, *Becoming Undone,* part III.
8. Cáceres was a leader of the four-hundred-thousand-member Lenca indigenous people. She was awarded the Goldman Environmental Prize in 2015 for leading a successful struggle to stop a huge Chinese hydroelectric project in Honduras. She was forty years old and a mother of four.

PRAISE FOR

ARTS OF LIVING

ON A DAMAGED PLANET

"What an inventive, fascinating book about landscapes in the Anthropocene! Between these book covers, rightside-up, upside-down, a concatenation of social science and natural science, artwork and natural science, ghosts of departed species and traces of our own human shrines to memory. . . . Not a horror-filled glimpse at destruction but also not a hymn to romantic wilderness. Here, guided by a remarkable and remarkably diverse set of guides, we enter into our planetary environments as they stand, sometimes battered, sometimes resilient, always riveting in their human—and nonhuman—richness. *Arts of Living on a Damaged Planet* is truly a book for our time."

—PETER GALISON, Harvard University

"Facing the perfect storm strangely named the Anthropocene, this book calls its readers to acknowledge and give praise to the many entangled arts of living that made this planet livable and that are now unraveling. Grandiose guilt will not do; we need to learn to notice what we were blind to, a humble but difficult art. The unique welding of scholarship and affect achieved by the texts assembled tells us that learning this art also means allowing oneself to be touched and induced to think and imagine by what touches us."

—ISABELLE STENGERS, author of *Cosmopolitics I* and *II*

"*Arts of Living on a Damaged Planet* exposes us to the active remnants of gigantic past human errors—the ghosts—that affect the daily lives of millions of people and their co-occurring other-than-human life forms. Challenging us to look at life in new and excitingly different ways, each part of this two-sided volume is informative, fascinating, and a source of stimulation to new thoughts and activisms. I have no doubt I will return to it many times."

—MICHAEL G. HADFIELD, University of Hawaiʻi at Mānoa

ARTS OF LIVING

ON A DAMAGED PLANET

ARTS OF LIVING ON A DAMAGED PLANET

MONSTERS OF THE ANTHROPOCENE

Anna Tsing
Heather Swanson
Elaine Gan
Nils Bubandt

Editors

University of Minnesota Press
MINNEAPOLIS • LONDON

The University of Minnesota Press gratefully acknowledges financial assistance for the publication of this book from the Aarhus University Research Fund.

Illustrations by Jesse Lopez, jesselopez.com.

Ursula K. Le Guin, "Deep in Admiration," was first presented at the Conference on Arts of Living on a Damaged Planet at UC Santa Cruz in 2014, then later appeared in *Late in the Day,* published by PM Press in 2015; copyright 2015 by Ursula K. Le Guin. "The Marrow" first appeared in *Hard Words,* published by Harper and Row in 1981; copyright 1981 by Ursula K. Le Guin. "Tao Song" first appeared in *Wild Angels,* published by Capra Press in 1974; copyright 1974 by Ursula K. Le Guin. "The Story" first appeared in *Late in the Day,* published by PM Press in 2015; copyright 2015 by Ursula K. Le Guin. "Kinship" first appeared in *Orion* magazine in 2014; copyright 2014 by Ursula K. Le Guin. "Whiteness" first appeared in *Los Angeles Review* published by Red Hen Press in 2014; copyright 2014 by Ursula K. Le Guin. "Infinitive" first appeared in *Sixty Odd,* published by Shambhala in 1999; copyright 1999 by Ursula K. Le Guin. "Futurology" first appeared in *Incredible Good Fortune,* published by Shambhala in 2006; copyright 2006 by Ursula K. Le Guin. "A Hymn to Aphrodite" first appeared in *Prairie Schooner* in 2014; copyright 2014 by Ursula K. Le Guin. All poetry reprinted by permission of Curtis Brown Ltd.

An earlier version of chapter 2 was previously published as Donna J. Haraway, "Sympoiesis: Symbiogenesis and the Lively Arts of Staying with the Trouble," in *Staying with the Trouble: Making Kin in the Chthulucene* (Durham, N.C.: Duke University Press, 2016); reprinted with permission of Duke University Press.

An earlier version of chapter 5 was previously published as Carla Freccero, "A Race of Wolves," *Yale French Studies* 127 (2015): 110–23; reprinted with permission.

Published by the University of Minnesota Press
111 Third Avenue South, Suite 290
Minneapolis, MN 55401-2520
http://www.upress.umn.edu

ISBN 978-1-5179-0236-0 (hc)
ISBN 978-1-5179-0237-7 (pb)
A Cataloging-in-Publication record for this book is available from the Library of Congress.

Printed in the United States of America on acid-free paper

22 10 9 8

Candice Lin, *5 Kingdoms*, 2015. Courtesy of the artist and Francois Ghebaly Gallery, Los Angeles.

Inspired by evolutionary biologist Lynn Margulis's theory of coevolution and symbiogenesis, Lin's etching shows a lively interplay of five kingdoms: plants, fungi, animals, and protists all radiate from bacteria. Filled with visual and narrative surprises about processes of digestion, distillation, and fermentation, Lin's drawings are invitations to look more closely at the wonders and terrors of the Anthropocene.

CONTENTS

MONSTERS AND THE ARTS OF LIVING

Acknowledgments *Mviii*

Introduction: Bodies Tumbled into Bodies *M1*
HEATHER SWANSON, ANNA TSING, NILS BUBANDT, AND ELAINE GAN

1 Deep in Admiration *M15*
URSULA K. LE GUIN

INHABITING MULTISPECIES BODIES *M23*

2 Symbiogenesis, Sympoiesis, and Art Science Activisms for Staying with the Trouble *M25*
DONNA HARAWAY

3 Noticing Microbial Worlds: The Postmodern Synthesis in Biology *M51*
MARGARET McFALL-NGAI

BEYOND INDIVIDUALS *M71*

4 Holobiont by Birth: Multilineage Individuals as the Concretion of Cooperative Processes *M73*
SCOTT F. GILBERT

5 Wolf, or *Homo Homini Lupus* *M91*
CARLA FRECCERO

6 Unruly Appetites: Salmon Domestication "All the Way Down" *M107*
MARIANNE ELISABETH LIEN

7 Without Planning: The Evolution of Collective Behavior in Ant Colonies *M125*
DEBORAH M. GORDON

AT THE EDGE OF EXTINCTION *M141*

8 Synchronies at Risk: The Intertwined Lives of Horseshoe Crabs and Red Knot Birds *M143*
PETER FUNCH

9 Remembering in Our Amnesia, Seeing in Our Blindness *M155*
INGRID M. PARKER

Coda. Beautiful Monsters: Terra in the Cyanocene *M169*
DORION SAGAN

ACKNOWLEDGMENTS

THIS BOOK WAS INCUBATED within the Aarhus University Research on the Anthropocene (AURA) project. Since 2013, coeditors Anna Tsing and Nils Bubandt have directed AURA, a transdisciplinary program crossing the humanities, arts, and sciences to examine co-species landscapes and global connections in a world shaped by anthropogenic disturbance. The project is supported by the Danish National Research Foundation as a Niels Bohr Professorship. AURA participants include the coeditors as well as Filippo Bertoni, Nathalia Brichet, Marilena Campos, Zachary Caple, Thiago Cardoso, Luz Cordoba, Rachel Cypher, Maria Dahm, Pierre du Plessis, Christine Fentz, Natalie Forssman, Peter Funch, Colin Hoag, Mathilde Højrup, Thomas Troels Kristensen, Jens Mogens Olesen, Ton Otto, Katy Overstreet, Pil Pedersen, Felix Riede, Meredith Root-Bernstein, Jens-Christian Svenning, Stine Vestbo, Thomas Schwartz Wentzer, and Rane Willerslev. Their modes of interdisciplinary engagement are integral to the fabric of this book. AURA's administrative assistants, Mia and Mai Korsbaek, have helped make this volume, and the gatherings that preceded it, possible.

All of the coeditors have relied on many interlocutors. In addition to those already mentioned, the coeditors thank Wanda Acosta, Ursula Biemann, Paulla Ebron, Michael Eilenberg, Jennifer A. González, Susan Harding, Adrienne Heijnen, Gail Hershatter, Rusten Hogness, John Law, Morten Axel Pedersen, Marean Pompidou, Danilyn Rutherford, Warren Sack, Shelley Stamp, Nina Holm Vohnsen, and Susan Wright.

We want to specially note the work of artist Jesse Lopez, whose illustrations enliven these pages. Ginger Clark helped secure the right to republish Ursula K. Le Guin's writing. Megan Martenyi and Filippo Bertoni helped with the final editing.

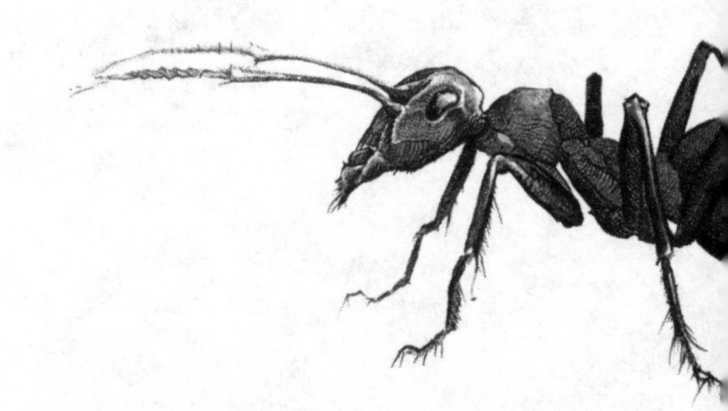

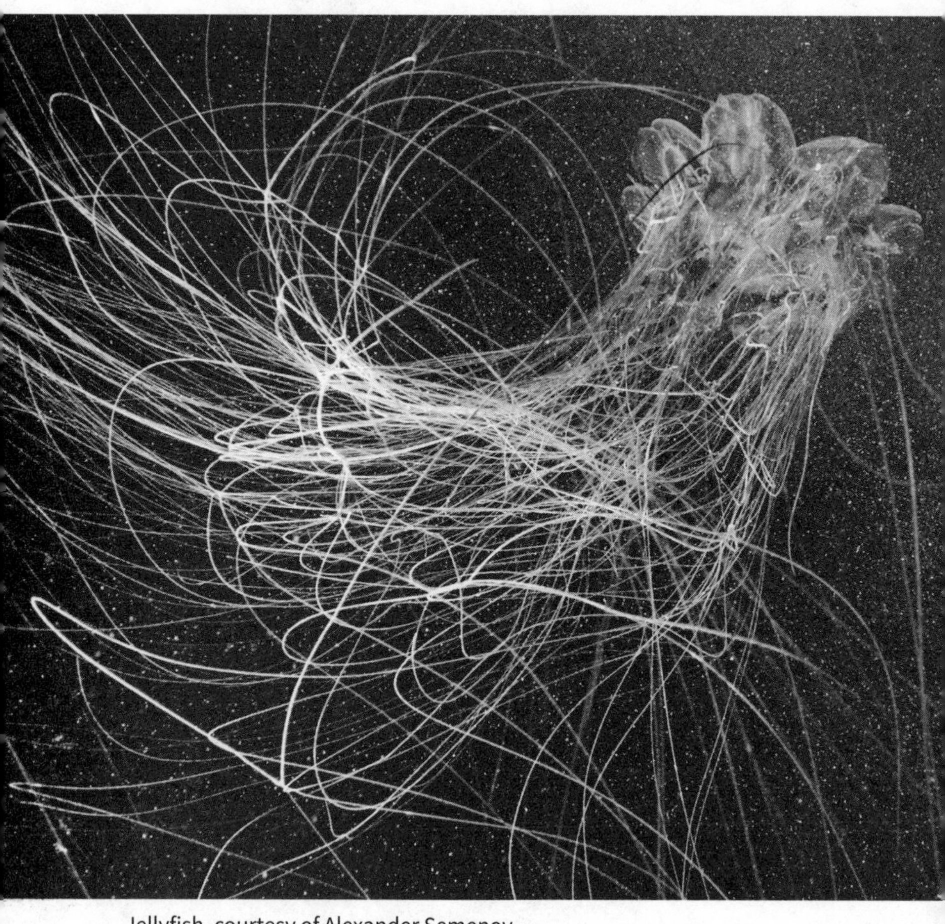

Jellyfish, courtesy of Alexander Semenov.

INTRODUCTION
BODIES TUMBLED INTO BODIES

Heather Swanson Anna Tsing Nils Bubandt Elaine Gan

What If All Organisms, Including Humans, Are Tangled Up with Each Other?

Jellyfish are monsters. Soft glass parasols as colorful as flowers, they blossom from watery depths with delicate grace. Yet woe to those tangled in their stinging tentacles. Along beaches in Australia, Florida, and the Philippines, jellies are becoming a greater threat than sharks, sending scores of swimmers to hospitals, some with fatal stings. Off the coast of Japan, 450-pound Nomura's jellies have capsized boats that have snared loads of them in their nets. In the Black Sea, comb jellyfish eat ten times their weight in a single day, destroying fish and fisheries. As jellyfish consume the small fish fry, emptying seas of other species, the waters fill up with jellies in fantastical numbers. The richness of earlier marine assemblages is overwhelmed. The ocean turns monstrous. Filling the seas with sloshing goo, jellyfish are nightmare creatures of a future in which only monsters can survive.

How did such monstrosity arise? Those Black Sea combs—so inspiring and so terrible—arrived in the ballast water of ships as recently as the 1980s. They took over too-warm seas emptied out by overfishing and polluted by the choking runoff of industrial farming. Under other conditions, jellies are capable of playing well with other species. If jellyfish are monsters, it is because of their entanglements—with us. Jellies become bullies through modern human shipping, overfishing,

pollution, and global warming. In all our heedless entanglements with more-than-human life, we humans too are monsters.

Coral reefs are monsters. Their polyps rise from reefs of their own making—but not just their own. Like the mythical chimeras of ancient Greece, beasts made up of the head of a lion, the body of a goat, and the tail of a snake, coral reefs are made of mismatched parts—animal, plant, and more—that hang together in fragile coordinations. In contrast to jellies, warming waters do not turn corals into bullies; rather, they drive off symbiotic dinoflagellates, weakening the corals. The necessity of working together makes coral life possible; indeed, symbiosis is essential to life on earth. But symbiosis is also vulnerable. Corals, like jellies, are tied to others in rapidly shifting worlds, but for them, disrupted relations lead not to riotous reproduction but to decline and death. In all our vulnerable entanglements with more-than-human life, we humans too are monsters.

Monsters are useful figures with which to think the Anthropocene, this time of massive human transformations of multispecies life and their uneven effects. Monsters are the wonders of symbiosis *and* the threats of ecological disruption. Modern human activities have unleashed new and terrifying threats: from invasive predators such as jellyfish to virulent new pathogens to out-of-control chemical processes. Modern human activities have also exposed the crucial and ancient forms of monstrosity that modernity tried to extinguish: the multispecies entanglements that make life across the earth, as in the coral reef, flourish. The monsters in this book, then, have a double  meaning: on one hand, they help us pay attention to ancient chimeric entanglements; on the other, they point us toward the monstrosities of modern Man. Monsters ask us to consider the wonders and terrors of symbiotic entanglement in the Anthropocene.

In the indeterminate conditions of environmental damage, nature is suddenly unfamiliar again. How shall we find our way? Perhaps sensibilities from folklore and science fiction—such as monsters and ghosts—will help. While ghosts (which appear in the other section of this book) help us read life's enmeshment in landscapes, monsters point us toward life's symbiotic entanglement across bodies. The double-sided format of *Arts of Living on a Damaged Planet* presents ghosts and monsters as two points of departure for characters, agencies, and stories that challenge the double conceit of modern Man. Against the fable of Progress, ghosts guide us through haunted lives

and landscapes. Against the conceit of the Individual, monsters highlight symbiosis, the enfolding of bodies within bodies in evolution and in every ecological niche. In dialectical fashion, ghosts and monsters unsettle *anthropos,* the Greek term for "human," from its presumed center stage in the Anthropocene by highlighting the webs of histories and bodies from which all life, including human life, emerges. Rather than imagining phantasms outside of natural history, the monsters and ghosts of this book are observable parts of the world. We learn them through multiple practices of knowing, from vernacular to official science, and draw inspiration from both the arts and sciences to work across genres of observation and storytelling.

The Art of Noticing Productive Crossings

It is unusual for natural scientists and humanists to have more than passing conversations about their work—yet learning about the conditions of livability in these dangerous times must surely be a common task. Consider, then, the excitement of the following exchange.

At the 2014 conference that forms the basis for this book, historian Kate Brown gave a talk about the sufferings of residents of the former plutonium-manufacturing district in Russia, where radioactive traces still course through soil and water. The residents' bodies were suffused with illness and unease. They complained of chronic fatigue, chronic pain, and digestive, circulatory, and immune disorders; they showed scars from multiple operations. Yet doctors could find no clear trace relating their multiple illnesses to radiation from the plutonium plant. The physicians checked for cancers traditionally associated with exposure to radioactivity; they did not find them. The patients' unspecific maladies did not fit standard diagnostic categories, and the doctors turned the residents away without treatment. The residents felt disregarded and betrayed.

Microbiologist Margaret McFall-Ngai listened to Brown's talk with interest: she recognized every symptom Brown listed from her own research, which focuses on how microbes affect the development of organisms, including humans. Rather than diffuse complaints, a product of bad living, as doctors had argued, McFall-Ngai thought all those ailments could easily arise from one cause: mutations in intestinal bacteria. Chronic doses of radiation that might not yet stimulate a human cell cancer could easily have caused bacterial mutations. It

seems likely that the plutonium-district residents were suffering from the ills of their microbial companions.

Suffering from the ills of another species: this is the condition of the Anthropocene, for humans and nonhumans alike. This suffering is a matter not just of empathy but also of material interdependence. We are mixed up with other species; we cannot live without them. Without intestinal bacteria, we cannot digest our food. Without endosymbiotic dinoflagellates, coral polyps lose their vitality. Yet such monstrosities have been anathema to the organization of modern industrial progress. Ironically, the denial of the monstrosity of entanglement has turned this life-making trait against us. Industrial campaigns exterminate impurities, undermining the coordinations that make life possible. Plantations grow monocultures, or single crops that deny the intimacies of companion species. Modern dairy and meat farms raise a handful of supercharged breeds. A new kind of monstrosity attacks us: our entanglements, blocked and concealed in these simplifications, return as virulent pathogens and spreading toxins. Industrial chemicals weave their way through our food webs; nuclear by-products sicken us not just through our human cells but also through our bacteria.

How shall we approach such blowback of the modern? Thinking together, a historian and a microbiologist found a new research problem, a problem both specific and of great import for our times. Their cross-disciplinary curiosity about the microbial worlds of the radiation-affected residents opens up questions about the multispecies mixes that make up our worlds. Brown and McFall-Ngai are both contributors to this book, and their dialogue is at once an example and a parable for the work that *Arts of Living on a Damaged Planet* seeks to do. We live on a human-damaged planet, contaminated by industrial pollution and losing more species every year—seemingly without possibilities for cleanup or replacement. Our continued survival demands that we learn something about how best to live and die within the entanglements we have. We need both senses of monstrosity: entanglement as life and as danger.

But Who, We Must Ask, Are "We"?

In the twentieth century, the natural and social sciences alike imagined the world as composed of individuals—with distinct bodies, genomes, and vested interests. Symbioses, when they were recognized,

were considered rare, anomalies in a world characterized by individual autonomy and relentless competition. It turns out, however, that such assumptions were wrong. Twenty-first-century research on organisms ranging from bacteria to insects to mammals has shown that symbiosis is a near-requirement for life—even for *Homo sapiens.* As developmental biologist Scott Gilbert explains in this side of the volume, our bodies contain more bacterial cells than human ones. Without bacteria, our immune systems do not develop properly. Even reproduction appears to be bacteria enabled. Life, put simply, is symbiosis "all the way down."

As Donna Haraway suggests, recognizing the importance of symbiotic makings *(sympoiesis)* is just the beginning of "staying with the trouble." Symbiotic relations must be constantly renewed and negotiated within life's entanglements. When conditions suddenly shift, once life-sustaining relations sometimes turn deadly. The case of low-dose chronic exposure to radioactivity shows us what can happen when symbiotic alliances are broken: essential gut microbes mutate into illness-causing enemies. Symbioses are vulnerable; the fate of one species can change whole ecosystems. As Ingrid Parker reminds us in her essay, the commercial hunting of sea otters off Pacific North America changed kelp forests to sea urchin barrens; without the otters, urchins took over. Because they were connected by common soil ecologies, whole suites of perennial grasses and wildflowers disappeared in California with the invasion of European annual grasses. This is one of the challenges of our times: entanglement with others makes life possible, but when one relationship goes awry, the repercussions ripple.

What Kinds of Monsters Are We Now?

Life has been monstrous almost from its beginnings. In ancient times, prokaryotes (bacteria and archaea) gave birth to monsters in which one organism engulfed others or joined immoderate liaisons, forming nucleated cells and multicellular organisms called eukaryotes. Ever since, we have muddled along in our mixes and messes. All eukaryotic life is monstrous.

Enlightenment Europe, however, tried to banish monsters. Monsters were identified with the irrational and the archaic. Category-crossing beings were abhorrent to Enlightenment ways of ordering

the world; sometimes they were classified as things of the devil, the antithesis of godly purity. Martin Luther, the Protestant reformer, identified the Catholic Church with monstrosity: in one vivid image, he offered a Papal Ass, a creature with the head of an ass and the breasts and belly of a woman. Luther helped forge what we think of as the modern world through his campaign against category-crossing monsters. But the forms of progress and rationalization that the Enlightenment and Reformation sparked have proved far scarier than the beasts they sought to banish. For later thinkers, rationalization meant individualization, the creation of distinct and alienated individuals, human and nonhuman. The landscape-making practices that followed from these new figures imagined the world as a space filled with autonomous entities and separable kinds, ones that could be easily aligned with capitalist fantasies of endless growth from alienated labor.

Ironically, the monstrosity of monocultures depends on the very multispecies relations that it denies. Anthropologist Marianne Lien provides a striking example of this logic of denial and dependence from Norwegian salmon farms. Commercial aquaculture aims to produce salmon exclusively, but this has proved impossible. When salmon are kept in close quarters, populations of sea lice—a naturally occurring but normally spatially scattered fish parasite—explode. Because the lice threaten fish health, farms first turned to chemical baths and medicated feeds, but the lice soon became resistant to the drugs. The situation forced the farms to turn to a multispecies intervention: putting wrasse, a lice-loving "cleaner fish," into the salmon pens to eat the parasites off the fishes' bodies. But wild wrasse populations were too small for the vast needs of the industry, so they had to begin farming wrasse. The wrasse had their own suite of relations: when young, wrasse require a diet of tiny crustaceans, served live. These copepods, however, proved hard to collect, so now they, too, must be cultivated.

The "simplifications" of industrial farming multiply beyond the original target species. Their multispecies modifications create ever more monsters—exploding numbers of parasites, drug-resistant bacteria, and more virulent diseases—by disrupting and torqueing the monsters that sustain life. The ecological simplifications of the modern world—products of the abhorrence of monsters—have turned monstrosity back against us, conjuring new threats to livability.

We Begin with Noticing

The seductive simplifications of industrial production threaten to render us blind to monstrosity in all its forms by covering over both lively and destructive connections. They bury once-vibrant rivers under urban concrete and obscure increasing inequalities beneath discourses of freedom and personal responsibility. Somehow, in the midst of ruins, we must maintain enough curiosity to notice the strange and wonderful as well as the terrible and terrifying. Natural history and ethnographic attentiveness—themselves products of modern projects—offer starting points for such curiosity, along with vernacular and indigenous knowledge practices. Such curiosity also means working against singular notions of modernity. How can we repurpose the tools of modernity against the terrors of Progress to make visible the other worlds it has ignored and damaged? Living in a time of planetary catastrophe thus begins with a practice at once humble and difficult: noticing the worlds around us.

Our monsters and ghosts help us notice landscapes of entanglement, bodies with other bodies, time with other times. They aid us in our call for a particular approach to noticing—one that draws inspiration from scientific observation alongside ethnography and critical theory. Ant expert Deborah Gordon embodies the forms of curiosity we hope to cultivate. Rather than be lulled by liberal economic theories, with their focus on individual determination of group outcomes, Gordon begins with questions about "collective behavior"—already in the realm of the monstrous. As a biologist committed to long-term fieldwork, Gordon has spent more than two decades observing ant interactions with the eye of a natural historian. Based on these observations, she has designed new kinds of experiments that show the flexible interdeterminacies of ant interactions with each other. Where other observers saw only rigid and mechanical "castes," Gordon was able to notice how ants are not individuals but shifting senses and signals that respond to situations of encounter as well as their environment. Mycologist Anne Pringle similarly enters the monstrous world of lichens, entanglements of algae and fungi. To study lichens, Pringle must begin by giving up modernist units of individuals and populations.

The modes of noticing we propose are purposefully promiscuous. The rigid segregation of the humanities and natural sciences was an

ideology for modern Man's conquest, but it is a poor tool for collaborative survival. Co-species survival requires arts of imagination as much as scientific specifications. But symbiotic scholarship takes time to evolve: many scholars in this book have spent decades in dialogue with others beyond their fields. Perhaps counterintuitively, slowing down to listen to the world—empirically and imaginatively at the same time—seems our only hope in a moment of crisis and urgency.

Our modes of noticing, however, are themselves monstrous in their connection to Man's conquest. Much of what we know about ecological connection comes from tracking the movement of radiation and other pollutants. Contamination often acts as a "tracer"—a way to see relations. We notice connections in part through their ruination; we see the importance of dinoflagellates to coral reefs only as the corals bleach and die. It is urgent that we start paying attention to more of our companions before we kill them off entirely.

We Listen for Modes of Storytelling

Some kinds of stories help us notice; others get in our way. Modern heroes—the guardians of progress across disciplines—are part of the problem. Thus, for example, McFall-Ngai has suggested that biologist Lynn Margulis, who first imagined symbiosis as the origin of cells, has not been accorded the preeminence she deserves because she is a woman and thus not eligible for hero status. Male scientists tend only to cite men, she explained, while women scientists tend to cite male and female scientists equally. Unless we learn to listen broadly, we may miss the biggest story of life on earth: symbiogenesis, the co-making of living things. Practices of storytelling matter.

Several forms of noticing and telling gather in *Arts of Living on a Damaged Planet*. We begin with creative writing, the necessary stimulus to imagining pasts, presents, and the yet-to-come. Ursula K. Le Guin starts off this half of the volume. She brings us into the craft of writing itself, always already part of other stories: "It's just part of a story, actually quite a lot of stories // if I'll only listen." There she teaches us quite properly to fear: "Whiteness crossed the continent / a poison fog where it went / villages were vacant / hearths and ways forsaken." And yet she shows us wonder, as the ocean "holds the whales as lightly / as the body holds the soul," even as it mixes the "slow swirl of pelagic polymers" and radioactive waste. Creative writing invites us to

imagine the world differently, to listen beyond newspaper headlines to hear those quiet stories about the Anthropocene whispered in small encounters. Imaginative writing draws us into what Donna Haraway, in her chapter, calls "art-science activisms," "sympoietic practices for living on a damaged planet."

To these imaginative frames we add the sciences of bodies tumbled into bodies, from developmental biology to ecology and from observation of ants to reflection on extinct elephants and rhinoceroses. They show us lichens, women in childbirth, strange sea creatures, missing wildflowers, and much more. Then, too, we need the environmental humanities and social sciences, which tell us of human and nonhuman histories, cultures and texts; they bring us into assemblages of power and meaning. We follow wolves, tentacular monsters, flying foxes, and stumps of chestnut trees. There are hybrid scholars, too, working across these lines, such as Donna Haraway, both biologist and cultural theorist; Karen Barad, a quantum field physicist and feminist philosopher; and Andrew Mathews, forester and anthropologist. They show us how to move beyond the exclusions that blocked our attention to cross-species entanglement. We follow kinds of stories that take us beyond the modern individual. Watching and writing: these, too, are arts of living.

Not all stories are equally useful in engaging us with collaborative survival, arts of living on a damaged planet. In her essay "The Carrier Bag Theory of Fiction," Ursula Le Guin quotes a Virginia Woolf glossary in which Woolf defines "heroism" as "botulism." This delightfully unexpected definition can again reframe the problem of livability in the Anthropocene. Woolf's "heroism" might stand in for the enactment of Man's conquest of Nature. This form of heroism has been a dream of modernity—and a cause of contemporary fears for life on earth. Heroic conquests, from big dams to mass relocations, have been dangerous acts, erasing many lives. Botulism is a form of food poisoning most commonly associated with canning; the anaerobic world inside the can may encourage the growth of toxin-producing *Clostridium botulinum* bacteria. These bacteria are common in soil and water, but they only produce toxins under special conditions, such as life inside a can. The aluminum can, a mid-twentieth-century invention, is a fitting icon of modern civilization and industrial distribution. The botulism in the can is similarly an icon of the monstrosities of the Anthropocene. Like radioactive contamination and proliferating sea lice, botulism is

produced from within the heart of modernity. Heroism—the story line of modern progress—is thus readable, indeed, as botulism. Livability in the Anthropocene is threatened by just those heroic story lines and practices that are thought to have made Man great.

Are there alternatives to heroism/botulism? Le Guin's essay suggests "carrier bags" as another way to tell a story. Collecting offers stories with more complex arcs of temporality, she argues; instead of a hero single-handedly making the future, there are entanglements and losses of many kinds.

Monsters are bodies tumbled into bodies; the art of telling monstrosity requires stories tumbled into stories. This is what literary critic Carla Freccero shows us through her attention to the jointly material and semiotic worlds of wolf–human relations. As she slips between literary tropes of the "lone wolf" and practices of wolf killing, forests tumble into fables tumble into politics. Material worlds and the stories we tell about them are bound up with each other. Meanwhile, biologist Andreas Hejnol shows us a dizzying range of body forms, from tapeworms to tunicates, in which each organism inherits the evolutionary solutions of its predecessors; old body plans are always mixed into contemporary ways of life. If we do not let progress "ladders" possess us, we are forced to recognize the monstrous in transformation. Classification systems are monster stories—and ghost stories—too. Nor can the question of monsters stop at the boundaries of life. An anthropogenic mud volcano, the subject of anthropologist Nils Bubandt's essay, is monstrous in just the ways we have been describing: both part of our natural connectedness and a threat to life. Spirits and stones emerge from the mud as our new sympoietic companions: they become part of us, and they urge us, as Haraway puts it, to stay with the trouble.

In this spirit, *Arts of Living on a Damaged Planet* is itself entangled. The volume seeks to draw out, rather than to simplify or banish, monsters and ghosts. It juxtaposes many genres to show how varied storytelling styles might inform each other both in learning about our challenged planet and in forging strategies for living with others in the yet-to-come.

While this introduction uses monsters to mix up bodies, challenging the rhetorical reign of the autonomous individual, the introduction to the other half of *Arts of Living on a Damaged Planet* uses ghosts. Ghosts show the layered temporalities of living and dying

that shape our landscapes, tripping up the forward march of progress. Ghosts, like monsters, are creatures of ambivalent entanglement. The landscape assemblages of multispecies living are possible because of ghosts; modern Man's singular timelines occlude our vision. Turn this book over and follow ghosts.

As an anthropologist with a lifelong interest in the worlds salmon and humans create together, **HEATHER SWANSON** explores the globe-spanning connections and comparisons of multispecies interactions. She is an assistant professor at Aarhus University and a postdoctoral researcher with the Aarhus University Research on the Anthropocene (AURA) project. She was a 2015–2016 fellow with the "Arctic Domestication in the Era of the Anthropocene" project, funded and hosted by the Centre for Advanced Study in Oslo, Norway.

ANNA TSING conceived of the conference from which this volume grew to spin common threads of curiosity across natural science, humanities, arts, and social science. She is a professor of anthropology at the University of California, Santa Cruz, and a Niels Bohr Professor with Aarhus University Research on the Anthropocene (AURA). Her most recent book, *The Mushroom at the End of the World: On the Possibility of Life in Capitalist Ruins,* received the Gregory Bateson Prize and the Victor Turner Prize in Ethnographic Writing.

An anthropologist, **NILS BUBANDT** is professor at Aarhus University. He is a co-convener of Aarhus University Research on the Anthropocene (AURA), with Anna Tsing, and editor in chief of the journal *Ethnos,* with Mark Graham. In his book *The Empty Seashell: Witchcraft and Doubt on an Indonesian Island,* he explores the relationship between monstrosity and uncertainty. He also takes plane rides to conferences about the Anthropocene with a certain amount of ambivalence.

ELAINE GAN makes clocks and time machines as speculative devices for sensing and mapping worlds otherwise. Working at the intersection of digital arts, environmental anthropology, and science and

technology studies, she traces coordinations with things as small as mycorrhizae and as large as rivers. A recent curatorial collaboration, *DUMP! Multispecies Making and Unmaking* at Kunsthal Aarhus (2015), gathered artists, scientists, and organisms to explore multispecies socialities that persist in the garbage dumps and rubble of modernity and to contest the celebration of technoscientific fixes and human exceptionalism that permeates contemporary discourse.

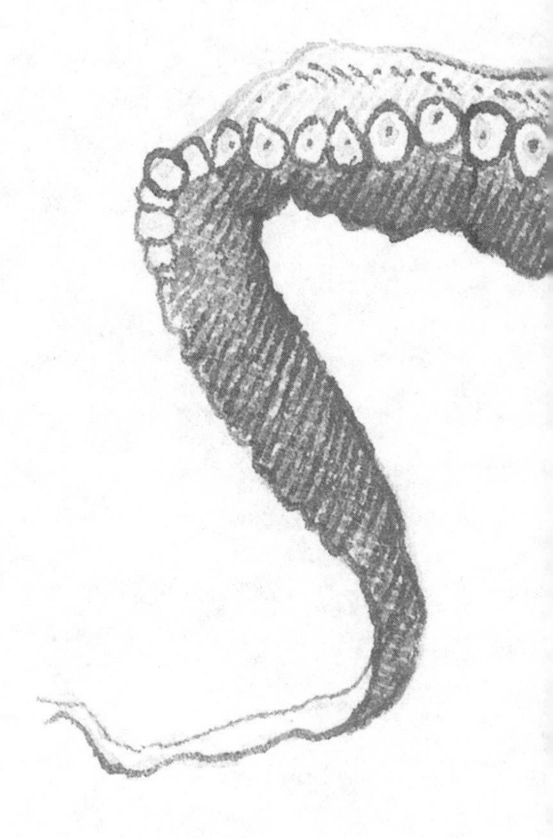

1

DEEP IN ADMIRATION

Ursula K. Le Guin

I HEARD THE POET BILL SIVERLY this week say that the essence of modern high technology is to consider the world as disposable: use it and throw it away. The people at this conference are here to think about how to get outside the mind-set that sees the technofix as the answer to all problems. It's easy to say we don't need more "high" technologies inescapably dependent on despoliation of the earth. It's easy to say we need recyclable, sustainable technologies, old and new—pottery making, bricklaying, sewing, weaving, carpentry, plumbing, solar power, farming, IT devices, whatever. But here, in the midst of our orgy of being lords of creation, texting as we drive, it's hard to put down the smartphone and stop looking for the next technofix. Changing our minds is going to be a big change. To use the world well, to be able to stop wasting it and our time in it, we need to relearn our being in it.

Skill in living, awareness of belonging to the world, delight in being part of the world, always tends to involve knowing our kinship as animals with animals. Darwin first gave that knowledge a scientific basis. And now, both poets and scientists are extending the rational aspect of our sense of relationship to creatures without nervous systems and to nonliving beings—our fellowship as creatures with other creatures, things with other things.

Relationship among all things appears to be complex and reciprocal—always at least two-way, back and forth. It seems that nothing is single in this universe, and nothing goes one way.

In this view, we humans appear as particularly lively, intense, aware nodes of relation in an infinite network of connections, simple or complicated, direct or hidden, strong or delicate, temporary or very long-lasting. A web of connections, infinite but locally fragile, with and among everything—all beings—including what we generally class as things, objects.

Descartes and the behaviorists willfully saw dogs as machines, without feeling. Is seeing plants as without feeling a similar arrogance?

One way to stop seeing trees, or rivers, or hills, only as "natural resources" is to class them as fellow beings—kinfolk.

I guess I'm trying to subjectify the universe, because look where objectifying it has gotten us. To subjectify is not necessarily to co-opt, colonize, exploit. Rather, it may involve a great reach outward of the mind and imagination.

What tools have we got to help us make that reach?

In *Romantic Things,* Mary Jacobus writes, "The regulated speech of poetry may be as close as we can get to such things—to the stilled voice of the inanimate object or insentient standing of trees."[1]

Poetry is the human language that can try to say what a tree or a rock or a river *is,* that is, to speak humanly *for it,* in both senses of the word "for." A poem can do so by relating the quality of an individual human relationship to a thing, a rock or river or tree, or simply by describing the thing as truthfully as possible.

Science describes accurately from outside; poetry describes accurately from inside. Science explicates; poetry implicates. Both celebrate what they describe. We need the languages of both science and poetry to save us from merely stockpiling endless "information" that fails to inform our ignorance or our irresponsibility.

By replacing unfounded, willful opinion, science can increase moral sensitivity; by demonstrating and performing aesthetic order or beauty, poetry can move minds to the sense of fellowship that prevents careless usage and exploitation of our fellow beings, waste and cruelty.

Poetry often serves religion; and the monotheistic religions, privileging humanity's relationship with the divine, encourage arrogance. Yet even in that hard soil, poetry will find the language of compassionate fellowship with our fellow beings.

The seventeenth-century Christian mystic Henry Vaughan wrote:

> So hills and valleys into singing break,
> And though poor stones have neither speech nor tongue,
> While active winds and streams both run and speak,
> Yet stones are deep in admiration.

By admiration, Vaughan meant reverence for God's sacred order of things, and joy in it, delight. By admiration, I understand reverence

for the infinite connectedness, the naturally sacred order of things, and joy in it, delight. So we admit stones to our holy communion; so the stones may admit us to theirs.

This talk was followed by a reading of a few of the author's poems.

THE MARROW

There was a word inside a stone.
I tried to pry it clear,
mallet and chisel, pick and gad,
until the stone was dropping blood,
but still I could not hear
the word the stone had said.

I threw it down beside the road
among a thousand stones
and as I turned away it cried
the word aloud within my ear
and the marrow of my bones
heard, and replied.

TAO SONG

O slow fish
show me the way
O green weed
grow me the way
The way you go
the way you grow
is the way
indeed
O bright Sun
light me the way
the right way
the one
no one can say
If one can choose it
it is wrong
Sing me the way
O song:
No one can lose it
for long

THE STORY

It's just part of a story, actually quite a lot of stories,
the part where the third son or the stepdaughter
sent on the impossible mission through the uncanny forest
comes across a fox with its paw caught in a trap
or little sparrows fallen from the nest
or some ants in trouble in a puddle of water.
He frees the fox, she puts the fledglings in the nest,
they get the ants safe to their ant-hill.
The little fox will come back later
and lead him to the castle where the princess is imprisoned,
the sparrow will fly before her to where the golden egg is hidden,
the ants will sort out every poppyseed for them
from the heap of sand before the fatal morning,
and I don't think I can add much to this story.
All my life it's been telling me
if I'll only listen who the hero is
and how to live happily ever after.

KINSHIP

Very slowly burning, the big forest tree
stands in the slight hollow of the snow
melted around it by the mild, long
heat of its being and its will to be
root, trunk, branch, leaf, and know
earth dark, sun light, wind touch, bird song.
Rootless and restless and warmblooded, we
blaze in the flare that blinds us to that slow,
tall, fraternal fire of life as strong
now as in the seedling two centuries ago.

WHITENESS

MEDITATIONS FOR MELVILLE

i

Whiteness crossed the continent
a poison fog where it went
villages were vacant
hearths and ways forsaken
Whiteness with greed and iron
makes the deep seas barren
Great migrations fly daylong
into whiteness and are gone

ii
Whiteness in its righteousness
bleaches creatures colorless
tolerates no
shadow

iii
People walk unseeing unseen
staring at a little screen
where the whiteness plays
an imitation of their days
Plugged in their ears white noise
drowns an ancient voice
murmuring to bless
darkness

INFINITIVE

We make too much history.

With or without us
there will be the silence
and the rocks and the far shining.
But what we need to be
is, oh, the small talk of swallows
in evening over
dull water under willows.
To be we need to know the river
holds the salmon and the ocean
holds the whales as lightly
as the body holds the soul
in the present tense, in the present tense.

FUTUROLOGY

I cannot break free from these iron stars.
I want the raspberry paw-pads of the fox,
but here are only claws, the Crab, the Scorpion,
great shining signs that slide across the sky.
I want the wisdom ignorant of wars
and the soft key that opens all the locks.
I want the touch of fur, the slant of sun
deep in a golden, slotted, changing eye.
O let there be no signs! Let fall the bars,
and walls be moss-grown, scattered rocks.
Let all the evil we have done be done
and minds lie still as sunlit meadows lie.

HYMN TO APHRODITE

Venus solis occasus orientisque, Dea pacifica,
foam-borne, implacable, tender:
war and storm serve you, and you wear
the fiery tiara of the volcanoes.
The young salmon swimming downriver
and the old upstream to breed and die
are yours, and the fog-drinking forests.
Yours are the scattered emerald half-circles
of islands, the lost islands. Yours
are the sunken warships of the Emperor.
Yours is each raindrop of the vast typhoon,
and the slow swirl of pelagic polymers.
The depths below all light are yours.
The moon is your hand-mirror.
Mother of Time and daughter of Destruction,
your feet are light upon the waters.
Death your dog follows you down the beaches
whining to see the breakers break
into blossom, into immortal
foam-flowers, where you have left
the bright track of your passing.
Pity your fearful, foolish children,
O Aphrodite of Fukushima.

As one of the most influential writers of our time, **URSULA K. LE GUIN** has stunned and stimulated many kinds of readers: from children to elders and from general readers to natural scientists, artists, humanists, and anthropologists. She is the author of many books, poems, and short stories, including *The Left Hand of Darkness* and *The Dispossessed.* Among many honors and awards, she holds the National Book Foundation Medal for Distinguished Contribution to American Letters. She is widely recognized for exploring the radical possibilities of society in her work—and the potential for varied ways humans might interact with the environment. She has consistently stretched Western environmental imaginations, inspiring what one scholar has called an "environmental paradigm shift."

Notes

Given at the conference "Anthropocene: Arts of Living on a Damaged Planet" at the University of California, Santa Cruz in May 2014, this short talk sums up ideas that many of my poems of the last few years have expressed or have been groping toward.

1. Mary Jacobus, *Romantic Things* (Chicago: University of Chicago Press, 2012).

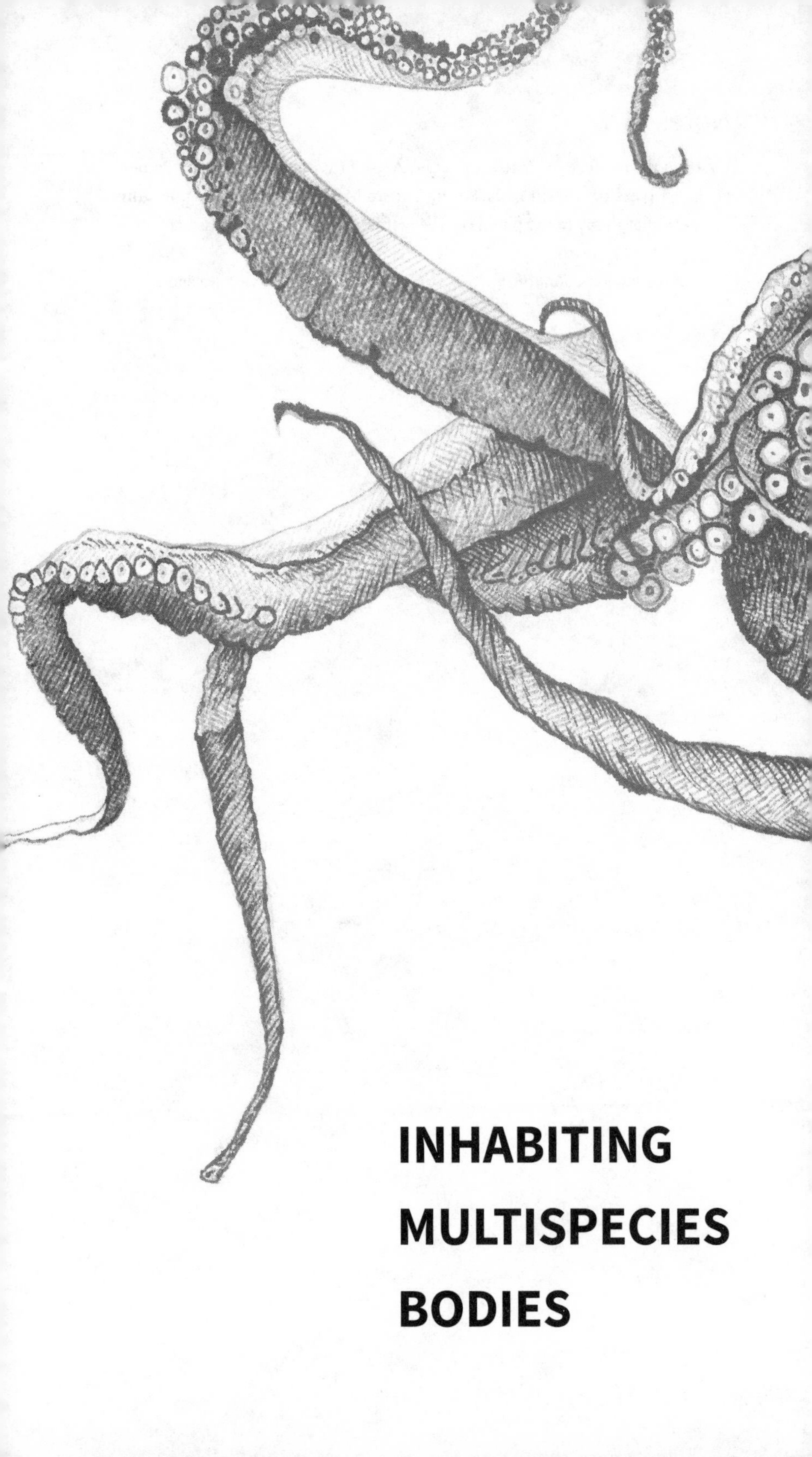

INHABITING MULTISPECIES BODIES

THERE IS A BIG NEW STORY IN BIOLOGY TODAY. Zoologist Margaret McFall-Ngai calls it the "postmodern synthesis." This postmodern is not the same as the *postmodernism* of the humanities. It requires readers to know that the "modern synthesis" in biology was a marriage of Darwinian evolutionary theory and Mendelian inheritance mechanisms; evolution worked through the passing of desirable traits to offspring. Every species evolved on its own. In the big new story, cross-species interaction has been shown as essential to development, evolution, and ecology. This volume offers articles by a number of the most prominent pioneers in this field, including Margaret McFall-Ngai, Scott Gilbert (see "Beyond Individuals"), and Donna Haraway. This trio has worked together, across distance and disciplinary boundaries, to move us toward new practices of research, inquiry, and storytelling, where natures and cultures, microorganisms and worlds, bodies and environments, are driven not by individuals but by symbiotic relationships all the way down.

McFall-Ngai explains the "nested ecosystems" that form our bodies in cross-species interaction. Her experimental organism, the bobtailed squid, has become a model for the whole field: bobtailed squid can only evade predators through their relation with *Vibrio* bacteria, with which they develop a light organ. The more we learn about microbes, the more we realize how integral they are to the lives of multicellular organisms.

Donna Haraway is a leading figure in the formation of radical interdisciplinarities, the combination of arts, sciences, and humanities into modes of "worlding," where McFall-Ngai's nested ecosystems and Gilbert's holobionts become our primary characters. This move challenges us to inhabit new ecologies, invent new words, and work across many knowledge practices; these demand new stories. "Worlding" is theorizing and storytelling that is rooted in the historical materialities of meetings between humans and nonhumans. In her chapter, she uses art–science activisms to craft new genres, what she calls "speculative fabulations."

For decades, McFall-Ngai, Haraway, and Gilbert have read and cited each other's work, diffracting their practices across disciplinary silos. Long-term dialogues across the science–humanities line are arts we need for living on a damaged planet. ●

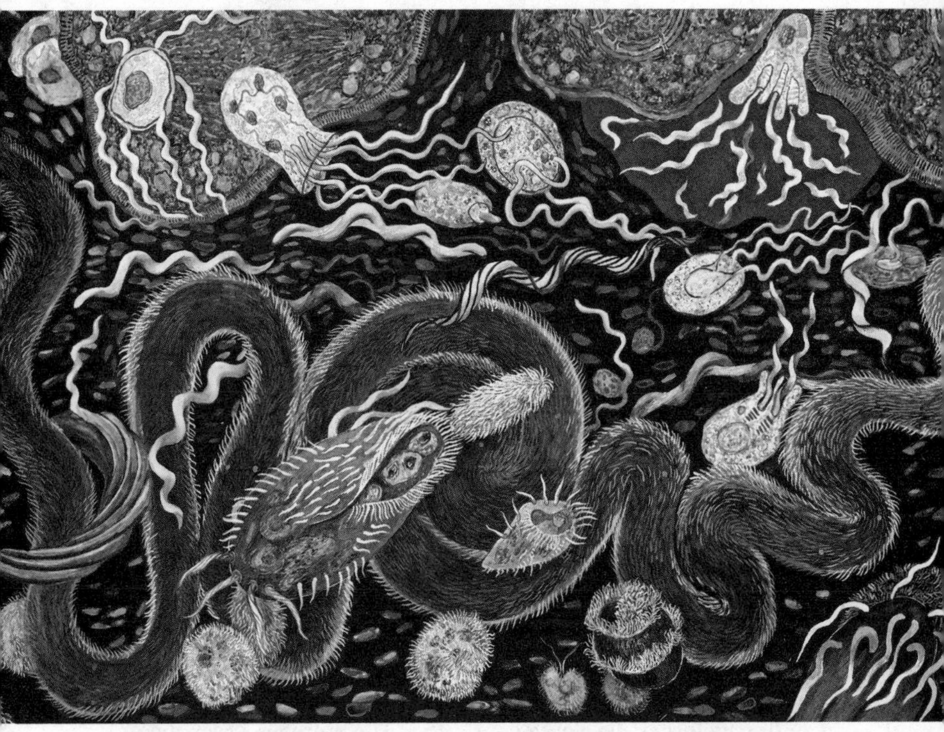

Figure M2.1. Shoshanah Dubiner, *Endosymbiosis: Homage to Lynn Margulis,* 2012. Gouache on watercolor paper, 23 × 35 inches. Reproduced for the Morrill Science Center, University of Massachusetts, Amherst, as a fine-art giclée on canvas, 48 × 72 inches. Courtesy of the artist. http://www.cybermuse.com/.

2

SYMBIOGENESIS, SYMPOIESIS, AND ART SCIENCE ACTIVISMS FOR STAYING WITH THE TROUBLE

Donna Haraway

Symbiogenesis

Shoshanah Dubiner's vivid painting called *Endosymbiosis* (Figure M2.1) hangs in the hallway joining the Departments of Geosciences and Biology at UMass Amherst, near the Life and Earth Café, a spatial clue to how critters become with each other.[1] Irresistible attraction toward enfolding each other is the vital motor of living and dying on earth. Critters interpenetrate one another, loop around and through one another, eat each another, get indigestion, and partially digest and partially assimilate one another, and thereby establish sympoietic arrangements that are otherwise known as cells, organisms, and ecological assemblages.

Sym-poiesis is a simple word; it means "making-with." Nothing makes itself; nothing is really auto-poietic or self-organizing. In the words of the Iñupiat computer "world game," earthlings are Never Alone.[2] That is the radical implication of sympoiesis. Sympoiesis is a word proper to complex, dynamic, responsive, situated, historical systems. It is a word for worlding.

Another word for these sympoietic entities is *holobionts,* or, etymologically, "entire beings" or "safe and sound beings."[3] That is decidedly not the same thing as One and Individual. Rather, in polytemporal, polyspatial knottings, holobionts hold together contingently and dynamically, engaging other holobionts in complex patternings. Critters do not precede their relatings; they make each other through semiotic material involution, out of the beings of previous such entanglements. Margulis (1938–2011) knew a great deal about "the co-opting of strangers," a phrase she proposed to describe the most fundamental practices of critters becoming with each other at every node of intra-action in earth history.[4]

Like Margulis, I use *holobiont* to mean symbiotic assemblages, at whatever scale of space or time, which are more like knots of diverse intra-active relatings in dynamic complex systems than like the entities of a biology made up of preexisting bounded units (genes, cells, organisms, etc.) in interactions that can be conceived only as competitive or cooperative. Like hers, my use of *holobiont* does not designate host + symbionts, because all the players are symbionts to each other, in diverse kinds of relationalities and with varying degrees of openness to attachments and assemblages with other holobionts. *Symbiosis* is not a synonym for "mutually beneficial." The array of names needed to designate the heterogeneous webbed patterns and processes of situated and dynamic dilemmas and advantages for the symbionts/holobionts is only beginning to surface as biologists let go of the dictates of methodological individualism and zero-sum games as the template for explanation.[5] I suggest we might also need a term like *holoent,* so as not to privilege only the living but to encompass the biotic and abiotic in dynamic sympoietic patterning.

An adept in microbiology, cell biology, chemistry, geology, and paleogeography, as well as a lover of languages, arts, stories, systems theories, and alarmingly generative critters, including human beings, Margulis was a radical evolutionary theorist. Her first and most intense loves were the bacteria and archaea of Terra and all their bumptious doings. The core of Margulis's view of life was that new *kinds* of cells, tissues, organs, and species evolve primarily through the long-lasting intimacy of strangers. The fusion of genomes in symbioses, followed by natural selection—with a very modest role for mutation as a motor of system-level change—leads to increasingly complex levels of good enough quasi-individuality to get through the day, or

the eon. Margulis called this basic and mortal life-making process *symbiogenesis.*

Bacteria and archaea did it first. My sense is that in her heart of hearts, Margulis felt that bacteria and archaea did it all, and there wasn't much left for so-called higher-order biological entities to do or invent. Eventually, however, by fusing with each other in stabilized, ongoing ways, archaea and bacteria invented the modern complex cell, with its nucleus full of ropy chromosomes made of DNA and proteins and diverse other sorts of extranuclear organelles, from undulating whips and spinning blades for locomotion to specialized vesicles and tubules for functions that work better kept separate.[6] Because she was a founder of Gaia theory with James Lovelock and a student of interlocked and multileveled systemic processes of nonreductionist organization and maintenance that make earth itself and earth's living beings unique, Margulis called these processes *autopoietic.*[7] I think she would have often—not always—preferred the term *sympoietic,* but the word and concept had not yet surfaced.[8] Autopoiesis and sympoiesis are in generative friction rather than opposition.

In 1998, M. Beth Dempster suggested the term *sympoiesis* for "collectively-producing systems that do not have self-defined spatial or temporal boundaries. Information and control are distributed among components. The systems are evolutionary and have the potential for surprising change." By contrast, autopoietic systems are "self-producing" autonomous units "with self defined spatial or temporal boundaries that tend to be centrally controlled, homeostatic, and predictable."[9] Symbiosis makes trouble for autopoiesis, and symbiogenesis is an even bigger trouble maker for self-organizing individual units. The more ubiquitous symbiogenesis seems to be in living beings' dynamic organizing processes, the more looped, braided, outreaching, involuted, and sympoietic is terran worlding.

Mixotricha paradoxa is everyone's favorite critter for explaining complex "individuality," symbiogenesis, and symbiosis. Margulis described this critter that is/are made up of at least five different taxonomic *kinds* of cells with their genomes this way: "Under low magnification, *M. paradoxa* looks like a single-celled swimming ciliate. With the electron microscope, however, it is seen to consist of five distinct kinds of creatures. Externally, it is most obviously the kind of one-celled organism that is classified as a protist. But inside each nucleated cell, where one would expect to find mitochondria, are many spherical

bacteria. On the surface, where cilia should be, are some 250,000 hair-like *Treponema spirochetes* (resembling the type that causes syphilis), as well as a contingent of large rod bacteria that is also 250,000 strong. In addition, we have redescribed 200 spirochetes of a larger type and named them *Canaleparolina darwiniensis.*"[10] Leaving out viruses, each *M. paradoxa* is not one, not five, not several hundred thousand, but a poster critter for holobionts. This holobiont lives in the gut of an Australian termite, *Mastotermes darwiniensis,* which has its own SF stories to tell about ones and manys.

Since Darwin's *On the Origin of Species* in 1859, biological evolutionary theory has become more and more essential to our ability to think, feel, and act well; and the interlinked Darwinian sciences that came together between the 1930s and 1950s into the "Modern Synthesis" or "New Synthesis" remain astonishing. How could one be a serious person without such works as Theodosius Dobzhansky's *Genetics and the Origin of Species* (1937), Ernst Mayr's *Systematics and the Origin of Species* (1942), George Gaylord Simpson's *Tempo and Mode in Evolution* (1944), and even Richard Dawkins's sociobiological formulations within the Modern Synthesis, *The Extended Phenotype* (1982)? However, bounded units (code fragments, genes, cells, organisms, populations, species) and relations described mathematically in competition equations are virtually the only actors and story formats of the Modern Synthesis. Evolutionary momentum, always verging on modernist notions of progress, is a constant theme, although teleology in the strict sense is not. Even as these sciences lay the groundwork for scientific conceptualization of the Anthropocene, they are undone in the very thinking of Anthropocenic systems that require both autopoietic and sympoietic analysis.

Rooted in units and relations, especially competitive relations, these sciences, for example population genetics, have a hard time with four key biological domains: embryology and development, symbiosis and collaborative entanglements of holobionts and holobiomes, the vast worldings of microbes, and exuberant critter biobehavioral inter- and intra-actions. Approaches tuned to "multispecies becoming-with" better sustain us in staying with the trouble on Terra. An emerging "new new synthesis" (or "extended synthesis") in transdisciplinary biologies and arts proposes string figures tying together human and nonhuman ecologies, evolution, development, history, affects, performances, technologies, and more.

Indebted to Margulis, I am undone and redone by the "New New Synthesis" unfolding in the early twenty-first century.[11] Formulations of symbiogenesis predate Margulis in the early-twentieth-century work of the Russian Konstantin Mereschkowsky and others.[12] However, Margulis, her successors, and her colleagues bring together symbiogenetic imaginations and materialities with all of the powerful cyborg tools of the late-twentieth-century molecular and ultrastructural biological revolutions. The strength of the "New New Synthesis" is precisely in the intellectual, cultural, and technical convergence that makes it possible to develop new model systems, concrete experimental practices, research collaborations, and both narrative and mathematical explanatory instruments. Such a convergence was impossible before the 1970s and after.

A model is a work object; a model is not the same *kind* of thing as a metaphor or analogy. A model is worked, and it does work. A model is like a miniature cosmos, in which a biologically curious Alice in Wonderland can have tea with the Red Queen and ask how this world works, even as she is worked by the complex-enough, simple-enough world. Models in biological research are stabilized systems that can be shared among colleagues to investigate questions experimentally and theoretically. Traditionally, biology has had a small set of hard-working living models, each shaped in knots of practices to be apt for some kinds of questions and not others. Listing seven model systems of developmental biology, namely, fruit flies, a nematode, the house mouse, a frog, the zebra fish, the chicken, and a mustard, Scott Gilbert wrote, "The recognition that one's organism is a model system . . . assures one of a community of like-minded researchers who have identified problems that the community thinks are important."[13]

Excellent for studying how parts fit together into cooperating and/or competing units, all seven of these individuated systems fail the researcher studying symbiosis and sympoiesis, in heterogeneous temporalities and spatialities. Holobionts require models tuned to an expandable number of quasi-collective/quasi-individual partners in constitutive relatings; these relationalities *are* the objects of study. The partners do not precede the relatings. Such models are emerging for the transformative processes of EcologicalEvolutionaryDevelopmental biology.

Every living thing has emerged and persevered (or not) bathed and swaddled in bacteria and archaea. Truly nothing is sterile; and

that reality is a terrific danger, basic fact of life, and critter-making opportunity. Margulis gave us dynamic multipartnered entities like *M. paradoxa* to study the symbiogenetic invention of eukaryotic cells from the entangling of bacteria and archaea. Nicole King's laboratory has proposed the clumping and subsequent tissuelike formations of choanoflagellates in the presence of specific bacteria as a new model system for studying the symbiogenetic origin of animal multicellularity.[14] Margaret McFall-Ngai and her colleagues have proposed the necessary infection of juvenile Hawaiian bobtail squid by specific vibrio bacteria as a symbiogenetic model system to study developmental patterning, in this case constructing the squid's ventral pouch to house light-emitting bacteria, so the moon cannot cast its shadow over the hunting squid, thus alerting the prey below.[15] Other emerging model systems tuned to symbiosis and EcoEvoDevo in mammals include both mouse brain and immune system development responding to signals from gut bacteria.[16] Coral reefs are an immense model for studying holobiome formation at the ecosystem level.

The collaborations of critters are matched by the string figures linking disciplines and methodologies, including genome sequencing, imaging technologies, functional genomics, and field biology, which make symbiogenesis such a powerful framework for twenty-first-century biology. Working on pea aphid symbiosis with *Buchnera,* Nancy Moran emphasizes this point: "The primary reason that symbiosis research is suddenly active, after decades at the margins of mainstream biology, is that DNA technology and genomics give us enormous new ability to discover symbiont diversity, and more significantly, to reveal how microbial metabolic capabilities contribute to the functioning of hosts and biological communities."[17] I add the necessity of asking how the multicellular partners in the symbioses affect the microbial symbionts. At whatever size, all the partners making up holobionts are symbionts to each other. They are holoents.

Two transformative papers embody for me the profound scientific changes afoot. Proclaiming "We Have Never Been Individuals," Gilbert, Sapp, and Tauber argue for holobionts and a symbiotic view of life by summarizing the evidence against bounded units from anatomy, physiology, genetics, evolution, immunology, and development.[18] In the second paper, signaling "A New Imperative for the Life Sciences," Margaret McFall-Ngai and Michael Hadfield, with twenty-four coauthors, present a vast range of animal-bacterial symbiotic

interactions at both ecosystem and intimate scales. They argue that this evidence should profoundly alter approaches to five questions: "how have bacteria facilitated the origin and evolution of animals; how do animals and bacteria affect each other's genomes; how does normal animal development depend on bacterial partners; how is homeostasis maintained between animals and their symbionts; and how can ecological approaches deepen our understanding of the multiple levels of animal-bacterial interaction?"[19]

Stories about worried colleagues at conferences, uncomprehending reviewers unused to so much evidential and disciplinary boundary crossing in one paper, or initially enthusiastic editors getting cold feet surround these papers. Such stories normally surround risky and generative syntheses and propositions. The critics are crucial to the holobiome of making science, and I am not a disinterested observer. Nonetheless, I think it matters that both of these papers were published in prominent places at a critical inflection point in the curve of research on, and explanation of, complex biological systems in the urgent times called the Anthropocene, when the arts for living on a damaged planet demand sympoietic thinking and action.

Interlacing Sciences and Arts with Involutionary Momentum

I am committed to art science activisms as sympoietic practices for living on a damaged planet. Carla Hustak and Natasha Myers gave us a beautiful paper titled "Involutionary Momentum" that is a hinge between symbiogenesis and science art activisms. These authors reread Darwin's own sensuous writing about his attention to absurdly sexual orchids and their pollinating insects. Hustak and Myers attend to the enfoldings and communications among bees, wasps, orchids, and scientists. The authors suggest that "involution" powers the "evolution" of living and dying on earth. Rolling inward enables rolling outward; the shape of life's motion traces a hyperbolic space, swooping and fluting like the folds of a frilled lettuce, coral reef, or bit of crocheting. Like EcoEvoDevo biologists, Hustak and Myers argue that a zero-sum game based on competing methodological individualists is a caricature of the sensuous, juicy, chemical, biological, material-semiotic, and science-making world. Counting "articulate plants and other loquacious organisms" among their number, living critters

practice the floridly repetitive mathematics of hyperbolic geometry.[20] "It is in encounters among orchids, insects, and scientists that we find openings for an ecology of interspecies intimacies and subtle propositions. What is at stake in this involutionary approach is a theory of ecological relationality that takes seriously organisms' practices, their inventions, and experiments crafting interspecies lives and worlds. This is an ecology inspired by a feminist ethic of 'response-ability' . . . in which questions of species difference are always conjugated with attentions to affect, entanglement, and rupture; an affective ecology in which creativity and curiosity characterize the experimental forms of life of all kinds of practitioners, not only the humans."[21]

Orchids are famous for their flowers looking like the genitals of the female insects of the particular species needed to pollinate them. The right sort of males seeking females of their own kind are drawn to the color, shape, and alluring insectlike pheromones of a particular orchid. These interactions have been explained (away) in neo-Darwinian orthodoxy as nothing but biological deception and exploitation of the insect by the flower, that is, an excellent example of the selfish gene in action. Even in this hard case of strong asymmetry of "costs and benefits," Hustak and Myers read aslant neo-Darwinism. The stories of mutation, adaptation, and natural selection are not silenced, but they do not deafen scientists, as if the evidence demanded it, when increasingly something more complex is audible in research across fields. "This requires reading with our senses attuned to stories told in otherwise muted registers. Working athwart the reductive, mechanistic, and adaptationist logics that ground the ecological sciences, we offer a reading that amplifies accounts of the creative, improvisational, and fleeting practices through which plants and insects *involve* themselves in one another's lives."[22]

But what happens when a partner involved critically in the life of another disappears from the earth? What happens when holobionts break apart? What happens when entire holobiomes crumble into the rubble of broken symbionts? This kind of question has to be asked in the urgencies of the Anthropocene if we are to nurture arts for living on a damaged planet. In his science fiction novel *The Speaker for the Dead,* Orson Scott Card explored how a boy, who had excelled in exterminationist technoscience in a cross-species war with an insectoid hive species, later in life took up responsibility for the dead, for collecting up the stories for those left behind when a being, or a way of

being, dies.[23] The man had to do what the boy, immersed only in cyber-realities and deadly virtual war, was never allowed to do; the man had to visit, to live with, to face the dead and the living—including the unexpected survivors—in all of their semiotic materialities. The task of the Speaker for the Dead is to bring the dead into the present so as to make more response-able living and dying possible in times yet to come.

My hinge to science art activisms turns on the ongoing performance of memory by an orchid for its extinct bee.

In "Bee Orchid" (Figure M2.2), we know a vanished insect once existed because a living flower still looks like the erotic organs of the avid female bee hungry for copulation. But the cartoon does something special; it does *not* mistake lures for identities; it does *not* say the flower is exactly like the extinct insect's genitals. Instead, the flower collects up the presence of the bee aslant, in desire and mortality. The shape of the flower is "an idea of what the female bee looked like to the male bee . . . as interpreted by a plant. . . . The only memory of the bee is a painting by a dying flower."[24] No longer embraced by living buzzing bees, the flower is a speaker for the dead. A stick figure promises to remember the bee flower when it comes time. The arts of memory enfold terran critters. That must be part of any possibility for resurgence!

Science Art Activisms for Staying with the Trouble

Consider two science art activisms committed to partial healing, modest rehabilitation, and still possible resurgence in the hard times of the Anthropocene and Capitalocene. I think of these science art activisms as stinger-endowed, unfurling tentacles of the ink-spurting, disguise-artist, hunting critters of an ongoing past, present, and future that I call the Chthulucene.[25] The Chthulucene is the time-space of the sym-chthonic ones, the symbiogenetic and sympoietic earthly ones, those now submerged and squashed in the tunnels, caves, remnants, edges, and crevices of damaged waters, airs, and lands. To live and die well as mortal critters in the Chthulucene is to join forces to reconstitute refuges, to make possible partial and robust biological-cultural-political-technological recuperation and recomposition, which must include mourning irreversible losses.

Each science art project cultivates robust response-ability for

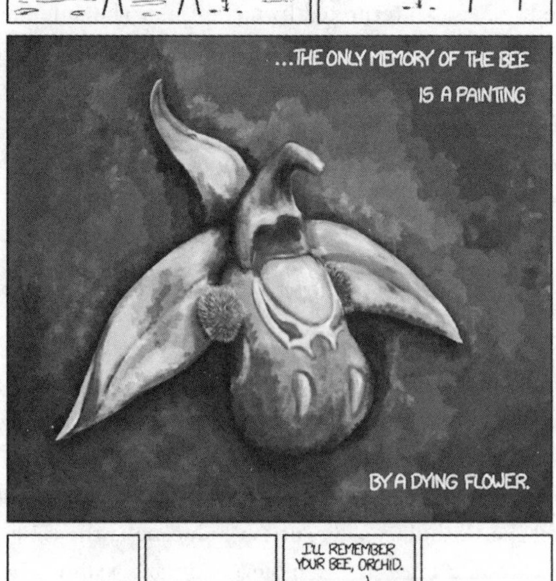

Figure M2.2. "Bee Orchid." Copyright Randall Munroe, http://xkcd.com/.

powerful and threatened places and beings. Each is located in a particularly sensitive place: the Great Barrier Reef and sister reefs, for the Crochet Coral Reef project, coordinated from the Institute for Figuring in Los Angeles, and the island Republic of Madagascar, for the Malagasy–English children's natural history book series called the Ako Project.[26] Each project is a case of noninnocent, risky, committed "becoming involved in one another's lives."[27] Each is a case of multispecies becoming-with, a model system in which scientists, artists, ordinary members of communities, and nonhuman beings become enfolded in each other's projects, in each other's lives. Each is an animating project in deadly times.

Like Anna Tsing's refugia in forests of the land, coral reefs are the forests of the sea, critical to resurgence for humans and nonhumans. "Resurgence is the work of many organisms, negotiating across differences, to forge assemblages of multispecies livability in the midst of disturbance."[28] Bathed in increasingly hot and acid oceans, coral holobiomes everywhere are threatened. Coral reefs have the highest biodiversity of any marine ecosystem. The symbiosis of cnidarian polyps, photosynthesizing dinoflagellates called zooanthellae living in coral tissues, and a hoard of microbes and viruses make up the keystone assemblage of the coral holobiome, home also to multitudes of other critters. Tens of millions of human beings, many of them very poor, depend directly on healthy coral ecosystems for their livelihoods. Recognition of bleached corals was crucial to advancing the terms *holobiont* and *holobiome* in the 1990s, *Anthropocene* in 2000, and *hologenome* in the 2000s. Corals, along with lichens, are the earliest instances of symbiosis recognized by biologists in the nineteenth century; these critters taught biologists to understand the parochialism of their ideas of individuals and collectives. These critters instruct people like me that we are all lichens, all coral. Besides all of this, coral reef worlds are achingly beautiful. I cannot imagine it is only people who know this beauty in their flesh.

A large island nation off the east coast of Africa, the Republic of Madagascar is home to complex, layered tapestries of historically situated peoples and other critters, including lemurs, close relatives of monkeys and apes. Nine out of ten kinds of Madagascar's nonhuman critters, including all species of lemurs, live nowhere else on earth. The rate of extinction and destruction of the many kinds of Madagascar's forests and watersheds vital for rural people (the large

majority of Madagascar's citizens), urban and town residents, and myriad nonhumans is almost beyond imagination, except that it is well advanced—but not uncontested locally and translocally. By 2015, only about 10 percent remained of the forests of Madagascar that were still thriving in the early twentieth century, despite a far from undisturbed history at that time. Forest well-being is one of the most urgent priorities for flourishing—indeed, survival—all over the earth. The contestations must matter; it's not a choice, it's a necessity.

The Crochet Coral Reef

In 1997, Daina Taimina, a Latvian mathematician at Cornell University, "finally worked out how to make a physical model of hyperbolic space that allows us to feel, and to tactilely explore the properties of this unique geometry. The method she used was crochet."[29] In 2005, after reading an article on coral bleaching, Christine Wertheim suggested to her twin sister Margaret, "We should crochet a coral reef" (Figure M2.3).[30] We can fight for the coral reefs that way, she intimated. The sisters were watching an episode of *Xena: Warrior Princess*; Xena's and her sidekick Gabrielle's fabulous fighting action inspired them.[31] The consequences have been utterly out of proportion to what the twin sisters imagined that night. So far, about eight thousand people, mostly women, in twenty-seven countries have come together to crochet in wool, cotton, plastic bags, discarded reel-to-reel tape, vinyl jelly yarn, plastic wrap, and anything else that can be induced to loop and whirl in the codes of crocheting.

The code is simple: crocheted models of hyperbolic planes achieve their ruffled forms by progressively increasing the number of stitches in each row. The emergent vitalities of this experimental life-form take diverse corporeal shape as crafters increase the numbers from row to row irregularly, strictly, or whimsically to see what forms they could make—not just any forms, but crenulated beings that take life as marine critters of vulnerable reefs.[32] "Every woolen form has its fibrous DNA."[33] But wool is hardly the only material. Plastic bottle anemone trees with trash tendrils and anemones made from *New York Times* blue plastic wrappers inhabit these reefs. Making fabulated, rarely mimetic, evocative models of coral reef ecosystems, the Crochet Coral Reef has morphed into what is probably the world's largest collaborative art project.

Figure M2.3. Beaded jellyfish made by Vonda N. McIntyre for the Crochet Coral Reef. From the collection of the Institute for Figuring. Copyright IFF.

The involutionary momentum of the Crochet Coral Reef powers the sympoietic knotting of mathematics, marine biology, environmental activism, ecological consciousness raising, women's handicrafts, fiber arts, museum display, and community art practices. A kind of hyperbolic embodied knowledge, the crochet reef stitches the materialities of global warming and toxic pollution. The makers of the reef practice multispecies becoming-with to cultivate the capacity to respond, response-ability. The crochet reef is the fruit of "algorithmic code, improvisational creativity, and community engagement."[34] The reef works not by representation but by open-ended, exploratory process. "Iterate, deviate, elaborate" are the principles of the process.[35]

The Crochet Coral Reef has core sets made for exhibitions, like the first ones at the Warhol Museum in Pittsburgh and the Chicago Cultural Center, both in 2007, to the Coral Forest exhibited in Abu Dhabi in 2014 and beyond. Morphing assemblages live at the Institute for Figuring and in the Wertheims' home. The IFF is the Wertheims' nonprofit organization, founded in 2003 and dedicated to "the aesthetic dimensions of mathematics, science, and engineering."[36] The core concept is material play, and the IFF proposes and enacts not think tanks or work tanks but play tanks, which I understand as arts for living on a damaged planet. The IFF and the Crochet Coral Reef are art science activisms, bringing people together to do string figures with math, sciences, and arts to make active attachments that might matter to resurgence in the Anthropocene and Capitalocene—that is, to make string figures in the Chthulucene. There are incarnations of a "biodiverse reef," "toxic reef," "bleached reef," "coral forest," "plastic midden," "bleached bone reef," "beaded coral garden," "coral forest medusa," and more, along with the satellite reefs made by collectives of crafters all over the world. Crafters make fabulated healthy reefs, but my sense is that most of the reefs bear the stigmata of plastic trash, bleaching, and toxic pollution. Crocheting with this trash feels to me like the looping of love and rage.

The skills and sensibilities of Margaret and Christine Wertheim, born in Brisbane near the Great Barrier Reef, are fundamental. With degrees in mathematics and physics, Margaret Wertheim is a science writer, curator, and artist. She has written extensively on the cultural history of theoretical physics. Over a million people have watched her 2009 TED talk on "The Beautiful Math of Coral."[37] With two books written in feminine feminist materialist poetics, Christine Wertheim

is a poet, performer, artist, critic, curator, crafter, and teacher. She aptly describes her work as "infesting fertile zones between cunning linguistics, psychoanalysis, poetry and gender studies."[38] These twin sisters were primed for sympoietic mergers.

Infecting each other and anyone who comes into contact with their fibrous critters, the thousands of crafters crochet psychological, material, and social attachments to biological reefs in the oceans, but not by practicing marine field biology, or by diving among the reefs, or by making some other direct contact. Rather, the crafters stitch "intimacy without proximity," a presence without disturbing the critters that animate the project, but with the power to confront the exterminationist, trashy, greedy practices of global industrial economies and cultures.[39] Intimacy without proximity is not "virtual" presence; it is "real" presence, in loopy materialities. The abstractions of the mathematics of crocheting are a lure to an affective cognitive ecology stitched in fiber arts. The crochet reef is a practice of caring without the neediness of touching by camera or hand in yet another travelogue of discovery. Material play builds caring publics. The result is a strong thread in the holobiome of the reef: we are all corals now.

The Madagascar Ako Project

As a Yale graduate student studying lemur behavior in 1962 in what is now the Berenty Primate Reserve, Alison Jolly fell into noninnocent love and knowledge in her first encounter with female-led, swaggering, opportunistic ringtail lemurs in the southern spiny forest. Transformed, this young six-foot-tall American white woman became a seeker of knowledge and well-being with and for the beings of Madagascar, especially the astonishing species of lemurs, the radically different forest ecosystems the length and breadth of the island, and the land's complex people and peoples. Author of many books and scientific papers and participant in numerous study and conservation teams, Jolly died in 2014. Her contributions to primatology, biodiversity conservation, and historically informed, passionate analyses of conservation conflicts and necessities were legion. But Jolly herself seemed especially to prize the sympoietic gift she helped craft, the Ako Project, which is tuned to practices for resurgence in vulnerable Malagasy worlds (Figure M2.4).[40] This is the part of her work I most love.[41]

Jolly understood well the terrible contradictions and frictions in

a Ohy. Niomana
ka. Tsy hiaraka
ndry vavy kely ry zareo.
R tamin'ilay zaza i
RR PRRRRR koa razandry
zana ary te- hitarika

xt day he set off again, alone.
sister. Maky Mazana said PURRRRRRR
zana wanted to stay home and become

Figure M2.4. Page from *Tik-Tik the Ringtailed Lemur/Tikitiki Ilay Maky.* UNICEF Madagascar and the Lemur Conservation Foundation. Text by Alison Jolly and Hanta Rasamimanana. Art by Deborah Ross. Courtesy Margaretta Jolly.

her embrace of *both* the rural people, who cut and burn the forests to make small agricultural plots called *tavy,* and her beloved prosimians with all their forest partners. Of course, she knew she was not Malagasy but at best a guest who might reciprocate appropriately and at worst another in a long line of colonizers, always taking land and giving advice for the best reasons. She knew better than almost any other Westerner and better than many Malagasy what made ongoing *tavy* burnings and other destructive practices so lethal to the future of the forests and all their critters, including the people who need them not just for their products (including lemurs for food) but to sustain fertility in phosphorous-poor, tropical, laterite soils. She also knew that *tavy* had once been part of the cycle of forest succession and biodiversity maintenance, with evidence in old stands in Ranamafuna Park.[42]

But not anymore. Nothing has time to regenerate anymore. Jolly knew in detail what the press of rapidly increasing human numbers means to the forests in the situated history of multiple land dispossessions, relocations, violent suppressions, a succession of failed national governments, huge solicited and imposed national debt, and broken development promises. She wrote vividly about local people's accurate assessment of the effects of generations of visiting experts, while the experts and visiting research scientists often knew little or nothing about the terrible history of land seizures, colonial and postcolonial search-and-destroy operations, rapacious extraction schemes, and the impact on villagers of the failed projects of usually well-intentioned but often ignorant foreign scientists and both local and foreign NGOs. She also knew what sustained committed work of real colleagues and friends could accomplish in Madagascar against the odds and across differences of all sorts. There are many possible examples and many important people, but I want to tell about one little project that might be considered a model system for sympoiesis.

Written in both English and Malagasy, each book in the Ako Project vividly narrates the adventures of a young Malagasy lemur from one of six species, from the tiny mouse lemur or *ny tsididy* to the queer-fingered aye-aye or *ny aiay,* to singing indri or *ilay babakoto.* The stories are rich natural histories, full of the empirical sensuous curiosity of that genre; and they are bumptious adventures of young lemurs living the joys and dangers of their habitats and their groups' social arrangements. Surrounding each lemur species with diverse plant and animal critters, the project provides teachers' guides in Malagasy and

beautifully crafted posters showing the unique regions of Madagascar where the stories take place. The books are *not* textbooks; they are stories, feasts for mind, heart, and body for children (and adults) who have no access to storybooks or to the critters of their own nation or even region. Most Malagasy never see a lemur in the wild, on television, or in a book. For generations, those privileged enough to go to a school with books saw pictures of French rabbits, a fact Alison Jolly told me with disgust in the 1980s when I interviewed her for *Primate Visions.*[43] Many villages are still without schools; and the formal curriculum remains conservative, modeled on French systems, with no place for local critters or ecologies in teacher training. In exciting, beautiful, funny, and scary stories distributed outside the school bureaucracies, the Ako Project nurtures empathy and knowledge about the extraordinary biodiversity of Madagascar *for the Malagasy.*

The Ako Project is the generative fruit of a colleagueship and friendship over decades.[44] In 1983, Alison Jolly met Hanta Rasamimanana, a scientist seventeen years her junior; they bonded as mothers doing fieldwork in challenging conditions, primatologists riveted by ringtail lemurs, lovers of Malagasy people and nature, and participants in global and local politics, with differently situated vulnerabilities and authority. Born in the capital and part of the generation sponsored by the Soviet Union under Didier Ratsiraka's socialism, Rasamimanana trained in animal husbandry at the Veterinary Academy in Moscow. She earned a PhD at the Museum d'Histoire Naturelle in Paris, and she has a master's in primate conservation. She is professor of zoology and scientific education at l'École Normale Supérieure of Antananarivo. Rasamimanana has published on ringtail feeding behavior, energy expenditure, and lemur female precedence and authority. Initiating a master's in primate conservation run in Mahajanga and the Comoros, her responsibilities in Madagascar's scientific academy have been multiple. An adviser on the Madagascar National Curriculum, she heads the Ako Project teacher support program, and she wrote the Malagasy teacher's guides based on workshops she ran in rural areas.

In all their attachments, together Jolly and Rasamimanana brought the Ako Project into the world. In their work and play across many crises in Madagascar and its conservation history, they have nurtured new generations of Malagasy naturalists and scientists, including small children, field station guides, and school and university

students. They have practiced the arts of living on a damaged planet; it matters.

Conclusion

Like coral reefs and forests, the arctic is profoundly vulnerable in the Anthropocene. Global warming is advancing at twice the rate in the arctic compared to anywhere else on earth. In the computer game *Never Alone* (Kisima Ingitchuna), a northern Alaskan Iñupiat girl and an arctic fox set out to find the source of a world-destroying blizzard (Figure M2.5). The idea that disaster will come is not new to indigenous peoples; genocidal disaster has already come, decades and centuries ago, and has not stopped, and the people have not ceased ongoing worlding either. No one acts alone; connections and corridors are practical and material, including in the spirit world. Stories for the Anthropocene must learn with these complex histories.

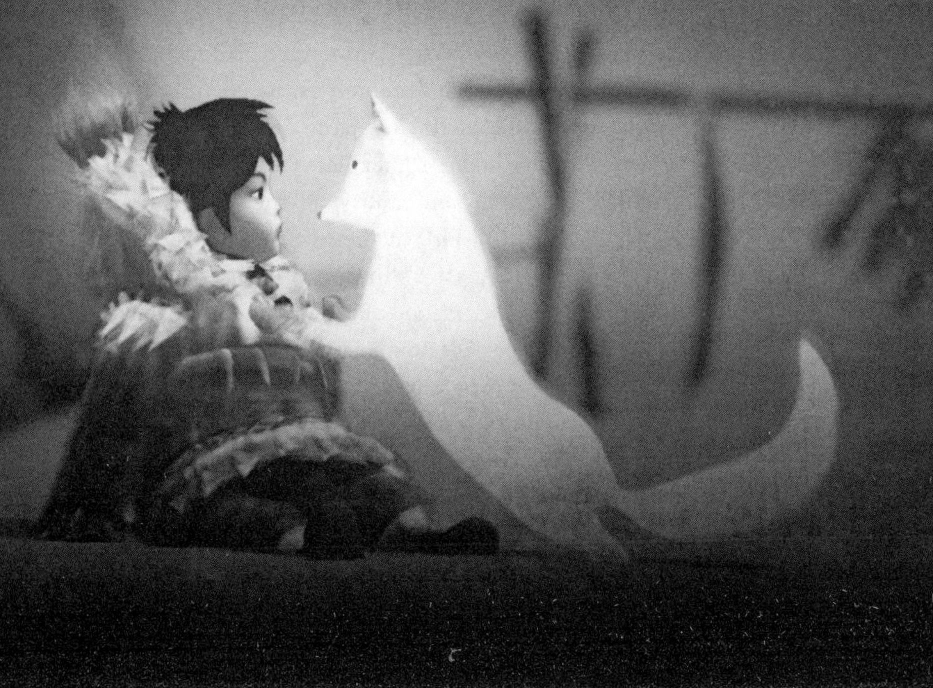

Figure M2.5. Cover image for *Never Alone* (Kisima Ingitchuna). Courtesy of E-line Media, in collaboration with Upper One Games and the Cook Inlet Tribal Council.

Taking place inside indigenous stories, world games invite contemporary sympoietic collaborations among designers of computer game platforms, indigenous storytellers, visual artists, carvers and puppet makers, digital-savvy youngsters, and community activists.[45] But the sympoiesis of *Never Alone* has another thread, too, namely, the spirit helpers crucial to the stories. *Never Alone* ties sym-anima-genic fibers into the string figure of this essay.

Working with Brazilian Amerindian hunters, with whom he learned to theorize the radical conceptual realignment he called multinaturalism and perspectivism, Eduardo Viveiros de Castro wrote, "Animism is the only *sensible* version of materialism."[46] Animism is not about "belief," a foreign Christian concept. Believing is not "sensible." I am talking about practices of worlding, about sympoiesis that is not only symbiogenetic but always a *sensible* materialism. The sensible materialisms of involutionary momentum are much more innovative than secular modernisms will allow. Stories for living in the Anthropocene demand a certain suspension of ontologies and epistemologies, holding them lightly, in favor of a more venturesome, experimental natural history. Without inhabiting symanimagenic sensible materialism, with all its pushes, pulls, affects, and attachments, one cannot play *Never Alone*; and the resurgence of this world might depend on learning to play.

We relate, know, think, world, and tell stories through and with other stories, worlds, knowledges, thinkings, yearnings. So do all the critters of Terra, in all our bumptious diversity and category-breaking compositions and decompositions. Words for this might be *materialism, evolution, ecology, sympoiesis, history, situated knowledges, animism,* and *science art activisms,* complete with the contaminations and infections conjured by each of these terms. Critters are at stake in each other in every mixing and turning of the terran compost pile. We are compost, not posthuman; we inhabit the humusities, not the humanities. Philosophically and materially, I am a compostist, not a posthumanist. Beings—human and not—become with each other, compose and decompose each other, in every scale and register of time and stuff in sympoietic tangling, in earthly worlding and unworlding. All of us must become more ontologically inventive and sensible within the bumptious holobiome that earth turns out to be, whether called Gaia or a Thousand Other Names.[47]

An original and pathbreaking scholar, **DONNA HARAWAY** has contributed to bringing many new fields into existence, including feminist science studies and multispecies storytelling. Distinguished Professor Emerita in the History of Consciousness program at the University of California, Santa Cruz, she is the author of many books that extend the scientific imagination, including *When Species Meet* (Minnesota, 2007), *Manifestly Haraway* (Minnesota, 2016), and *Staying with the Trouble.*

Notes

For an extended development of this chapter, see Donna J. Haraway, *Staying with the Trouble: Making Kin in the Chthulucene* (Durham, N.C.: Duke University Press, 2016).

1. Shoshanah Dubiner, "New Painting in Honor of Lynn Margulis," *Science in Service to Society,* no. 3 (October 2012), https://www.cns.umass.edu/about/newsletter/october-2012/memorial-painting-in-honor-of-lynn-margulis.
2. http://neveralonegame.com/game/.
3. In 1991, "Margulis proposed any physical association between individuals of different species for significant portions of their lifetime constitutes a 'symbiosis' and that all participants are bionts, such that the resulting association is a holobiont." Sarah Walters, "Holobionts and the Hologenome Theory," *Investigate: A Research and Science Blog,* September 4, 2013, http://www.intellectualventureslab.com/investigate/holobionts-and-the-hologenome-theory; Lynn Margulis, "Symbiogenesis and Symbionticism," in *Symbiosis as a Source of Evolutionary Innovation: Speciation and Morphogenesis,* ed. L. Margulis and R. Fester, 1–14 (Boston: MIT Press, 1991).
4. Lynn Margulis and Dorion Sagan, *Acquiring Genomes: A Theory of the Origin of Species* (New York: Basic Books, 2002), 205.
5. It would be hard to summarize the symbiotic/holobionic view better than this statement does: "Life is sustained by symbioses between nitrogen-fixing rhizobial bacteria and legumes, sulphide-oxidizing bacteria and clams in tidal seagrass communities, algae and reef-building corals, and protective mycorrhizal or endophytic fungi and plants. In addition to these grand symbioses are the nodes of symbiosis called organisms." Scott F. Gilbert, Thomas C. G. Bosch, and Cristina Lédon-Rettig, "Eco-Evo-Devo: Developmental Symbiosis and Developmental Plasticity as Evolutionary Agents," *Nature Reviews Genetics* 16 (October 2015): 612.

6. Like many paradigm-setting papers, Margulis's theory of the origin of the nucleated cell was rejected several times before being published. Lynn Sagan, "On the Origin of Mitosing Cells," *Journal of Theoretical Biology* 14, no. 3 (1967): 225–74.
7. James E. Lovelock, "Gaia as Seen through the Atmosphere," *Atmospheric Environment* 6, no. 8 (1967): 579–80; James E. Lovelock and Lynn Margulis, "Atmospheric Homeostasis by and for the Biosphere: The Gaia Hypothesis," *Tellus, Series A* 26, no. 1–2 (1974): 2–10.
8. Autopoietic systems theory was crucial to formulating the concept of the Anthropocene. Lovelock and Margulis's Gaia describes complex nonlinear couplings between processes that compose and sustain entwined but nonadditive subsystems as a partially cohering systemic whole. As Stengers stresses, Gaia is not a nurturing mother but an intrusive event that undoes thinking as usual. Gaia can flip out, in system collapse after system collapse. Complexity can unravel; earth can die. Isabelle Stengers, in conversation with Heather Davis and Etienne Turpin, "Matters of Cosmopolitics: On the Provocations of Gaïa," in *Architecture in the Anthropocene: Encounters among Design, Deep Time, Science, and Philosophy,* ed. Etienne Turpin, 171–82 (London: Open Humanities Press, 2013).
9. M. Beth Dempster, "A Self-Organizing Systems Perspective on Planning for Sustainability" (MA thesis, University of Waterloo, 1998), http://www.bethd.ca/pubs/mesthe.pdf.
10. Lynn Margulis and Dorion Sagan, "The Beast with Five Genomes," *Natural History,* June 2001, http://www.naturalhistorymag.com/htmlsite/master.html?http://www.naturalhistorymag.com/htmlsite/0601/0601_feature.html.
11. Margaret McFall-Ngai suggested "Postmodern Synthesis" for the revolutionary changes to the "Modern Synthesis" coming from EcoEvoDevo. McFall-Ngai, "Divining the Essence of Symbiosis: Insights from the Squid-Vibrio Model," *PLOS Biology* 12, no. 2 (February 2014): e1001783, doi:10.1371/journal.pbio.10017833. Allergic to "post-" prefixes, I suggest the ungainly "New New Synthesis." Scott Gilbert prefers "extended synthesis." Scott F. Gilbert and David Epel, *Ecological Developmental Biology,* 2nd ed. (Sunderland, Mass.: Sinauer Associates, 2015).
12. Konstantin Mereschkowsky, "Theorie der zwei Plasmaarten als Grundlage der Symbiogenesis, einer neuen Lehre von der Entstehung der Organismen," *Biologisches Zentralblatt* 30 (1910): 353–67.
13. Scott F. Gilbert, "The Adequacy of Model Systems for Evo-Devo," in *Mapping the Future of Biology,* ed. A. Barberousse, T. Pradeu, and M. Morange (New York: Springer, 2009), 57.
14. Nicole King, "King Lab: Choanoflagellates and the Origin of Animals," https://kinglab.berkeley.edu/.

15. Margaret McFall-Ngai, "Pacific Biosciences Research Center at the University of Hawai'i at Manoa," http://www.pbrc.hawaii.edu/index.php/margaret-mcfall-ngai.
16. Jeffrey Gordon, "Gordon Lab," https://gordonlab.wustl.edu/; Sarkis Mazmanian, "Sarkis Mazmanian Lab Site," https://www.sarkis.caltech.edu/.
17. Statement from Nancy Moran, "Nancy Moran's Lab," http://web.biosci.utexas.edu/moran/.
18. Scott F. Gilbert, Jan Sapp, and Alfred Tauber, "A Symbiotic View of Life: We Have Never Been Individuals," *Quarterly Review of Biology* 87, no. 4 (2012): 325–41. Gilbert cowrote a separate paper because of unresolved disagreements at the National Evolutionary Synthesis Center workshop concerning whether the holobiont can be considered a level of selection. Gilbert maintains that it must be so considered and that the immune system has evolved to manage a dialogue with (and not merely exterminate or exclude) potential symbionts and to block cooperation-destroying cheaters.
19. Margaret McFall-Ngai, Michael G. Hadfield, T. C. Bosch, H. V. Carey, T. Domazet-Lošo, A. E. Douglas, N. Dubilier, et al., "Animals in a Bacterial World: A New Imperative for the Life Sciences," *Proceedings of the National Academy of Sciences of the United States of America* 110, no. 9 (2013): 3229. This paper is the result of a workshop supported by the National Evolutionary Synthesis Center.
20. Margaret Wertheim, *A Field Guide to Hyperbolic Space* (Los Angeles, Calif.: Institute for Figuring, 2007).
21. Carla Hustak and Natasha Myers, "Involutionary Momentum," *differences* 23, no. 3 (2012): 79, 97, 106.
22. Ibid., 77.
23. Orson Scott Card, *The Speaker for the Dead* (New York: Tor Books, 1986).
24. http://www.explainxkcd.com/wiki/index.php/1259:_Bee_Orchid. This orchid, *Ophrys apifera,* mimics the not-quite-extinct solitary bee *Eucera.*
25. Donna Haraway, "Anthropocene, Capitalocene, Plantationocene, Chthulucene: Making Kin," *Environmental Humanities* 6 (2015): 159–65.
26. Crochet Coral Reef, http://crochetcoralreef.org/; Ako Project, http://www.lemurreserve.org/akoproject2012.html.
27. Hustak and Myers, "Involutionary Momentum," 77.
28. Anna Tsing, "A Threat to Holocene Resurgence Is a Threat to Livability" (unpublished manuscript, 2015). Tsing argues that the Holocene was/is the long period when refugia, places of refuge, still existed. Veronique Greenwood, "Hope from the Deep," *Nova Next,* March 4, 2015, http://www.pbs.org/wgbh/nova/next/earth/deep-coral-refugia/.
29. Wertheim, *Field Guide to Hyperbolic Space,* 35.

30. Margaret Wertheim and Christine Wertheim, *Crochet Coral Reef: A Project by the Institute for Figuring* (Los Angeles, Calif.: IFF, 2015), 17.

31. Rob Tapert and John Schulian, "Dreamworker," *Xena: Warrior Princess,* season 1, episode 3, dir. Bruce Seth Green, aired September 18, 1995 (United States: Renaissance Pictures, 1995).

32. On the crochet coral reefs as experimental life-forms, see Sophia Roosth, "Evolutionary Yarns in Seahorse Valley," *differences* 25, no. 5 (2012): 9–41.

33. Wertheim and Wertheim, *Crochet Coral Reef,* 21.

34. Ibid., 23.

35. Ibid., 17.

36. Ibid., 202.

37. Margaret Wertheim, "The Beautiful Math of Coral," February 2009, http://www.ted.com/talks/margaret_wertheim_crochets_the_coral_reef?language=en.

38. Christine Wertheim, "CalArts Faculty Staff Directory," https://directory.calarts.edu/directory/christine-wertheim.

39. Jacob Metcalf, "Intimacy without Proximity," *Environmental Philosophy* 5, no. 2 (2008): 99–128.

40. "Ako Project: The Books," http://www.lemurreserve.org/akobooks.html. Written by Alison Jolly, illustrated by Deborah Ross, Malagasy text by Hantanirina Rasamimanana. Published by the Lemur Conservation Foundation in the United States and Canada and by UNICEF in Madagascar. Unilingual books are available in English and Chinese, with more translations planned. Book artist Deborah Ross has works in major magazines, zoos, and botanical gardens, plus watercolor workshops for Walt Disney Studios, DreamWorks, Pixar, and Cal Arts and rural art workshops for Malagasy villagers. Poster artist Janet Mary Robinson has degrees in scientific illustration and ecology and environment.

41. See Alison Jolly, *Thank You, Madagascar* (London: Zed Books, 2015), for an astute, quirky, gorgeously written, often tragic account of tangles in the history of Malagasy-Western conservation encounters and projects over the late twentieth and early twenty-first centuries, all of which Jolly participated in. Thanks to Margaretta Jolly for documents on the Ako Project.

42. Ranomafana National Park could not have existed without the multistranded commitment of primatologist and conservationist Patricia Wright, whose work Jolly praised in *Thank You, Madagascar.* See Patricia C. Wright and B. A. Andriamihaja, "Making a Rain Forest National Park Work in Madagascar: Ranomafana National Park and Its Long-Term Commitment," in *Making Parks Work: Strategies for Preserving Tropical Nature,* ed. J. Terborgh et al., 112–36 (Washington, D.C.: Island Press, 2002).

43. Donna J. Haraway, *Primate Visions: Gender, Race, and Nature in the World of Modern Science* (New York: Routledge, 1989).
44. Margaretta Jolly, "Alison Jolly and Hantanirina Rasamimanana: The Story of a Friendship," *Madagascar Conservation and Development* 5, no. 2 (2010): 45.
45. Dean Takahashi, "After *Never Alone,* E-Line Media and Alaska Native Group See Big Opportunity in 'World Games,'" *GamesBeat,* February 5, 2015, http://venturebeat.com/2015/02/05/after-never-alone-e-line-media-and-alaska-native-group-see-big-opportunity-in-world-games/. Thanks to Marco Harding for the reference and for teaching me to play.
46. Eduardo Viveiros de Castro, pers. comm., October 2, 2014. In conversations and in "Secular Trouble," an unpublished paper for the conference on Religion and Politics in Anxious States, University of Kentucky, April 4, 2014, Susan Harding teaches me about the cultural and historical roots of "belief" in colonial and Christian practices. See Déborah Danowski and Eduardo Viveiros de Castro, "L'Arret du monde," in *De l'univers clos au monde infini,* ed. Émilie Hache, Bruno Latour, Christophe Bonneuil, and Pierre de Jouvancourt, 221–339 (Bellevaux, France: Dehors, 2014).
47. "The Thousand Names of Gaia/Os Mil Nomes de Gaia: From the Anthropocene to the Age of the Earth," conference in Rio de Janeiro, September 15–19, 2014, https://thethousandnamesofgaia.wordpress.com/.

3

NOTICING MICROBIAL WORLDS

THE POSTMODERN SYNTHESIS IN BIOLOGY

Margaret McFall-Ngai

WE ARE IN THE MIDST OF YET ANOTHER REVOLUTION IN BIOLOGY. Just over 150 years ago, Charles Darwin published his *Origin of Species,* sparking one revolution in biology by providing compelling evidence for evolution through natural selection. Just over sixty years ago, James Watson and Francis Crick described the "double helix" structure of DNA, another major revolution in biology. Today, genomics research is driving still another revolution, one that is changing entire disciplines as well as the ways we understand the world.

The genomics revolution is ushering in new modes of noticing—modes that allow us to better see the complex relationships that humankind is disturbing. The Anthropocene, genomics research shows us, is indeed an epoch loss—but it is not just the loss of individual organisms or species or even macrobiomes, such as coral reefs. Rather, it is the loss of complex microbial worlds both within and beyond organismal bodies—worlds that make nearly all life possible. My goal in this essay is to illustrate how the new approaches to biology that are emerging out of genomics give us a new sense of the multispecies relationships that we are damaging.

One of the wondrous parts of biology is that it shows us that living beings—including ourselves—are not what we thought. This is nothing

new for the field. In its early days, biology focused on unexpected developmental transformations, such as how creatures like humans begin as eggs. Later, biologists illustrated that kinds, or species, are not static but changing, through processes that act on genes. Both the insights of early developmental biology and the "Modern Synthesis"—the combination of Darwin's natural selection and Mendel's genetics—opened up our senses to the surprising processes around us. However, those key observations are turning out to highlight only a very small part of who and what organisms are. We are now beginning to realize that "individuals" aren't particularly individual at all. The organisms of developmental biology, along with Darwin's species, all turn out to be complex assemblages, typically made up of more cells of others than of their "own."

Consider a phenomenon called *microchimerism,* which occurs when cells from a fetus pass through the placenta and take up residence in the mother's body, and vice versa.[1] Doctors and scientists have long thought that *something* transfers from babies to mothers, often with lasting effects on mothers' immune systems. However, they hypothesized that some kind of molecules were on the move. Very recently, biologists have learned that one feature that migrates are *cells.* The cells of the mother infiltrate across the placenta into the child, and the cells of the child infiltrate into the mother. Thus each one of us is a chimera of sorts, our bodies containing cell lines of others. If you are a first-born child, you will have a set of cells that come from your mother, including cells that she acquired from her own mother in the same way. If you are a youngest child, not only will you receive your mother's cells, but you will also receive all of your siblings' cells. We are thus not what we thought: every "I" is also a "we." And, while the preceding example is highly illustrative, our bodies are not only—not even primarily—composed of cells from our close kin. Indeed, we are more microbe than human. Our fundamental microbial-ness is at the heart of this essay—and it is precisely what the genomics revolution is allowing us to notice.

Technologies Facilitate Novel Insights

The genomics revolution is, in large part, a technical one. New technologies are changing our ideas about the relationship between the micro and macro worlds by changing *what we can see.* Our understanding of

the biological world has always been fundamentally linked to how we are able to perceive it, and what we can perceive is tied to the technologies we have for seeing. This has always been the case. During the earliest periods of biological classification, people organized things into categories such as animals and plants, because these were the kinds of organisms that they could see. Then, in the seventeenth century, Anton van Leeuwenhoek became the first person actually to see microbes. He scraped the inside of his cheek and looked at what he found there on an early microscope that he made himself. He saw what he called "animalcules," because—with his low-resolution visual technology—he thought he was seeing tiny animals. Yet his discovery piqued curiosity about the legions of living things that cannot be seen by unaided human eyes. By the mid-nineteenth century, with improved microscopes, biologists recognized microorganisms as distinct from animals, and they began to divide living things into three groups: animals, plants, and microbes.

Biology changed again with the invention of the transmission electron microscope, which offered powers of magnification about five thousand times greater than those of conventional light microscopes. Although the earliest electron microscope became commercially available in 1939, improved models began to transform biology in the 1960s and 1970s, when researchers used them to look closely at the characteristics of cells. This technology enabled biologists to see that some cells were more complex than others, and these observations further complexified classifications of life-forms. The result was a five kingdom model:[2] animals, plants, fungi, protists ("complex small things"), and monera ("not-so-complex small things"). Electron microscopes also allowed Lynn Margulis to look closely at eukaryotic cells and theorize about their origins. Based on what she could see, Margulis hypothesized that the organelles of complex cells arose from endosymbiosis—that is, that the coordination and cooperation of simple bacteria were the foundation of more elaborate forms of life. By improving the power and resolution of visual analyses, transmission electron microscopes clearly sparked many new biological insights. However, they remained *visual* modes of "seeing" the world.

Then, in the late 1970s, a whole new observational practice arrived on the scene. Carl Woese developed better practices for sequencing DNA, and through them, he was able to probe evolutionary relationships among organisms via genetics rather than comparative

anatomy.[3] In doing so, his work illuminated the diversity and wildness of microbial worlds. Woese's DNA methods, however, remained slow. Genetics research took off after Kary Mullis refined the polymerase chain reaction (PCR), allowing DNA analyses to be done much more quickly. PCR also allowed for easier isolation, amplification, replication, and quantification of particular DNA sequences. These new technologies lent support to earlier hypotheses, such as those that Lynn Margulis first developed from visual observations. Her work on endosymbiosis finally garnered widespread support in the 1980s, after new techniques revealed that cell organelles had genetic signatures that differed from those found in cell nuclei.

Changing Technology, Changing Conceptual Models

The technological advancements that have occurred since the 1970s have once again transformed the tree of life. Although Robert Whittaker's five kingdom model of 1969 was a mainstay of biology textbooks and lingers in the minds of many biologists, phylogenetic diagrams have continued to rapidly shift.

In 1990, Woese, along with his colleagues, drew on the most recent DNA-based research to craft a new phylogenetic tree that radically decentered plants, animals, and fungi.[4] The new diagram depicted three domains of life—the Archaea, the Bacteria, and the Eukarya—and moved the long-standing groups of plants, animals, and fungi in one tiny corner of the tree of life. The reordering reflected what new technologies highlighted: that the earth's biological diversity is far more microbial than ever imagined. Since the early 1990s, Woese and colleagues' model has garnered great attention—and it has become the new "standard" with major implications for how biologists view the world. The animals, plants, and fungi—once assumed to encompass the majority of biological life—are now a set of very tiny branches on a very big tree.

Yet, recently biologists are even beginning to question the very notion of a tree of life, as such. Eugene Koonin, a biologist who focuses on comparative genomics, is among those who have argued that the tree needs to go because classic notions of evolutionary descent and reproductive transmission of genes no longer hold.[5] Recent research appears to demonstrate that the paradigmatic Modern Synthesis,

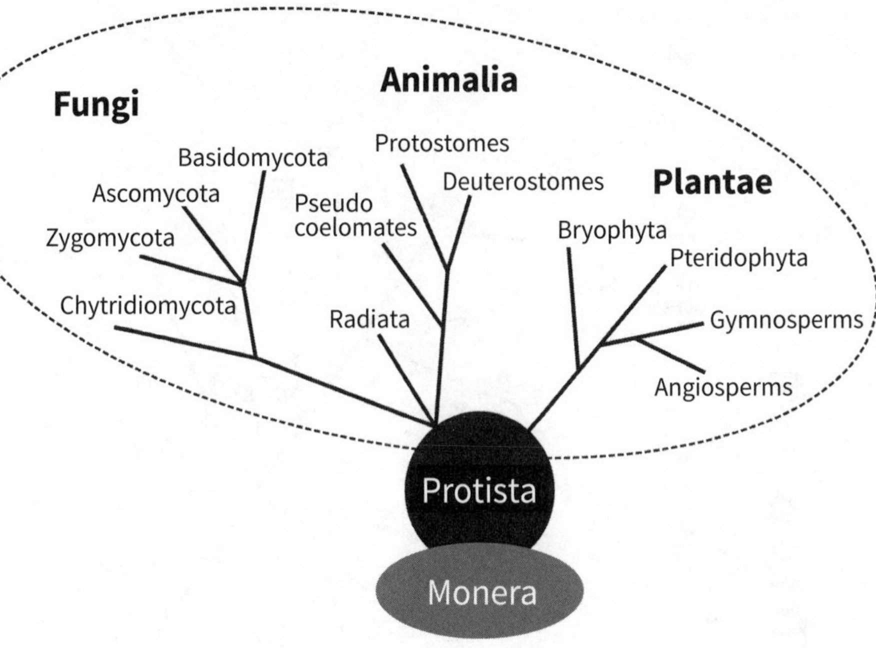

Figure M3.1. A version of the Whittaker diagram. By Ib Jensen, Department of Cartography, Aarhus University.

the focus on the vertical transmission of genetic traits that emerged from the integration of Darwinian selection and Mendelian genetics, applies only to a very small portion of the biological world. Most of the world's living things—particularly its microbes—are busy engaging in horizontal gene transfer, the lateral movement of genetic traits, even among very distantly related organisms from different domains. Horizontal DNA swapping has upended what we thought we knew about gene transfer: we now know that genetic material from bacteria sometimes ends up in the bodies of beetles, that of fungi in aphids, and that of humans in malaria protozoa. For bacteria, at least, such transfers are not the stuff of science fiction but of everyday evolution. Koonin asserts that the new findings related to endosymbiosis, horizontal gene transfer, and bacterial phylogeny demand a new biologic paradigm—one he has called the "Postmodern Synthesis." In the Postmodern Synthesis, organisms are not individuals and Mendelian modes of evolution are moved to the margins. With all of the horizontal

Phylogenetic Tree of Life

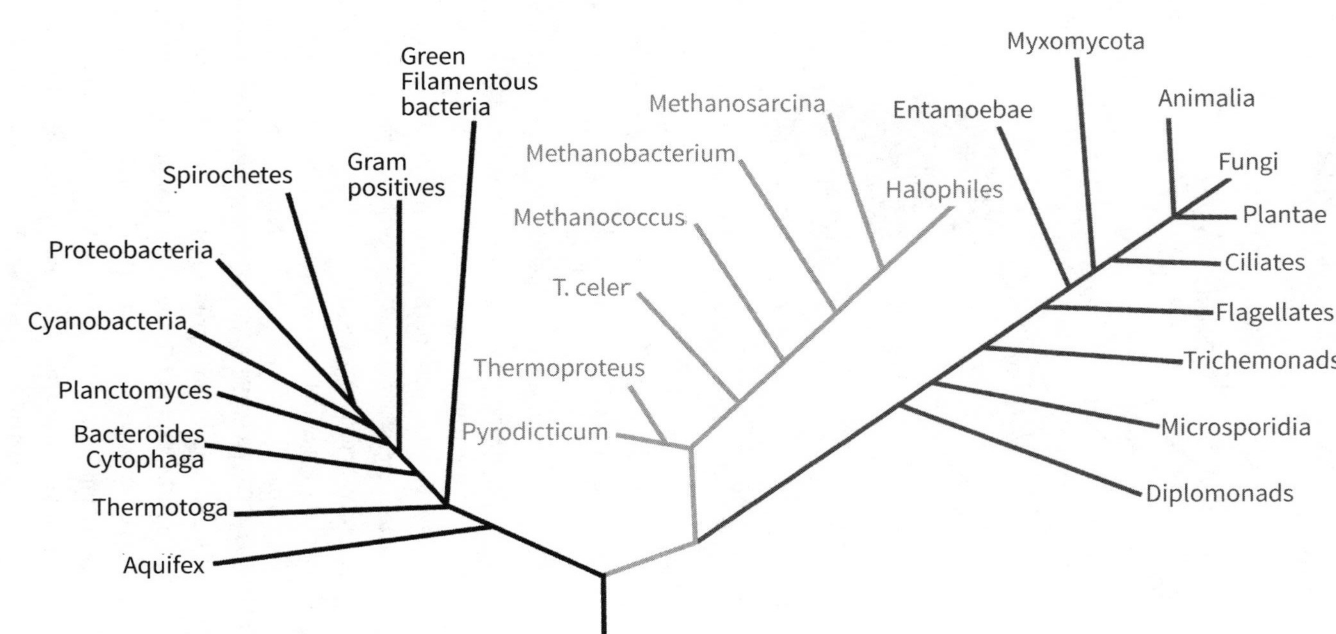

Bacteria
Archaea
Eucarya
Green Filamentous bacteria
Gram positives
Spirochetes
Proteobacteria
Cyanobacteria
Planctomyces
Bacteroides Cytophaga
Thermotoga
Aquifex
Methanosarcina
Methanobacterium
Methanococcus
T. celer
Thermoproteus
Pyrodicticum
Halophiles
Entamoebae
Myxomycota
Animalia
Fungi
Plantae
Ciliates
Flagellates
Trichemonads
Microsporidia
Diplomonads

exchange of genes, Koonin draws phylogenetic diagrams that depict a "web of life" rather than a tree.

Koonin's diagram may not be the final answer. Ideas about how the biological world is put together remain very much in flux. For example, one ongoing debate is about whether large viruses should be considered a fourth domain of life. This and many other debates are emerging with and through improved PCR processes—and PCR processes made cheaper. In the last six years, molecular technology has revolutionized biology because it has become dramatically less expensive. In 2006, it cost $6,000 per megabase to sequence DNA. Today, it costs a mere three to four cents. Suddenly, it is possible to conduct genomic analysis on a scale that was unfathomable only a few years ago. Biologists are now in future shock: too much information, too short a time. It is truly remarkable. It is as if we have suddenly found ourselves in the science fiction world depicted in the 1997 film *Gattaca,* where genetic technologies pervade nearly every facet of everyday life.

Today, we are collecting more and more full genome sequences, as prices drop even lower. Although we are not there yet, biology is headed toward what many refer to as the "one dollar genome," a future that seems sure to arrive soon. It used to be that if someone wanted to know what kind of genetic change had occurred in a bacterium to make it behave differently, she would have to do years' worth of experiments in the lab. Now, one sends it out for complete gene sequencing, at a cost of only a couple hundred dollars, then simply looks for the genetic change. The same kind of revolution is happening in relation to gene products, that is, proteins. Cyrus Chothia, a well-respected microbiologist, predicted that there would be no more than a total of one thousand protein families.[6] However, Pfam, a protein family database run by the European Molecular Biology Laboratory and Wellcome Trust, shows that as of May 2015, 16,230 protein families have been documented, and the ongoing discovery rate is about two to three per day. We are clearly in a moment when biological knowledge is in a period of change that is at once surprising, thrilling, and overwhelming.

Figure M3.2. Woese and colleagues' diagram. By Ib Jensen, Department of Cartography, Aarhus University. Redrawn from https://en.wikipedia.org/wiki/File:Phylogenetic_tree.svg.

Importance of Horizontal Gene Transfer

Koonin & Wolf, 2012

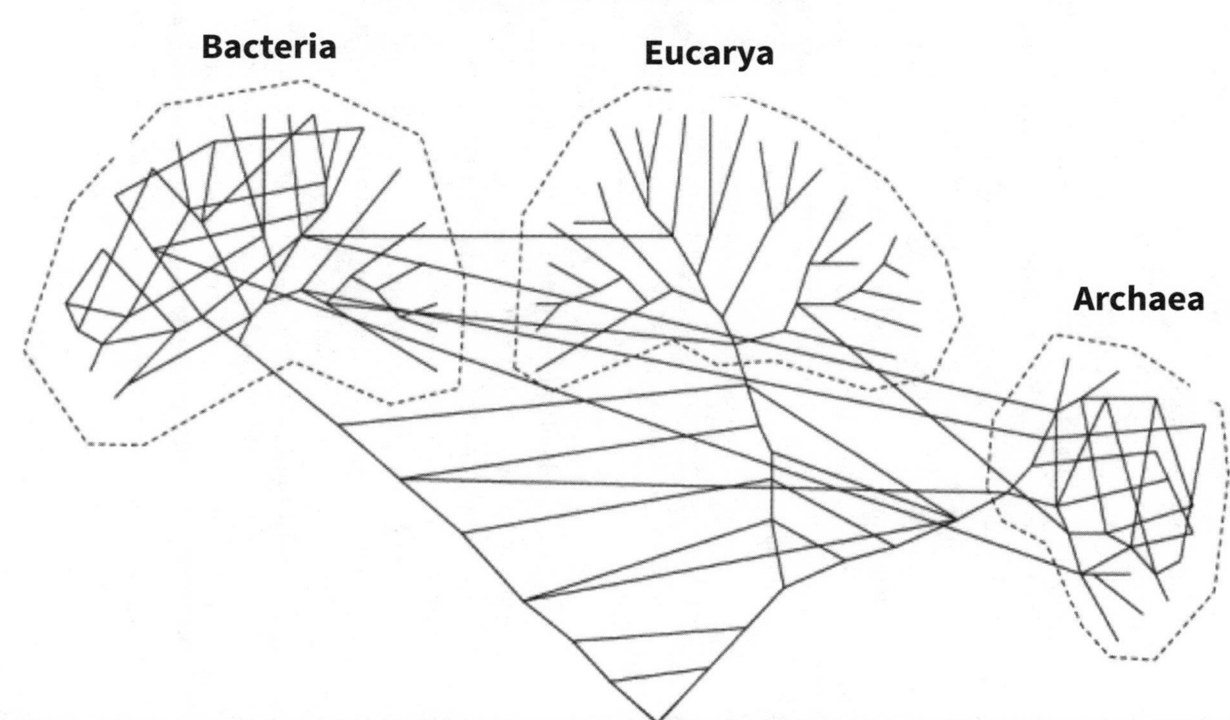

Mainly Microbes

Investigations into microbial life haven't just changed a once-simplistic tree of life into a complex web. They have also shown us that single-celled organisms are among the most important creatures for all life on earth.[7] Indeed, the vast diversity of the biosphere appears dependent on the microbial world. Microbes don't just "rule" the world: they make every life-form possible, and they have been doing so since the beginning of evolutionary time. How are such insights changing biology as we know it? When we realize that plants and animals are deeply embedded in the microbial world, it changes all aspects of plant and animal biology. It turns out that most animals and plants have coevolved in and with microbial-rich environments and have relied on the microbial world for their own evolution and health.

The work of Michael Hadfield at the University of Hawai'i offers but one illustration.[8] Hadfield's research shows that many, if not most, of the marine invertebrates that have larval stages require interaction with bacterial biofilms to develop. Larvae generally drift or swim in ocean waters until they are able to find a place to settle. If they do not, they are not able to metamorphose into their next life stage and soon perish. For biologists, it was no surprise that various invertebrate larvae require particular physical conditions (i.e., particular temperatures, salinity levels, etc.) to settle and continue their development. But what Hadfield revealed is that they also require particular *microbial* conditions. If the right microbes, which typically occur in assemblages called biofilms, aren't on a rock or other substrate, larvae do not—and perhaps cannot—settle there. The microbes seem to both cue the larvae to settle and stimulate their development. This is highly important, because should biofilms change due to ocean warming or acidification, some invertebrates might lose their bacterial cues. The effects of such possible disruptions of biofilm-invertebrate interactions hold the potential to substantially change ocean ecosystems. In short, bacteria matter not only in themselves but also in relation to other living beings, who depend on them for processes as basic as bodily development.

Figure M3.3. Is there a tree of life? Koonin's web of life diagram. Reprinted with permission from Eugene V. Koonin.

In contrast to Hadfield's research, my own work focuses on bacterial symbioses—on bacteria that take up residence *on* or *inside* bodies rather than provide cues from the outside. I have been deeply inspired by the work of Lynn Margulis, the scholar who first drew attention to symbiosis as a major developmental and evolutionary force.[9] Margulis, far ahead of her time, crafted entirely new biological paradigms and founded the entire field of symbiosis. Unfortunately, too many recent works barely refer to her and her pioneering accomplishments. I prominently cite her here, because I do not want the accomplishments of women scientists to continue to be dismissed and forgotten. Although Margulis made symbiosis into a field of study, she did not invent symbiosis as such. By the late nineteenth century, biologists were already aware of a handful of symbiotic relationships, such as those that made up lichens and algae that lived inside sea anemones. But before Margulis's work, symbiosis was seen as a rare exception in a world dominated by unmitigated competition. Margulis showed, instead, that symbiosis was the "norm"—and a core form of relationality. Importantly, she was able to do so by focusing on microbes.

Microbes are key symbiotic partners for plants, animals, and fungi, because they are far more diverse and versatile than larger organisms. Members of Eukarya, the phylogenetic domain that includes all multicellular organisms, have very limited metabolic diversity. Their strategy is to be tied to oxygen. Either they are photosynthetic or they use oxygen as a terminal electron acceptor during cellular respiration. Oxygen is, indeed, an excellent terminal electron acceptor, and that is why oxygen-centric metabolism has led to the evolution of so many interesting and evolutionarily successful organisms. But when it comes to metabolic strategies, that is all Eukarya have: they have only the limited ability to exchange electrons with oxygen. Conversely, bacteria have phenomenally greater metabolic diversity, including metabolic pathways that use such diverse terminal electron receptors as nitrate, sulfate, and ferric iron.

In addition, microbes have very short generation times compared to many eukaryotes, particularly animals, plants, and fungi. They also tend to have large population sizes and a propensity for exchanging DNA. To be sure, eukaryotes also swap genetic material. Remember the example of cellular exchange between the human fetus and mother described earlier in this essay. But bacteria are the experts when it comes to genetic transfers. Multicellular organisms—animals,

plants, and fungi—tend to have long generation times, comparatively small population sizes, and a propensity against horizontal gene transfer. All of these traits slow the ability of eukaryotes to rapidly adapt to changing conditions. Thus, as one can imagine, it makes sense for eukaryotes in particular to partner with microbes to increase their metabolic scope and their capacities to adjust to fluctuating conditions. This is what symbiosis is all about.

Squid–Vibrio Symbiosis

In 1988, I began a collaborative project with Edward Ruby that explores symbiotic relationships between Hawaiian bobtailed squid *Euprymna scolopes* and a bacteria called *Vibrio fischeri.*[10] At its inception, the project seemed like an opportunity to study a rare host-microbe alliance. In the center of their translucent bodies, these nocturnal squid have a two-chambered light organ filled with bacteria that luminesce, helping to camouflage the squid from predators swimming below. The bacterial glow hides the squid's silhouette, making these animals less visible against the background of light that filters down from the water's surface. Hawaiian bobtailed squid are not born with these bacteria; rather, they must quickly acquire them by secreting a sticky mucus. Once a tiny handful of *V. fischeri* bacteria arrive, they rapidly outcompete other microbes while colonizing crypts within the squid's light organ. Only hours after the squid have hatched, the newly acquired vibrio are already munching away on sugars and amino acids provided by their host, and the squid's light organs are set to shine.

The vibrio bacteria, however, do not merely inhabit a preexisting niche in a squid's mantle. Instead, they are essential in bringing these squid into being. Vibrio cells initiate changes in squid gene expression, shaping the development of squid bodies and immune systems. The presence of these bacteria causes morphological changes in squid light organs, shifts the functions of certain tissues, and alters protein production. These observations challenged what we thought we knew about organismal development, namely, that it was driven primarily by inherited genetic codes. In contrast, our research showed us that squid develop, in part, through relations with microbes, not exclusively through inherited genetic scripts.[11] What we have learned about the squid-vibrio symbiosis has certainly been interesting in and of itself. However, in the years since we began this research, we have

Figure M3.4. The Hawaiian bobtailed squid, *E. scolopes,* harbors a luminous bacterial symbiont, the gram-negative marine bacterium *V. fischeri.* The animal, which is a night-active predator, uses the light of the bacterial partner in a camouflage behavior. It emits ventral luminescence to match downwelling moonlight and starlight so that it does not cast a shadow. Photograph by C. Frazee for University of Wisconsin/Margaret McFall-Ngai.

discovered that what we thought was a rare symbiosis is not particularly rare at all. As we find that the organ function and developmental sequences of more and more animals are intertwined with bacterial relations, the squid-vibrio symbiosis is shifting from a quirky curiosity to one of a number of "model organisms" for biological investigations in the era of the Postmodern Synthesis.

Importance of Microbes for Vertebrates

When it comes to symbiotic relations with microbes, humans turn out to resemble Hawaiian bobtailed squid far more than one might expect. Vertebrate body plans include ten organ systems, and eight of those ten organ systems have components that associate with microbes. Microbes are so abundant in the human body that the number of nonhuman cells is at least equal to that of the number of human cells. This amazing number of bacteria within us has profound effects on our biology. Bacteria are not only changing the way our guts behave; their metabolic products interact with our entire bodies in complicated ways that we are just beginning to explore. For example, we are finding out that gut bacteria have significant impacts on our brains, affecting the ways we think and feel.[12]

One thing that gut microbes do is regulate the third wave of the mammalian sleep cycle. Thus, if one takes lots of antibiotics, it is very likely that one's sleep cycle will be disrupted. Furthermore, there is growing evidence that the presence or absence of certain microbial strains is linked to depression, anxiety, and autism.[13] Some researchers think that many developmental and psychiatric disorders may one day be treated through the adjustment and management of gut bacteria.

Such approaches to bacteria mark a huge conceptual shift in microbiology and medicine. From the nineteenth century onward, bacteria have been seen primarily as pathogens. Robert Koch, considered to be the founder of modern bacteriology, worked on the microbes that cause anthrax, cholera, and tuberculosis. With his focus on infectious diseases, Koch saw bacteria as bad—as the entities that make people sick. Although some environmental microbiologists pointed out the beneficial roles of bacteria in ecological processes, such as soil formation, their insights did not infect other parts of microbiology, where bacteria continued to be classified predominately as a menace. Today, biologists face the conceptual challenge of rethinking bacteria as

partners in health. Human bodies can no longer be seen as fortresses to defend against microbial onslaught but must be reenvisioned as nested ecosystems.

A Need for New Institutional Structures

Research on bodies-as-ecosystems struggles to find a place in current disciplinary configurations. They sit awkwardly in a liminal between-space, at once part microbiology, part developmental biology, part ecology, yet not fully at home within any of those fields. I have had the privilege of being both chair and vice-chair of the American Society for Microbiology general meeting for a number of years, and I have seen how the strong subdisciplinary silos remain. Even though more than ten thousand microbiologists gather in the same place, the environmental microbiologists and the pathogenic microbiologists still don't talk very much to each other. Their respective ways of conceptualizing microbes—as disease-causing entities within bodies or as ecological actors outside of them—largely remain on separate tracks. Academic departments and research institutes suffer from similar structural divisions. In a moment when the Postmodern Synthesis demands increasing cross-field collaboration, forms of biological knowledge remain specialized and segregated. Today, one can get a degree in biology without knowing any ecology or molecular biology. A typical thirteen-hundred-page introductory biology book might have a small, nonintegrated, twenty- to thirty-page section in the middle on prokaryotes (bacteria), continuing to present basic information in a way that runs counter to our current knowledge about how the biological world is put together. We need more attention to microbiology across the entire discipline. The challenges extend to professional organizations and funding agencies, too. Indeed, I am probably one of the only zoologists among the ten thousand scholars who attend the American Society for Microbiology meetings. While there are certainly some small steps toward more integrative biological approaches, macrobiologists and ecologists need to be bolder in building bridges with microbiologists, while microbiologists need to be more willing to move away from biomedical approaches that, from the outset, cast bacteria as disease-causing invaders rather than potential symbiotic partners.[14]

Symbioses in the Anthropocene

There is a real urgency for the exploration of microbial worlds that calls out for better scientific collaboration. That urgency is the disappearance of many of the symbiotic relationships of which microbes are a part.[15] All animals, plants, and fungi have their own unique microbiota—suites of tiny beings with which they live. A few eukaryote-prokaryote symbioses have already garnered scientific attention—those in the light organs of marine organisms and the human gut mentioned earlier but also those found in termites, aphids, and nematodes, just to name a few. The vast majority of symbiotic worlds, however, remain unexplored.

At the same time, though, they are likely slipping away. We are in a moment when animal and plant extinctions are occurring at an elevated rate. When one of these larger organisms disappears, its suite of coevolved microbes likely disappears, too.[16] Once we recognize that individuals are ecosystems, it shows us that the loss of a single species probably entails the loss of many kinds, not just one. Attention to microbial life raises the specter that our extinction crisis may be even more serious than we thought. It also poses a plethora of additional questions about how we might best live on a damaged planet. So far, microbes have been almost entirely ignored by conservation biologists. But might native microbial communities also be increasingly distressed? In the midst of global climate change, one worries about how higher temperatures might disrupt symbiotic alliances. If one symbiotic partner is more sensitive than the other, might all those dependent on that relation be at risk?

Invasive species are another pressing problem in our world of intensive human movement and global trade. But might disruptive species introductions be microbial as well as animal, vegetal, and fungal? While the travel of a pathogen occasionally draws attention, countless microbial movements are undoubtedly going on under the radar. Perhaps it would it be wise for conservation projects to pay more attention to their possible consequences for ecological relations. When an invasive species colonizes a new region, it is likely to bring a new host of microorganisms with it. What effects might these new microbes be having on endemic ones—and how might those effects matter to all entwined with microbial life?

In the era of the Anthropocene, noticing microbial worlds seems

more important than ever. Simply put, sustaining life requires sustaining symbioses. Noticing those symbioses—and their wondrousness—is a critical step. New technologies are clearly a great help for seeing transdomain partnerships, but we need more: we also need new collaborations among the different branches of biology—and beyond. The Postmodern Synthesis is a theoretical frame that demands such disciplinary practices. However, it is also something more: it is also a general call for all of us to be more curious about the microbes that make up our worlds.

MARGARET MCFALL-NGAI's research on the role of vibrio bacteria in the development of the light organs of Hawaiian bobtailed squid revolutionized modern biology by drawing attention to the constitutive role of interspecies interactions. She is director of the Pacific Biosciences Research Center, University of Hawai'i at Manoa, and holds appointments as an AD White Professor-at-Large at Cornell University and a EU Marie Curie ITN Professor.

Notes

1. Ashley Moffett and Charlie Loke, "Immunology of Placentation in Eutherian Mammals," *Nature Reviews Immunology* 6, no. 8 (2006): 584–94, doi:10.1038/nri1897.
2. Lynn Margulis, Karlene V. Schwartz, and Michael Dolan, *The Illustrated Five Kingdoms: A Guide to the Diversity of Life on Earth* (New York: HarperCollins, 1994).
3. Norman R. Pace, Jan Sapp, and Nigel Goldenfeld, "Phylogeny and Beyond: Scientific, Historical, and Conceptual Significance of the First Tree of Life," *Proceedings of the National Academy of Sciences of the United States of America* 109, no. 4 (2012): 1011–18, doi:10.1073/pnas.1109716109.
4. Carl R. Woese, Otto Kandler, and Mark L. Wheelis, "Towards a Natural System of Organisms: Proposal for the Domains Archaea, Bacteria, and Eucarya," *Proceedings of the National Academy of Sciences of the United States of America* 87, no. 12 (1990): 4576–79.
5. Eugene Koonin, "The Origin at 150: Is a New Evolutionary Synthesis in Sight?," *Trends in Genetics* 25, no. 11 (2009): 473–75, doi:10.1016/j.tig.2009.09.007.
6. Cyrus Chothia, "Proteins: One Thousand Families for the Molecular Biologist," *Nature* 357 (1992): 543–44, doi:10.1038/357543a0.

7. Margaret McFall-Ngai et al., "Animals in a Bacterial World: A New Imperative for the Life Sciences," *Proceedings of the National Academy of Sciences of the United States of America* 110, no. 9 (2013): 3229–36, doi:10.1073/pnas.1218525110.
8. Michael G. Hadfield, "Biofilms and Marine Invertebrate Larvae: What Bacteria Produce That Larvae Use to Choose Settlement Sites," *Annual Reviews of Marine Science* 3 (2011): 453–70, doi:10.1146/annurev-marine-120709-142753.
9. Lynn Margulis and René Fester, *Symbiosis as a Source of Evolutionary Innovation: Speciation and Morphogenesis* (Boston: MIT Press, 1991).
10. Margaret McFall-Ngai, "Divining the Essence of Symbiosis: Insights from the Squid-Vibrio Model," *PLOS Biology* 12, no. 2 (2014a): e1001783, doi:10.1371/journal.pbio.1001783.
11. Margaret McFall-Ngai, "The Importance of Microbes in Animal Development: Lessons from the Squid-Vibrio Symbiosis," *Annual Review in Microbiology* 68 (2014b): 177–94, doi:10.1146/annurev-micro-091313-103654.
12. For review, see Timothy R. Sampson and Sarkis K. Mazmanian, "Control of Brain Development, Function, and Behavior by the Microbiome," *Cell Host and Microbe* 17, no. 5 (2015): 565–76, doi:10.1016/j.chom.2015.04.011.
13. Yang Wang and Lloyd H. Kasper, "The Role of the Microbiome in Central Nervous System Disorders," *Brain, Behavior, and Immunity* 38 (2014): 1–12, doi:10.1016/j.bbi.2013.12.015.
14. In late fall 2011, twenty-seven scientists, from six countries on three continents, gathered at the National Evolutionary Synthesis Center in Durham, North Carolina, for a workshop titled "The Origin and Evolution of Animal-Microbe Interactions." The purpose of this adventure was to discuss a looming revolution in biology, specifically the emerging recognition that every animal requires interactions with the microbial world for all aspects of its biology. The group was led by two biologists, Michael Hadfield and Margaret McFall-Ngai; the remaining twenty-five individuals were divided into five groups based on expertise. The five areas of focus included the following aspects of animal biology: origin and evolution, genomics, development, physiology, and ecology. Each group was given three days to produce a seven-hundred-word summary of current knowledge in their area with a supporting figure. The groups interacted to integrate their contributions into a single voice, and the resulting contributions were published in the *Proceedings of the National Academy of Sciences*

just over a year following the workshop. McFall-Ngai et al., "Animals in a Bacterial World."

15. E.g., see Martin J. Blaser, *Missing Microbes* (New York: Henry Holt, 2014).

16. James Staley, "Biodiversity: Are Microbial Species Threatened?," *Current Opinion in Biotechnology* 8, no. 3 (1997): 340–45.

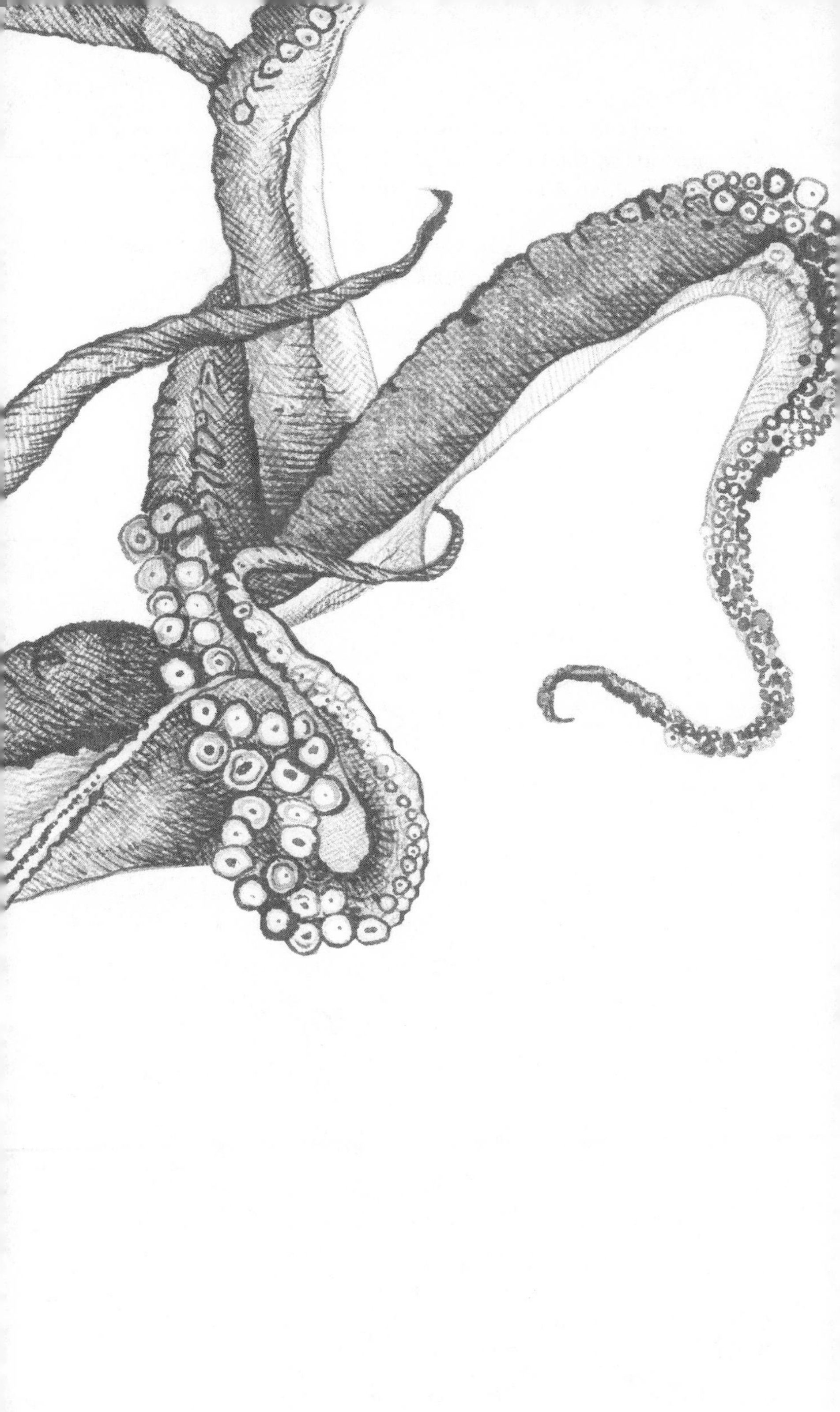

BEYOND INDIVIDUALS

THE TWENTIETH CENTURY WAS A POWERFUL TIME for thinking through individuals. Individuals were the ideological unit of political "freedom." They also became the analytic motor of influential sciences, from economics to population biology. "Imagine individuals," both scholars and pundits told us, "and you can conjure the world."

The imagined autonomy of the individual was tied to the autonomy of the species. Each species was thought to rise or fall on its own merits, that is, through the fitness of the individuals it produced. Individuals were just one kind of self-contained unit that could be summed up or divided like building blocks, from genes to populations to species—and sometimes even to nations, religions, or civilizations.

Today, the autonomy of all these units has come under question, and each question works to undermine the edifice built from the segregation of each from each. As biologist Scott Gilbert tells us, "we have never been individuals." His "we" refers to all life; his "individuals" are autonomous species as well as single organisms. If most of the cells in the human body are microbes, which "individual" are we? We can't segregate our species nor claim distinctive status—as a body, a genome, or an immune system. And what if evolution selects for relations among species rather than "individuals"?

Gilbert's refusal of species autonomy is joined in this section by related refusals from literary criticism and anthropology. Carla Freccero introduces us to the ancient figure of the *cynanthrope,* the merger of dog and man. Species autonomy, she shows, is neither common sense nor timeless. Wolves and dogs lope through our histories as both human and not human. In the stories we tell each other, wolves raise children; wolves become men and men wolves. These stories will guide us as we question who we are in a more-than-human world.

Anthropologist Marianne Lien investigates these issues through material practices. It is impossible to raise farmed salmon in Norwegian fjords, she finds, without the help of wrasse, a fish that cleans sea lice from the salmon. In working to domesticate salmon, fish farmers find themselves learning to domesticate wrasse. In the context of fish farming, these species cannot live without each other: so much for the autonomous individual and species.

Bringing us back to biology, Deborah Gordon investigates the ant colony, perhaps the best-known example of an entity in which no individual acts alone. Should the colony then be seen as the individual, in which the ants are merely mechanical parts? Gordon suggests that we change the question. Instead of trying to identify autonomous individuals, ant or colony, she asks how collective effects are created from the relations of ants to each other. Collective behavior begins not as an internally generated algorithm but rather in the midst of encounter. "When one ant touches another with its antennae, it decides whether the other ant is a nestmate," she explains. From tiny interactions such as these, colonies develop foraging strategies and reproductive trajectories, not as "individuals," but as encounters in motion.

Imagining individuals as the motor of analysis made computations possible without field observations, because all individuals were posited to behave in the same internally generated way. Without this simplification, we must return to the field to watch relationships, encounters, and histories. A world of wonder beckons. Perhaps such observations will help us learn to live on a damaged planet. ●

4

HOLOBIONT BY BIRTH
MULTILINEAGE INDIVIDUALS AS THE CONCRETION OF COOPERATIVE PROCESSES

Scott F. Gilbert

WHEN YOU THINK OF A COW, you probably envision an animal grazing, eating grass, and perhaps producing methane at her other end. However, cows cannot do this. Their bovine genome does not encode proteins with the enzymatic activity needed to digest cellulose. What the cow does is chew the grass and maintain a symbiotic community of microorganisms in her gut. It is this population of gut symbionts that digests the grass and makes the cow possible.

The cow is an obvious example of what is called a *holobiont,* an organism plus its persistent communities of symbionts.[1] The notion of the holobiont is important both within and beyond biology because it shows a radically new way of conceptualizing "individuals." Recognizing the holobiont as a critical unit of life highlights process and reciprocal interactions, while challenging notions of genomic purity.

In one sense, though, the holobiont concept is not completely new. For decades, plant-focused scientists have been attentive to symbioses and have recognized that plants are composite organisms. But those studying animals have been relatively resistant to such ideas. Lynn Margulis and René Fester[2] mooted the term *holobiont* for such

a complete organism. As further studies demonstrated the critical roles symbionts play for the host, Eugene Rosenberg and Ilana Zilber-Rosenberg utilized the term *holobiont* to indicate the obligatory nature of symbiosis for the life of the organism.

Animal-focused biologists may have struggled to see organisms as holobionts because the holobiont concept undermines the classic definitions of animal individuality.[3] Those definitions have included the following:

1. *Anatomical individuality.* Anatomical boundaries are what separate us from the environment and from each other. When we look at each other, we appear to be anatomically distinct individuals.
2. *Genetic individuality.* This is the notion that each of us has a single genome that allows forensic scientists to say, "Ah-ha, he's the perpetrator!" This genomic individuality has largely superseded others and is now often considered prior to other forms of individuality.
3. *Developmental individuality.* This is the concept that each of us comes from the fertilized egg and that we are defined as individuals by the common origin of all our cells.[4]
4. *Immune individuality.* If I were to put my skin onto you, you would reject it because is not you. In this definition, our immune systems exist to recognize that which is nonself and to protect us against a hostile outside world waiting to infiltrate and destroy the individual.
5. *Physiological individuality.* This is the individuality wherein the different parts of the body come together for a common end. It is a body defined by a harmonious division of labor.
6. *Evolutionary individuality.* This form of individuality focuses on the individual that gets selected within evolutionary processes, be it a genome or an organism.

These categories are usually what we think of when we think of biological ideas associated with individuality. The concept of the holobiont, however, upsets all of them.

Consider corals, which have been very important in studies of holobionts. Many species of coral have green algal symbionts, *Zooxanthellae,* which live inside their cells and provide them with oxygen and sugars. The coral with symbionts is a functional organism and is the

basis of a reef ecosystem; without the symbionts, the coral whitens and dies. Thermal stress and pollutant stress can cause these symbionts to exit the coral or die, leaving the coral without oxygen and without sugars. It is an obligate relationship. The coral simply doesn't exist without its symbiont.[5]

Symbioses are equally important for the termite *Mastotermes darwiniensis,* one of the poster organisms for holobionts. The termite eats wood. It eats trees. It eats houses. It is a major agricultural pest. Only, it cannot eat wood. It does not have a genome that allows it to eat wood. What it has inside its gut is a symbiotic protist, *Mixotricha paradoxica,* that eats the wood. Only, it doesn't. *Mixotricha* is a composite organism containing a protist and at least four different types of bacteria. Termites are thus composite organisms all the way down. Bacteria and protists act together to make *M. paradoxica,* which is essential to the functioning of the gut of a termite, which itself lives in a termite community. So what is the individual? A so-called individual worker termite cannot live without its symbionts or its colony.[6] Clearly individuality is being questioned here at many levels. In this chapter, I want to focus specifically on the holobiont, the organism plus its persistent microbial communities, and the ways that this concept disrupts the tenets of individualism that have structured dominant lines of thought not only within biology but also in fields as diverse as economics, politics, and philosophy. The holobiont is powerful, in part, because it is not limited to nonhuman organisms. It also changes what it means to be a person.

So let me begin with the first in a series of disruptions, namely, the disruption of the notion of the anatomical individual. If you look at a human body, you will find that for every eukaryotic cell derived from the zygote, there is a bacterial symbiont. Only about half the cells in our bodies contain a "human genome." The other cells include about 160 different bacterial genomes. We have about 160 major species of bacteria in our bodies, and they all form complex ecosystems. Human bodies are and contain a plurality of ecosystems. Our mouths are different ecosystems than our intestines, or our skin, or our airways. The volume of the microbial organisms in our bodies is about the same as the volume of our brain, and the metabolic activity of those microbes is about equivalent to that of our liver. The microbiome *is* another organ; so we are not *anatomically* individuals at all.[7]

But what about the notion of the *genetic individual*? Supposedly,

each of our cells has the same genome established at fertilization. This is a concept that is used increasingly to define who we are. Today, genetic individuality is the individuality that is promulgated as the core of individual selves. *Life* magazine, when it describes "the first days of creation," that is, our embryonic development, tells us that

> the result of fertilization is a single nucleus that contains an entire biological blueprint for a new individual. Genetic information governing everything from the length of the nose to the diseases that will be inherited.[8]

DNA is constantly being represented to us as the secular version of our soul. It is depicted as that which is our essence and that which determines our behaviors, namely, our real selves. Dorothy Nelkin and Susan Lindee have called this our "sacred DNA."[9] This genomic notion of individuality is the one that's being used by anti-abortion lobbyists, because if your genome is formed at fertilization, and if DNA is your essence, then fertilization has become the equivalent of ensoulment. Several anti-abortion websites each tell us, "and even more amazingly, intelligence and personality, the way you look and feel, were already in place in your genetic code. At the moment of conception, you were essentially and uniquely you." It was the message of the Republican presidential primary contenders in 2015 as well as the message of conservative philosophers:[10] your DNA defines who you are.

However, this alleged genetic basis of individuality is scientifically wrong.[11] The symbionts are another mode of inheritance. Indeed, while humans have about twenty-two thousand different genes, the bacteria in us bring approximately 8 million more genes to the scene. We get our symbionts primarily by infection from the mother as we pass through the birth canal after the amnion breaks. These bacteria are supplemented by those from the mother's skin and from the environment.[12] In many arthropods, the bacteria come packaged, like organelles, in the egg cytoplasm. In *Drosophila* (fruit flies), for instance, the *Wolbachia* bacteria that are important for their immune system are transmitted from their mothers, inside the oocytes from which the flies develop.[13] Animals are not monogenomic organisms.

Moreover, symbiont genes play important roles, even selectable roles, in the lives of holobionts. Pea aphids provide a beautiful illustration. They are parthenogenetic and reproduce exclusively through altering meiosis. There are no males in the species and no sexual recombination of genes.[14] As a result, there should not be any diversity

among these aphids, except what arises from rare mutations. Yet, these organisms are indeed diverse. This is because pea aphids can and do acquire different sets of symbionts.

One of the pea aphid's common symbiotic bacteria, *Buchnera,* can provide thermotolerance to the holobiont. This bacterium has a gene capable of encoding a heat-tolerating chaperonin protein. A bacterium can have a long form of this gene, which produces a full chaperonin protein, or it could have a short form of this gene, which produces a short, nonfunctional protein. If you are a pea aphid, having the short form of the gene is usually a little bit better because it allows you to produce more offspring under most temperatures. But under relatively high temperatures, you lose fecundity if you don't have the full chaperonin gene. In other words, here is a selectable trait: the ability to have offspring at high temperatures. This trait, however, is not from an allele of the host genome of the pea aphid. It is from an allele of the pea aphids' symbiont. The symbiont is giving a property to the holobiont that can shape evolutionary processes.[15]

Pea aphids often have another different symbiont, *Hamiltonella,* that prevents parasitoid wasp infection. A parasitoid wasp can insert its egg into a juvenile pea aphid. The egg of the wasp will develop inside the pea aphid, forming a small larva that eats the aphid from within, being very careful not to kill it. Eventually, when the aphid is about to die, the wasp larva digests a hole through the cuticle of the aphid. If pea aphids do not have *Hamiltonella* symbionts, about 80 percent of the aphids become parasitized in this way by wasps. But with the bacteria, the aphid is basically immune to wasp infection—yet not quite. Not any *Hamiltonella* bacteria can do this job. The bacteria have to have a certain allele to be effective. Only, it's not really an allele. It is the *bacteriophage* of the bacteria that confers parasitoid resistance to the holobiont. In short, it is the symbiont of the symbiont that is doing the work of giving immunity to the entire organism.[16]

Do the symbionts of humans have different, potentially selectable alleles? The answer is perhaps. We know that our symbiotic bacteria have different alleles that can potentially give the holobiont different properties. A paper that came out in *Nature* a few years ago demonstrated that people in Japan have a different type of *Bacteroides plebius* in their guts than do people in America. The genomes that are found in *B. plebius* in Japan contain two genes gained by horizontal gene transfer from a related organism that lives on red algae. These are genes

that allow the organism to digest the complex sugars found in seaweed. Having gut bacteria with these genes may allow the Japanese population to get more calories from sushi.[17]

These examples illustrate that we are not genetically individuals. But what about developmental criteria for individuality? Recent evidence shows that we are not individuals by developmental criteria, either. Organisms need their symbionts to construct "themselves." In the case of *Asobara,* another parasitoid wasp, *Wolbachia* bacteria are essential for reproduction. The reproductive tracts of *Asobara* females are dependent on the presence of the *Wolbachia.* If the *Wolbachia* are removed, the cells of the wasps' ovaries die.[18]

For mammals, bacteria are also critical for body development. Without its normal gut symbionts, a mouse cannot form its gut capillary system nor the gut-associated lymphoid tissue. How do bacteria play such an important role in development? Researchers have found that the bacteria induce normal gene expression in the Paneth cells of the mouse intestine. These intestinal epithelial cells are being told what to do by chemical signals coming from the bacteria! When mice are raised to be "germ-free," their guts do not develop properly. Normal bacteria induce a sixfold increase in the mRNA for colipase, a protein involved in lipid metabolism. The expression of angiogenin-4, which is important for forming gut capillaries, is increased tenfold when normal gut bacteria are present. In other words, a germ-free mouse is like a mutant with only 10 percent of the angiogenin mRNA production of normal mice.[19] Without such proteins, guts simply do not develop and function well. Bacterial symbionts are needed for normal development.

We are learning more and more about how microbes can be critically important in development. A couple of recent papers indicate that a certain strain of *Lactobacillus* is able to reverse the experimental osteoporosis that develops in female mice whose ovaries have been removed. Without estrogen, the mice lose bone density, as is also seen in postmenopausal women. However, if one adds a particular strain of *Lactobacillus* to ovariectomized mice, the mice develop higher bone density again. What are these bacteria doing? They are inhibiting osteoclast development, just like estrogen does. Osteoclasts are the bone-digesting cells. The balance between osteoblasts (which make bone) and osteoclasts (which degrade bone) is controlled by estrogen and testosterone hormones, and when one loses estrogen,

the osteoclasts start increasing at the expense of osteoblasts. What these bacteria do is stop the differentiation of the osteoclasts and preserve bone.[20]

There is growing evidence that in mice, and possibly in humans, bacteria are also partly responsible for normal brain development. One large study[21] showed that if one virally stresses a pregnant mouse, the resulting pups have symptoms similar to those of the autism spectrum in humans. For instance, these mice spend a lot of time self-grooming, and they are not interested in other mice. They prefer solitary cages to cages with other mice in them. This study found that the autistic-like mice act more like normal mice when you alter their bacteria. If one adds certain bacteria to their guts, especially certain species of *Bacteroides,* this alters the community of the bacterial symbionts, and it increases the integrity of the gut epithelium. This simple procedure stops the leaking of bacterial products into the gut and normalizes several of the autism-like behavioral abnormalities. One of these products, 4EPS, is made by bacteria and causes anxiety-like symptoms in mice. In the "autistic" mice, this product is seen in relatively high amounts in the circulation. In the mice without these symptoms—and in the "autistic" mice that were treated with the bacteria—4EPS can hardly be detected. This study opens up a new area that investigates cognitive and emotional situations as products of bacterial metabolism.[22]

Understanding the developmental roles of bacterial symbionts is important in fields such as medicine and in agriculture. For example, *Mansonella*—a parasitic worm that infects people, giving them symptoms such as lethargy and edema—has evolved resistance to the helminthicide drugs, which are usually used to kill off parasites of this kind. However, researchers have found an alternate treatment; they have discovered that doxycycline, a good old-fashioned antibacterial drug, kills the worms indirectly. Without its bacterial symbionts, the worm can't molt, and its cells die. One kills the worm by killing its symbiont.[23]

Unfortunately, we might be doing the same thing in the American Midwest to the spotted salamander. The female spotted salamander lays a huge egg mass, and within these clumps of eggs, the inside eggs cannot receive enough oxygen or food on their own. However, the eggs are normally coated with a particular symbiotic algae. The algae eat the jelly around the eggs, while also undergoing photosynthesis. In

so doing, they produce the highly oxygenated environment that the eggs—particularly the inner ones—need to survive.[24] However, in U.S. agricultural areas, farmers are using herbicides, such as Atrazine, to control their weeds. These herbicides move through waterways and into the ponds where these salamanders breed. The herbicides kill the algae, and when they kill the algae, few salamander eggs survive.

What, then, of *physiological* individuality? In theory, the organism is an individual whose component parts cooperate for the betterment of the whole. But the parts that cooperate can be other organisms living with it. Let us consider the milkweed bug *Planococcus.* This insect has a symbiont, *Tremblaya,* a bacterium residing within its cells. In turn, those bacteria have another bacterial symbiont, *Moranella,* inside of them. This set of symbionts, nested one inside the other like a set of Russian *matryoshka* dolls, is necessary for the insect to synthesize several amino acids. For instance, *Planococcus,* alone, cannot make phenylalanine. Its genome does not contain the genes encoding the enzymes involved in its synthesis. The symbionts do. Phenylalanine synthesis starts in the symbiont, then it goes into the symbiont's symbiont. The product made by the symbiont's symbiont returns into the symbiont. Only the last step of phenylalanine synthesis is done by the enzyme encoded by the genome of the insect itself.[25] The production of a single compound thus requires a threefold symbiosis.

This co-metabolism—the physiological integration of the host and the symbiont—is seen in mammals, too. Research on mice indicates that as much as one-third of a mammal's metabolome, the diversity of molecules carried inside its blood, has a microbial origin. The circulatory system extends the chemical impact of the microbiota throughout the body. For instance, 95 percent of the serotonin in mammal blood appears to be made, not only by the eukaryotic cells, but by induction from the bacteria that dwell within us.[26]

Co-metabolism matters for humans. Kwashiorkor was long considered a serious protein deficiency disease. However, when researchers went to Africa to study identical twins wherein one of the twins had kwashiorkor and the other did not, they found that the twins had the same genome and nearly identical low-protein diets. What, then, was causing one twin to be healthy and the other twin to be sick? It turned out to be their bacteria. When the researchers gave a healthy twin's gut bacteria to the twin with kwashiokor, the kwashiorkor was alleviated. Scientists replicated this experiment in mice, giving some mice the

bacteria from the kwashiokor twin and another set of mice the bacteria from the healthy twin. What happened when both sets of mice were given protein-deficient diets was that the mice with the first set of bacteria (from the sick children) developed kwashiorkor-like symptoms. The mice with the bacteria from the healthy children did not get the disease. This physiological cooperation between symbiont and host was called *microbe-host co-metabolism.*[27] It is the metabolism of the holobiont.

Human pregnancy, too, is an amazing co-metabolic situation. The bacteria that are in a woman's reproductive and digestive tracts at the end of pregnancy are different than those that are usually present. This matters because these are the bacteria that are going to be seen by the fetus as it is leaving the birth canal.[28] It appears that the hormones of the mother are changing the bacteria so as to shape the bacterial population that her baby acquires during birth. Moreover, the mother's body is going even one step further to promote health in her offspring. A mother has two sets of nutrients in her milk—one set for the newborn and one set for the bacteria that will help finish the construction of its gut capillaries and lymphoid tissues.[29] A human mother's milk contains several oligosaccharides, complex sugars, that cannot be digested by the baby. These are not sugars for the baby; these are sugars for bacteria such as *Bifidobacteria,* which has genes that encode enzymes capable of digesting those special milk sugars. Through her symbionts and through her milk, the mother is causing developmental changes in her infant even after its birth. Indeed, "birth" is not the birth of a so-called individual. Birth is the continuation of the holobiont community.[30]

The grounds for immune individuality are as shaky as those for developmental and physiological individuality. When I was a postdoctoral fellow in immunology, and when I taught immunology in the 1980s, we thought that the immune system was the defense network of our bodies, an amazing set of weaponry to protect us against a hostile environment. Now, it seems that such protective functions probably constitute a relatively minor part of the work that an immune system does. The immune system, rather than being imagined as a force of protective soldiers made by the host, can be thought of as a group of passport control agents and bouncers. They know who to let in and who to keep out—and they have learned this through millions of years of evolution. The immune system is a composite product of

the holobiont, and it is not simply fighting anything that is "not-self." Rather, it knows that there are some bacteria that are supposed to be welcomed into our bodies because, as the many examples so far have shown, the bacteria are needed for completing our development and for our physiological functioning.[31]

This is not the immune system I had learned about. It is a much more interesting immune system—one that allows countless microbes to become part of our bodies. Indeed, even the immune system itself is built by microbes. Without the proper microbial symbionts, important subsets of immune cells fail to form. Germ-free mice have an immunodeficiency syndrome.[32] We are thus not individuals by immune criteria.

Lastly, I want to focus on holobiont evolution. Evolution may be the evolution of holobionts, not monogenomic individuals. Lynn Margulis said, "In short, I believe that most evolutionary novelty arose and still arises directly from symbiosis."[33] We are seeing this being played out in many fascinating ways. First, as mentioned earlier, in our discussion of pea aphids, the symbionts can provide selectable variation to the holobiont. Second, symbionts can effect reproductive isolation. Seth Bordenstein's laboratory has demonstrated that symbiotic bacteria can cause reproductive isolation through cytoplasmic incompatibility. Here certain symbionts can function only in the presence of certain nuclei, but not others. The hybrids between individuals of the same species having different sets of symbionts either fail to form or are sterile.[34] Another mechanism for reproductive isolation is mate selection. In Eugene Rosenberg's lab, Sharon and colleagues have shown that bacteria on larval food changed the contact pheromones of the adult.[35] *Drosophila* that grew up on starch prefer other *Drosophila* that grew up on starch. This is not a property of the starch itself. Rather, it is a property of the *Lactobacillus* bacteria on the starch. Once within the body, *Lactobaccillus* changes the pheromones on the cuticles. When the bacteria are eliminated from the food, there is no sexual preference.

Third, research suggests that the ability to become a metazoan, a multicellular animal, is probably a result of bacteria-eukaryotic interaction among choanoflagellates.[36] One has to remember that the bacteria were here first. They had a 2 billion year head start on eukaryotes. When eukaryotes came into being, they found themselves in a rich microbial environment. We talk about the Anthropocene. We talk

about the age of fishes, and we talk about the age of reptiles, and the age of mammals. No. It is the age of bacteria, always was, and always will be.[37] We evolve as teams, as consortia—and we likely always have.

It appears that there is no individuality in the classical biological sense. We have no anatomical individuality: most of our cells are microbial. No physiological individuality: we are joined in co-metabolism with our microbes. No developmental individuality: the microbes help build our guts and our immune systems. They help build the light organ of squids (see McFall-Ngai, this volume), and they prevent ovarian cell death in parasitoid wasps. We are not individuals by immune criteria—the microbes actually help make our immune systems, and the immune system helps make niches for the microbes in our bodies. Indeed, recent studies provide evidence for convergent evolution between symbiotic microbes and immune receptors, where both are under selective pressure to mimic the molecules characteristic of the mammalian species.[38] Genetic individuality falls apart, too: we have more than 150 genomes in our bodies beside our eukaryotic inheritance, and these bacteria collectively have many more different genes than we have as human eukaryotes. We are multilineage organisms. Evolutionarily, we are not individuals either: symbionts can provide selectable variation. They can cause genetic isolation, and by acquiring genomes, as Lynn Margulis said, bacteria might actually cause huge evolutionary transitions, such as the one to multicellularity.

Because this chapter was first presented at a meeting where biologists, sociologists, and anthropologists were speaking to one another, let me mention some of the political ramifications of literally "becoming with the other." In a social sense, symbionts play havoc with the notion of a pure body politic. We are definitely not monogenomic individuals. We do not share the same lineage. So what are symbionts? If one thinks of an animal organism in the classical sense of being an individual, then the symbionts are *Gastarbeiter*, guest workers who do the work that the stable members of the population won't dirty themselves with. (One can think of such places as Saudi Arabia and Yemen, where certain lineages have citizenship and most of the population in the country are not citizens but rather temporary residents.) If one thinks of an animal as having porous borders, then the symbionts are legal resident aliens, like green card holders in the United States. Only if one thinks of the animal or any other organism as a holobiont are the symbionts full citizens of an evolving and heterogeneous

community. The body and the body politic reflect each other's awareness and anxieties.

Symbiosis is the strategy that supports life on earth. Rhizomal bacteria interact with legumes, allowing nitrogen fixation, the basis of terrestrial life. The coral reef ecosystem and the tidal sea grass ecosystem depend on the symbionts of corals and clams. These major symbiotic webs rule the planet, and within these big symbioses are the smaller symbiotic webs of things we call organisms. And within organisms are the products of even more ancient symbioses called cells, and the products of other ancient symbioses, which we call genomes. Symbiosis is the way of life on earth; we are all holobionts by birth.

A renowned scholar of developmental genetics and embryology, **SCOTT F. GILBERT** has helped to build a new subdiscipline of biology that brings together ecology, evolution, and organismal development. A true transdisciplinary thinker, he holds degrees in religion and history of science. He is the Howard A. Schneiderman Professor of Biology (Emeritus) at Swarthmore College and a Finland Distinguished Professor at the University of Helsinki.

Notes

1. Ilana Zilber-Rosenberg and Eugene Rosenberg, "Role of Microorganisms in the Evolution of Animals and Plants: The Hologenome Theory of Evolution," *FEMS Microbiology Reviews* 32, no. 5 (2008): 723–35, doi:10.1111/j.1574-6976.2008.00123.x.
2. Lynn Margulis and René Fester, *Symbiosis as a Source of Evolutionary Innovation: Speciation and Morphogenesis* (Boston: MIT Press, 1991).
3. Scott F. Gilbert, Jan Sapp, and Alfred I. Tauber, "A Symbiotic View of Life: We Have Never Been Individuals," *Quarterly Review of Biology* 87, no. 4 (2012): 325–41, doi:10.1086/668166; Margaret McFall-Ngai, Michael G. Hadfield, Thomas C. G. Bosch, Hannah V. Carey, Tomislav Domazet-Lošo, Angela E. Douglas, Nicole Dubilier, et al., "Animals in a Bacterial World: A New Imperative for the Life Sciences," *Proceedings of the National Academy of Sciences of the United States of America* 110, no. 9 (2013): 3229–36, doi:10.1073/pnas.1218525110; Scott F. Gilbert and David Epel, *Ecological Developmental Biology: The Environmental Regulation of Development, Health, and Evolution* (Sunderland, Mass.: Sinauer Associates, 2015).

4. Thomas H. Huxley, "Upon Animal Individuality," *Edinburgh New Philosophical Journal* 53 (1852): 172–77.

5. Leonard Muscatine, Paul G. Falkowski, James W. Porter, and Zvy Dubinsky, "Fate of Photosynthetic Fixed Carbon in Light- and Shade-Adapted Colonies of the Symbiotic Coral *Stylophora pistillata*," *Proceedings of the Royal Society, Series B* 222, no. 1227 (1984): 181–202, doi:10.1098/rspb.1984.0058; Eugene Rosenberg, Omry Koren, Leah Reshef, Rotem Efrony, and Ilana Zilber-Rosenberg, "The Role of Microorganisms in Coral Health, Disease, and Evolution," *Nature Reviews in Microbiology* 5 (2007): 355–62, doi: 10.1038/nrmicro1635.

6. Lynn Margulis and Dorion Sagan, "The Beast with Five Genomes," *Natural History* 110, no. 5 (2001): 38.

7. Ruth E. Ley, Daniel A. Peterson, and Jeffrey I. Gordon, "Ecological and Evolutionary Forces Shaping Microbial Diversity in the Human Intestine," *Cell* 124, no. 4 (2006): 837–48, doi:10.1016/j.cell.2006.02.017; Yun Kyung Lee and Sarkis K. Mazmanian, "Has the Microbiota Played a Critical Role in the Evolution of the Adaptive Immune System?," *Science* 330, no. 6012 (2010): 1768–73, doi:10.1126/science.1195568; McFall-Ngai et al., "Animals in a Bacterial World."; Ron Sender, Shai Fuchs, and Ron Milo, "Revised Estimates for the Number of Human and Bacterial Cells in the Body," *PLoS Biology* 14, no. 8 (2016): e1002533.

8. "The First Days of Creation," *Life* 13, no. 10 (1990): 26–46.

9. Dorothy Nelkin and M. Susan Lindee, *The DNA Mystique: The Gene as a Cultural Icon* (New York: W. H. Freeman, 1996).

10. Robert P. George and Patrick Lee, "Back to Science Class for the Science Guy," *National Review,* September 28, 2015, http://www.nationalreview.com/article/424721/bill-nye-youtube-abortion.

11. Scott F. Gilbert, "When 'Personhood' Begins in the Embryo: Avoiding a Syllabus of Errors," *Birth Defects Research, Part C* 84, no. 2 (2008): 164–73, doi:10.1002/bdrc.20123; Scott F. Gilbert and Rebecca Howes-Mischel, "'Show Me Your Original Face before You Were Born': The Convergence of Public Fetuses and Sacred DNA," *History and Philosophy of the Life Sciences* 26, no. 3-4 (2004): 377–94.

12. Lisa J. Funkhauser and Seth R. Bordenstein, "Mom Knows Best: The Universality of Maternal Microbial Transmission," *PLOS Biology* 11, no. 8 (2013): e1001631, doi:10.1371/journal.pbio.1001631; Scott F. Gilbert, "A Holobiont Birth Narrative: The Epigenetic Transmission of the Human Microbiome," *Frontiers in Genetics* 5 (2014): 282, doi:10.3389/fgene.2014.00282.

13. Patrick M. Ferree, Horacio M. Frydman, Jennifer M. Li, Jian Cao, Eric Wieschaus, and William Sullivan, "*Wolbachia* Utilizes Host Microtubules and Dynein for Anterior Localization in the *Drosophila* Oocyte," *PLOS Pathogens* 1, no. 2 (2005): e14, doi:10.1371/journal.ppat.0010014.

14. Jan Sapp, *The New Foundations of Evolution: On the Tree of Life* (New York: Oxford University Press, 2009); Junjie Qin, Ruiqiang Li, Jeroen Raes, Manimozhiyan Arumugam, Kristoffer Solvsten Burgdorf, Chaysavanh Manichanh, and Trine Nielsen, "A Human Gut Microbial Gene Catalogue Established by Metagenomic Sequencing," *Nature* 464 (2010): 59–65, doi:10.1038/nature08821.
15. Helen E. Dunbar, Alex C. C. Wilson, Nicole R. Ferguson, and Nancy A. Moran, "Aphid Thermal Tolerance Is Governed by a Point Mutation in Bacterial Symbionts," *PLOS Biology* 5, no. 5 (2007): e96, doi:10.1371/journal.pbio.0050096; Scott F. Gilbert, Emily McDonald, Nicole Boyle, Nicholas Buttino, Lin Gyi, Mark Mai, Neelakantan Prakash, and James Robinson, "Symbiosis as a Source of Selectable Epigenetic Variation: Taking the Heat for the Big Guy," *Philosophical Transactions of the Royal Society, Series B* 365, no. 1540 (2010): 671–78, doi:10.1098/rstb.2009.0245; Nancy A. Moran and Yueli Yun, "Experimental Replacement of an Obligate Insect Symbiont," *Proceedings of the National Academy of Sciences of the United States of America* 112, no. 7 (2015): 2093–96, doi:10.1073/pnas.1420037112.
16. Kerry M. Oliver, Patrick H. Degnan, Martha S. Hunter, and Nancy A. Moran, "Bacteriophages Encode Factors Required for Protection in a Symbiotic Mutualism," *Science* 325, no. 5943 (2009): 992–94, doi:10.1126/science.1174463.
17. Jan-Hendrik Hehemann, Gaëlle Correc, Tristan Barbeyron, William Helbert, Mirjam Czjzek, and Gurvan Michel, "Transfer of Carbohydrate-Active Enzymes from Marine Bacteria to Japanese Gut Microbiota," *Nature* 464 (2010): 908–12, doi:10.1038/nature08937; Jan-Hendrik Hehemann, Amelia G. Kelly, Nicholas A. Pudlo, Eric C. Martens, and Alisdair B. Boraston, "Bacteria of the Human Gut Microbiome Catabolize Red Seaweed Glycans with Carbohydrate Active Enzyme: Updates from Extrinsic Microbes," *Proceedings of the National Academy of Sciences of the United States of America* 109, no. 48 (2012): 19786–91, doi:10.1073/pnas.1211002109.
18. Margaret J. McFall-Ngai, "Unseen Forces: The Influence of Bacteria on Animal Development," *Developmental Biology* 242, no. 1 (2002): 1–14, doi:10.1006/dbio.2001.0522; Bart A. Pannebakker, Benjamin Loppin, Coen P. H. Elemans, Lionel Humblot, and Fabrice Vavre, "Parasitic Inhibition of Cell Death Facilitates Symbiosis," *Proceedings of the National Academy of Sciences of the United States of America* 104, no. 1 (2007): 213–15, doi:10.1073/pnas.0607845104.
19. Lora V. Hooper, Melissa H. Wong, Anders Thelin, Lennart Hansson, Per G. Falk, and Jeffrey I. Gordon, "Molecular Analysis of Commensal Host-Microbial Relationships in the Intestine," *Science* 291, no. 5505 (2001): 881–84, doi:10.1126/science.291.5505.881; Thaddeus S. Stap-

penbeck, Lora V. Hooper, and Jeffrey I. Gordon, "Developmental Regulation of Intestinal Angiogenesis by Indigenous Microbes via Paneth Cells," *Proceedings of the National Academy of Sciences of the United States of America* 99, no. 24 (2002): 15451–55, doi:10.1073/pnas.202604299.

20. Robert A. Britton, Regina Irwin, Darin Quach, Laura Schaefer, Jing Zhang, Taehyung Lee, Narayanan Parameswaran, and Laura R. McCabe, "Probiotic *L. reuteri* Treatment Prevents Bone Loss in a Menopausal Ovariectomized Mouse Model," *Journal of Cell Physiology* 229, no. 11 (2014): 1822–30, doi:10.1002/jcp.24636; Claes Ohlsson, Cecilia Engdahl, Frida Fåk, Annica Andersson, Sara H. Windahl, Helen H. Farman, Sofia Movérare-Skrtic, Ulrika Islander, and Klara Sjögren, "Probiotics Protect Mice from Ovariectomy-Induced Cortical Bone Loss," *PLOS ONE* 9, no. 3 (2014): e92368, doi:10.1371/journal.pone.0092368.

21. Elaine Y. Hsaio, Sara W. McBride, Sophia Hsien, Gil Sharon, Embriette R. Hyde, Tyler McCue, Julian A. Codelli, et al., "Microbiota Modulate Behavioral and Physiological Abnormalities Associated with Neurodevelopmental Disorders," *Cell* 155, no. 7 (2013): 1451–63, doi:10.1016/j.cell.2013.11.024.

22. Ibid.; Lieve Desbonnet, Gerard Clarke, Fiona Shanahan, Timothy G. Dinan, and John F. Cryan, "Microbiota Is Essential for Social Development in the Mouse," *Molecular Psychiatry* 19, no. 2 (2014): 146–48, doi:10.1038/mp.2013.65; McFall-Ngai et al., "Animals in a Bacterial World."

23. Frederic Landmann, Denis Voronin, William Sullivan, and Mark J. Taylor, "Anti-filarial Activity of Antibiotic Therapy Is due to Extensive Apoptosis after *Wolbachia* Depletion from Filarial Nematodes," *PLOS Pathogens* 7, no. 11 (2011): e1002351, doi:10.1371/journal.ppat.1002351.

24. Heather M. Olivier and Brad R. Moon, "The Effects of Atrazine on Spotted Salamander Embryos and Their Symbiotic Algae," *Ecotoxicology* 19, no. 4 (2010): 654–61, doi:10.1007/10646-009-0437-8; Ryan Kerney, Eunsoo Kim, Roger P. Hangarter, Aaron A. Heiss, Cory D. Bishop, and Brian K. Hall, "Intracellular Invasion of Green Algae in a Salamander Host," *Proceedings of the National Academy of Sciences of the United States of America* 108, no. 16 (2011): 6497–502, doi:10.1073/pnas.1018259108; Erin R. Graham, Scott A. Fay, Adam Davey, and Robert W. Sanders, "Intracapsular Algae Provide Fixed Carbon to Developing Embryos of the Salamander *Ambystoma maculatum*," *Journal of Experimental Biology* 216 (2013): 452–59, doi:10.1242/jeb.076711.

25. John P. McCutcheon and Carol D. von Dohlen, "An Interdependent Metabolic Patchwork in the Nested Symbiosis of Mealybugs," *Current Biology* 21, no. 16 (2011): 1366–72, doi:10.1016/j.cub.2011.06.051.

26. Fredrik Bäckhed, Hao Ding, Ting Wang, Lora V. Hooper, Gou Young Koh, Andras Nagy, Clay F. Semenkovich, and Jeffrey I. Gordon, "The

Gut Microbiota as an Environmental Factor That Regulates Fat Storage," *Proceedings of the National Academy of Sciences of the United States of America* 101, no. 44 (2004): 15718–23, doi:10.1073/pnas.0407076101; Margaret McFall-Ngai, Elizabeth A. C. Heath-Heckman, Amani A. Gillette, Suzanne M. Peyer, and Elizabeth A. Harvie, "The Secret Languages of Coevolved Symbioses: Insights from the *Euprymna scolopes-Vibrio fischeri* Symbiosis," *Seminars in Immunology* 24, no. 1 (2012): 3–8, doi:10.1016/j.smim.2011.11.006; Jessica M. Yano, Kristie Yu, Gregory P. Donaldson, Gauri G. Shastri, Phoebe Ann, Liang Ma, Cathryn R. Nagler, Rustem F. Ismagilov, Sarkis K. Mazmanian, and Elaine Y. Hsiao, "Indigenous Bacteria from the Gut Microbiota Regulate Host Serotonin Biosynthesis," *Cell* 161, no. 2 (2015): 264–76, doi:10.1016/j.cell.2015.02.047.

27. Michelle I. Smith, Tanya Yatsunenko, Mark J. Manary, Indi Trehan, Rajhab Mkakosya, Jiye Cheng, and Andrew L. Kau, "Gut Microbiomes of Malawian Twin Pairs Discordant for Kwashiorkor," *Science* 339, no. 6119 (2013): 548–54, doi:10.1126/science.1229000.
28. Omry Koren, Julia K. Goodrich, Tyler C. Cullender, Aymé Spor, Kirsi Laitinen, Helene Kling Bäckhed, and Antonio Gonzalez, "Host Remodeling of the Gut Microbiome and Metabolic Changes during Pregnancy," *Cell* 150, no. 3 (2012): 470–80, doi:10.1016/j.cell.2012.07.008.
29. Angela M. Zivkovic, J. Bruce German, Carlito B. Lebrilla, and David A. Mills, "Human Milk Glycobiome and Its Impact on the Infant Gastrointestinal Microbiota," *Proceedings of the National Academy of Sciences of the United States of America* 108 (2011): 4653–58, doi:10.1073/pnas.1000083107; Hiroshi Makino, Akira Kushiro, Eiji Ishikawa, Hiroyuki Kubota, Agata Gawad, Takafumi Sakai, Kenji Oishi, et al., "Mother-to-Infant Transmission of Intestinal Bifidobacterial Strains Has an Impact on the Early Development of Vaginally Delivered Infant's Microbiota," *PLOS ONE* 8, no. 11 (2013): e78331, doi:10.1371/journal.pone.0078331.
30. Gilbert, "A Holobiont Birth Narrative"; Lynn Chiu and Scott F. Gilbert, "The Birth of the Holobiont: Multi-species Birthing through Scaffolding and Niche Construction," *Biosemiotics* 8, no. 2 (2015): 191–210, doi:10.1007/s12304-015-9232-5.
31. Takashi Obata, Yoshiyuki Goto, Jun Kunisawa, Shintaro Sato, Mitsuo Sakamoto, Hiromi Setoyama, Takahiro Matsuki, et al., "Indigenous Opportunistic Bacteria Inhabit Mammalian Gut-Associated Lymphoid Tissues and Share a Mucosal Antibody-Mediated Symbiosis," *Proceedings of the National Academy of Sciences of the United States of America* 107, no. 16 (2010): 7419–24, doi:10.1073/pnas.1001061107; June L. Round, S. Melanie Lee, Jennifer Li, Gloria Tran, Bana Jabri, Talal A. Chatila, and Sarkis K. Mazmanian, "The Toll-like Receptor 2 Pathway

Establishes Colonization by a Commensal of the Human Microbiota," *Science* 332, no. 6032 (2011): 974–77, doi:10.1126/science.1206095.

32. Alfred I. Tauber, "Expanding Immunology: Defense versus Ecological Perspectives," *Perspectives in Biology and Medicine* 51, no. 2 (2008): 270–84, doi:10.1353/pbm.0.0000; Torsten Olszak, Dingding An, Sebastian Zeissig, Miguel Pinilla Vera, Julia Richter, Andre Franke, Jonathan N. Glickman, et al., "Microbial Exposure during Early Life Has Persistent Effects on Natural Killer T Cell Function," *Science* 336, no. 6080 (2012): 489–93, doi:10.1126/science.121932; Duane Wesemann, Andrew J. Portuguese, Robin M. Meyers, Michael P. Gallagher, Kendra Cluff-Jones, Jennifer M. Magee, Rohit A. Panchakshari, et al., "Microbial Colonization Influences Early B-Lineage Development in the Gut Lamina Propria," *Nature* 501, no. 7465 (2013): 112–15, doi:10.1038/nature12496.

33. Lynn Margulis, *Symbiotic Planet: A New Look at Evolution* (New York: Basic Books, 1999), 33.

34. Robert M. Brucker and Seth R. Bordenstein, "The Hologenomic Basis of Speciation: Gut Bacteria Cause Hybrid Lethality in the Genus *Nasonia*," *Science* 341, no. 6146 (2013): 667–69, doi:10.1126/science.1240659.

35. Gil Sharon, Daniel Segal, John M. Ringo, Abraham Hefetz, Ilana Zilber-Rosenberg, and Eugene Rosenberg, "Commensal Bacteria Play a Role in Mating Preference of *Drosophila melanogaster*," *Proceedings of the National Academy of Sciences of the United States of America* 107, no. 46 (2010): 20051–56, doi:10.1073/pnas.1009906107.

36. Rosanna A. Alegado, Laura W. Brown, Shugeng Cao, Renee K. Dermenjian, Richard Zuzow, Stephen R. Fairclough, Jon Clardy, and Nicole King, "A Bacterial Sulfonolipid Triggers Multicellular Development in the Closest Living Relatives of Animals," *eLife* 1 (2012): e00013, doi:10.7554/eLife.00013; Mark J. Dayel, Rosanna A. Alegado, Stephen R. Fairclough, Tera C. Levin, Scott A. Nichols, Kent McDonald, and Nicole King, "Cell Differentiation and Morphogenesis in the Colony-Forming Choanoflagellate *Salpingoeca rosetta*," *Developmental Biology* 357, no. 1 (2011): 73–82, doi:10.1016/j.ydbio.2011.06.003.

37. Stephen Jay Gould, "Planet of the Bacteria," *Washington Post Horizon* 119, no. 344 (1996): H1.

38. Leonard Moise, Sarah Beseme, Ryan Tassone, Rui Liu, Fazana Kibnria, Frances Terry, William Martin, and Anne S. De Groot, "T Cell Epitope Redundancy: Cross-Conservation of the TCR Face between Pathogens and Self and Its Implications for Vaccines and Autoimmunity," *Expert Review of Vaccines* 15, no. 5 (2016): 607–17; Robert Root-Bernstein, "Autoimmunity and the Microbiome: T-Cell Receptor Mimicry of 'Self' and Microbial Antigens Mediates Self Tolerance in Holobionts," *Bioessays* 38 (2016), doi:10.1002/bies.201600083.

5

WOLF, OR *HOMO HOMINI LUPUS*

Carla Freccero

WOLF IS EVERYWHERE IN THE WESTERN IMAGINATION. From werewolf trials and fairy tales to postmodern wolf-human hybrids, the wolves that prowl through moral fables are clothed in political allegories, dragged unwittingly into juridical encounters, and, most recently, stand in for the queerness of transpecies becomings. Wolves and werewolves are ancient and familiar enemies and monsters, and they are emergent figures in the Anthropocene, both through their disappearance and "reintroduction" in discussions of vanishing species and habitats, and through the recent imaginative refigurations of wolf-human becoming in the age of the holobiont, spawning new imaginaries for communities of the living.

I am interested in wolves and humans and human-wolf encounters and becomings as a result of my work on the genealogy of the cynanthrope, the ancient merger of dog and man. Cynanthropes were thought to live on the edge of civilization and to be intelligent and rational like humans but also ferocious and hostile toward strangers, devouring their enemies. They are examples of what I call, after Jacques Derrida, "carnivorous virility," a cultural fantasy that the merger between dog and human restores to human men a measure of primitive strength, virility, and savagery.[1] The figure of the cynanthrope is a "material-semiotic" figure: material because it was thought to actually exist as an entity—there are depictions of it—and it persists in forms of masculine-canine becoming, and "semiotic," that is, meaningful (it persists in part because it is meaningful). What happens when wolf replaces dog in such becomings? Werewolves mark

the place where the monstrous announces itself: something that cannot be domesticated or mastered inheres in the human and makes it other than human. Certainly the imagined and actual history of humans-against-wolves and wolves-against-humans would indicate a certain intractability, nonnegotiability, in this relation. But if, as Derrida suggests, monsters also announce futurity, then perhaps the persistence of the becoming-wolf of humans in the popular imagination has something to tell us about the survival, the living-on, of humans and wolves on a damaged planet in apocalyptic times.[2]

Here I would like to do a queer figural historiography of the semiotic-material nexus embodied in figures of wolf-human interactions. How, where, and why do wolves signify, and what are the material histories, cultures, and encounters that make lupine figures ubiquitous in human oral and scriptural cultures? If "stories" is what there is to offer about the arts of living in the Anthropocene, then figural historiography provides a way of understanding what is "old" and persistent in narratives of human-wolf being and encounter, and what, in the figural and imaginative domain, has been accreted in the archive of the long togetherness, the "intimacy without proximity" of humans and wolves.[3] The fantasies, especially the fantasies of monstrosity that affect the interrelation of species, continue to exert their force in the present; exploring them can teach us, perhaps, to prepare ourselves to be open to the future. "All experience open to the future is prepared or prepares itself to welcome the monstrous *arrivant*," Derrida reminds us, an arrival that is monstrous because it cannot be known in advance and may, or may not, include "us" in a form we recognize.[4] This figural historiography is queer because it follows twisted paths, denatures temporal chronologies and ideologies of what Lee Edelman calls reproductive futurism (the future is the fulfillment of the promise of the present, the present begets the future), and because the encounter and sometimes merger between human and wolf is itself a sign of queer couplings, especially in the hybridized being of werewolves and cynanthropes.[5] Animal theory more generally (and I include humans here) is also queer in the sense that it opens up questions of nonnormative subjectivities, sexualities, and desire. It denormativizes or decenters the human by showing how the human is one subject position among others. Such theory—and, I hope, my story—helps untangle the long and enmeshed interimplications of sex, species, and race toward a queered future for the living

such that the normative categories for "knowing in advance" might no longer be available. Finally, insofar as it follows figures that cannot be said to be "individuals" but are rather beings-always-in-relation, this figural historiography also queers the tyranny—and the myth—of the one (see Gilbert and Haraway in this volume).

I am also tracking wolves in their spectrality; however fictionally and allegorically ubiquitous, material wolves (I originally wrote "in their integrity," but what could I have meant? Am I in search of the authentic wolf, the never-before-eradicated and reseeded wolf? Am I looking for "wild" wolves? Indeed, wolves are also asked to stand in for a nostalgia for the wholeness of the human and the natural, a nostalgia that is both spatial and temporal) are largely absent from most people's lives (not all, but most), except where they are protected or where, as in parts of Eastern Europe and Asia, there hasn't been as much systematic eradication (the so-called southern wolves, in Africa, South America, the Middle East, and South Asia, are considerably smaller and often hybridized with jackals, thus posing a diminished perceived threat). There is a work to be done here, a work of transpecies mourning that also seeks to come to terms with the spectral returns of lupine being within the cultures that have expelled and eradicated it.

How does the material history of the competition between humans and wolves in Europe that results in wolf eradication continue to haunt the figure of the human-as-animal in literature, political theory, and popular culture, and what (identificational? compensatory? melancholic?) work does this figure of the merger of human and wolf do? How might thinking about the long genealogy of both hostile and proximate human–wolf togetherness reformulate notions of "wild" and "domestic," "civilized" and "savage"? How might an understanding of wolfly being as thoroughly entangled with human being change the relationship between us?

I begin with a proverbial wolf; I am looking for the *lupus in fabula* (the wolf in the story), which is a way of saying "speak of the devil" . . .

In a 1963 Warner Brothers cartoon featuring Ralph E. Wolf and Sam Sheepdog—originally created by Chuck Jones (in 1953) then remade by ex-Jones animators Phil Monroe and Richard Thompson—a wolf and a sheepdog share a companionable dailiness and friendship involving coffee together in the morning and a return home arm in arm at the end of the day.[6] In between, of course, they assume their roles as enemies: the sheepdog guards the flock, while the wolf devises numerous

stratagems to steal the sheep, foiled at every turn by a seemingly dopey yet powerful and alert guard dog. The cartoon, with its reference to the workaday world, cleverly points to the human cultural roles within which dog and wolf are forced to play out their opposed roles and marks as capital the framework for their opposition. There is an invisible boss and a system within which they must perform: someone—presumably human—owns the flock, and both are employed in its maintenance and devastation. The cartoon is knowing and innovative in that it remarks on the "insiderness" to human culture of the wolf—he is *supposed* to try to steal the sheep, although, in this *fort-da* of mastery and triumph over trauma, he will never succeed. Instead, his actions confirm and reconfirm the superior agency of the human creation: the sheepdog. No matter how much the thief tries to bring down the empire, he is foiled. Part of the critique this cartoon offers is to assert that the economic system of private property and primitive accumulation requires an enemy.

And "enemy" is most often what wolf is, especially in the economic arena. The archive of wolves and humans in intimate naturecultures is a record of economic competition, top-of-the-food-chain predators finding themselves side by side, the one in the space of the city-state (the polis), the other on its borders (wolves in literary records are always in the forest, the space of romance; wolves occupy the genre of romance, or they are *unheimlich,* uncanny, "homelike" yet not, and thus also occupy the genre of horror). Both human and wolf are interested in the flesh of ungulates, whether they be the domesticated sheep and cattle whose accumulation furnished primitive wealth or the "wild" deer whose abundance furnished royalty with hunting grounds throughout the Middle Ages and early modernity in Europe. And they do, or did, populate the world. Of wolves, Garry Marvin says that they're "the most widely distributed of all land mammals, apart from humans," while Aleksander Pluskowski notes that they have adaptive success "in being able to survive in virtually any environment."[7] They're both (humans and wolves) found in almost every corner of the earth, although now the one is far more widespread than the other (as a direct consequence of one side's intervention). And if *homo homini lupus* (a man is a wolf to/for other men)—a phrase whose originating context is economic—then wolves too are often wolves for other wolves, because in reserves and parks, it would seem that half if not more of their fatalities are due to predation by other wolves

in territorial disputes.[8] Their sociality—both human and wolf—is the nuclear family, sometimes extended to clans or packs, and when rival children are born, it has been a customary strategy to kill them, wolves wolves, and humans humans. A wolf pack's females will practice infanticide on the subordinates' litters when they are born simultaneously, while, with brilliant economy of phrasing, in John Webster's play *The Duchess of Malfi,* Ferdinand declares, after having had the Duchess's children killed, that "the death of young wolves is never to be pitied."[9] There is cross-species infanticide too: humans kill wolf cubs and wolves kill human children. One of the systematic practices for eradicating wolves in the Middle Ages and early modernity was to find the den and kill all the cubs in spring or summer, while wolf attacks on humans primarily target children.

Do wolves (and wolves and humans) have a history? And what have we learned? In the history of mutual predation by humans and wolves, there are a few glimpses of cultural histories to be had. Like humans, wolves excel in visual observational learning. They have learned to fear firepower and to calibrate the distances they need to keep from guns. Pluskowski writes that "modern wolves have had many generations of experience with firearms and their general timidity may be related to this. But this shyness is conditional and wolves have been known to overcome their fear of people in a number of situations. . . . It is likely that wolves in medieval northern Europe were even more fearless."[10] It also seems to be the case, from documenting human-wolf encounters, that there are "no examples of humans being incorporated into long-term predation strategies" on the part of wolves.[11] The obverse is certainly not true, as the example of the Luparii attests. They were designated wolf-hunting royal officials receiving bounties for killed wolves from the ninth century on in Europe. Wolves prefer wild ungulates to tame ones, given the choice, and they only kill tame ones in surplus (which has led, among humans, to their reputation as greedy, vindictive, and wantonly destructive). A longer cultural history would explore the many microdecisions, genetically selective and conscious as well as unconscious, that led eventually to dog, for, as Donna Haraway and others have argued, dog is a naturecultural history of mutual domestication, wolf-for-human, human-for-wolf, the results being (for the wolf-become-dog) smaller brains, smaller teeth, neoteny, and an ability to read humans visually and vocally—a kind of language acquisition.[12] Wolves, for the most part, recognize dogs and

recognize the human and wolf allegiances of dogs (they can trick and ambush dogs, and they can also have sex with them—female wolves occasionally solicit male dogs). There seems to be no wholesale strategy to become dog on the part of wolves, while it is unclear, at least for most of the history we can discern, whether humans have practiced a systematic strategy of the becoming-dog of wolf.[13] For a long time humans have intended—as far as one can tell—genocide for wolves. Where wolves have survived, it is because they found places to live that were inaccessible to most human hunters.

The historically and politically loaded refrain *homo homini lupus* is from Plautus's *Asinaria,* and it is the phrase the merchant in the text utters: "lupus est homo homini, non homo, quom qualis sit non novit" (One man to another is a wolf, not a man, when he doesn't know what sort he is).[14] But this phrase's more famous future is political, not economic, and it takes out the qualifying phrase: Hobbes's *homo homini lupus,* which is the evil twin of the other Latin adage, commented on by Erasmus and Hobbes, *homo homini deus.* Man is wolf and god, god and wolf are man's possibilities, man is somewhere between wolf and god, if he is man. This is the subject of Jacques Derrida's 2001–2002 seminar *The Beast and the Sovereign, LA* (feminine) *bête et LE* (masculine) *souverain,* which is also a homonym for *The Beast Is the Sovereign*—a kind of queer heterosexual couple or a hermaphroditic being.[15] Indeed, Derrida performs a *gene*lycology of sovereignty, noting the ways wolf and sovereign mirror each other, become each other, and raise questions of reason and force in the polis. And because, for Aristotle, humankind *(anthropos)* is by nature *(physei)* a political animal (to digress for a moment: man is the animal of the polis, that is, man is by nature a political living being, *politikon zoon*, therefore not, as Giorgio Agamben would have it, *bios,* not the exception but part of the *zoe,* life, the living), one might also suggest that humankind is the sheep, the flock that is so favored in the Christian metaphor, surrounded, as it were, by two sovereigns (or one?), two exceptions to the law, two bodies of the King.[16] And in the long and ancient *gene*lycology of political animals, in the naturecultural formations that give rise to wolf-with/against/for-human and human-with/against/for-wolf, we find many places where humans, most often *male* humans, derive their heroic, exceptional, savage, strong, noble, ferocious status from the wolf beside them: (1) as their twin brother—Derrida cites an Ojibwe hero legend of fraternal rivalry between brothers, one human, one wolf;

(2) as their wolf mother (the one who suckles Romulus and Remus, here again the theme of fraternal rivalry); (3) as their wolf father and brothers (Rudyard Kipling's Akela, the adoptive father of Mowgli the man-cub and Mowgli's brothers the wolf pack); and (4) as the wolf who is preserved in their names, nicknames, and totems.[17] Derrida wants us to think through

> this becoming-beast, this becoming-animal of a sovereign who is above all a war chief, and is determined as sovereign or as animal faced with the enemy. He is instituted as sovereign by the possibility of the enemy, by that hostility in which Carl Schmitt claimed to recognize, along with the possibility of the political, the very possibility of the sovereign, of sovereign decision and exception.[18]

Schmitt, in "The Concept of the Political," bases his conceptual realm of state sovereignty and autonomy upon the distinction between *friend* and *enemy*.[19] An enemy establishes the very notion of the political. The enemy is a hostile equal, an other like the self. Marvin notes, in his study of Albanian shepherds, that the wolf is accorded the subjectivity of enemy in precisely Schmittian terms, that is, as a brother/other with agency and intention, an equal, if you will. "The wolf is a stranger, an Other, the wild outsider who continues to be wild and does not succumb to domestication and incorporation," he writes, and, although in Norway, the shepherds are economically compensated for wolf depredations, farmers nevertheless want revenge, as from an enemy with purpose and intent.[20]

This sovereign/tyrant who is a wolf lives in the literature of fables and popular stories as well, and he (for he is often, if not always, a male wolf) is also a noble animal, unlike the degraded servant, the dog, whose collar of servitude, famously in Aesop, Marie de France, and La Fontaine, among others, will not be traded by the wolf for an easier life. The hungry wolf encounters a dog who explains that he, the dog, is well fed because he works for a living. The wolf expresses interest until he sees a raw ring around the dog's neck from his master's collar, at which point the wolf in Aesop's tale declares that freedom, even under conditions of starvation, is preferable to slavery.[21] Marie's knight-wolf (in the *lai* or narrative poem "Bisclavret," the story of a werewolf who is not like the others) demonstrates his nobility (and his doglike nature) through his recognition of the king.[22] (And it is odd that Derrida neglects to comment on this text in what is otherwise an

amazing archive of wolves in literary history.) For Marie, there must be two wolves, a bad one and a good one, to rehabilitate wolfishness: Bisclavret is not a *loup garou/garwaf* (the French and Norman terms, respectively, for "werewolf"): "Le Loup-garou est une bête sauvage; / tant qu'il est possedé de cette rage, / il dévore les gens, leur cause grand mal" (A werewolf is a savage beast; / while his fury is on him / he eats men, does much harm). The difference in the two contradictory valences of the wolf-man hybrid is marked by a linguistic difference: "Such [Bisclavret] is its name in Breton, while the Normans call it *Garwaf*."[23] And yet there must be something of the wolf in the man, the tyrant in the sovereign—in Marie's tale, the wolf is asked to stand in for something particularly "savage" about sovereign power, for Bisclavret uses his savagery righteously to punish the Lady (his adulterous wife) and her usurping husband (he attacks them both and tears off his former Lady's nose). The King, in turn, tortures the wife to elicit a confession. Peggy McCracken, analyzing this tale from the point of view of translation, notes that for Marie, translation frequently occurs at the site of an animal's name, as here, thus signaling "a translatability between human and animal forms," insofar as translation is a figure for the transformation that also occurs thematically between humans and animals in the *Lais*.[24] As the tale bears out, and the twinning of beast and sovereign suggests, the difference asserted between *Bisclavret* and *Garwaf* may be "merely" a turn of phrase. The wolf is the sovereign turned tyrant or he is the tyrant in the sovereign. The beast is the sovereign; the sovereign is the beast.[25]

La Fontaine nicely encapsulates this "fabular" or "fabulous" dimension of the political when he comments on problems of power and subordination in the fable of the wolf and the lamb, whose first line serves as a refrain for Derrida's seminar: "La raison du plus fort est toujours la meilleure" ("The reason of the strongest is always the best," or "might makes right").[26] The wolf, in a fabulous demonstration of *ressentiment*, feels wronged from the outset and seeks to blame the lamb, who very reasonably—that is, using rational faculties (La Fontaine knew, and disagreed with, Descartes's theory of the animal machine)—explains that he could not possibly be the culprit (because he drinks downstream from the wolf, because he was not born when the wolf was insulted/slandered last year). The lamb addresses him as "Your Majesty," and it is clear that he is dealing with a powerful and arbitrary ruler. That ruler (both plaintiff and judge) always already feels wronged in advance and

seeks vengeance. The story ends as one might expect, the wolf carrying off and eating the lamb, "sans autre forme de procès," without any (other) form of trial. Ultimately, this framing of wolf-lamb relations that pits an arbitrary, ruthless, and unjust power against innocence and reason will binaristically inform all future representations of power and subordination in Western narratives of tyrants and their people, and of wolves and humans.

In more secular, though equally political, Western European contexts, the wolf is the father, from the examples I have already mentioned, to Freud's Wolf-Man, where Freud diagnoses the wolf as father (but Gilles Deleuze and Félix Guattari object that Freud doesn't account for the wolves' multiplicity, for the wolves as the non-oedipalized, nondomesticable pack or swarm—more on that in a moment).[27] And in psychological/psychoanalytic readings of Little Red Riding Hood, from Bruno Bettleheim to the 2004 film *The Woodsman,* the wolf is the sexualized father-lover, though he can either be sexualized by the little girl herself or be a predator.[28] Bettleheim in fact makes the point that the wolf is both the seductive heterosexual projection onto an object of the girl's budding sexuality *and* the danger represented to her innocence and youth by male strangers. It is also interesting that there is specificity to the gender of the child—she's female—which adds a (perhaps unconscious) dimension of genetic and reproductive competition to the fantasy of wolf-human antagonism. Current adaptations of the tale also sometimes introduce queer species competition through the thematics of incest and pedophilia.

In *The Woodsman,* the wolf, whom we are led to believe throughout the film is Walter (a convicted sex offender, a "pedophile," released after twelve years in prison), turns out to be the girl Robin's father; in other words, the father is already there in the place of the wolf, the wolf is already inside. The film thus brilliantly deconstructs the work of displacement and projection in a culture that assigns (rapacious, devouring, predatory) "wolf" status to the outlaw/outcast/pervert in order not to confront the wolf within. (Remember the "wolf pack" of Central Park, the young Black men wrongfully accused of attacking a jogger? This too was a remake of Little Red Riding Hood, racialized by turning the "lone wolf" into a teeming multiplicity, an undifferentiated mass of wolves.)

There are other ideological valences—on the one hand masculinist

and traditional, on the other new-feminist and agential—to the strange intimacy between human women and individual (father) wolves: women love the wolf, as they love the sovereign/tyrant—this is Bettleheim's oedipal/Freudian reading. As Sylvia Plath's furious poem "Daddy" declares, "Every woman adores a Fascist, / The boot in the face, the brute / Brute heart of a brute like you." (Brute, of course, is an explicitly animal/animalized reference.)[29] Her poem is also replete with references to the color black, reminding us of the racialization of the primitive and dangerous sexualized father/sovereign, a very weak—and positively valorized—remnant of which is still operative in Stephenie Myers's Twilight series, where the werewolf is a Native American man.[30] The wolf, like the racialized other of a white European cultural imagination, connotes sexual potency, vigor, a carnality that supplements—with sexualized embodiedness—civilization. It remains to be seen whether this refashioning of feminine agency and sexualized transpecies (but nevertheless heterosexual) proximity to wolves—wolves of color, it seems, specifically—has any kind of subversive role to play in reworking cultural fantasies about the wolf in the man.

Wolf-women have a different life in Western cultural fantasies. Wolves are not gendered qua "empirical, actual" animals for the Western imagination; but when wolves *are* gendered, maternal wolves confer salutary savagery. I understand this fantasy as consonant with early modern (Italian) practices of sending children of the nobility to be wet-nursed by rural peasant women, despite the high rates of mortality thereby entailed. The wet nurse was thought to confer the sturdiness and vigor of peasant rurality and "naturalness" through her breast milk.[31] Maternal wolves are important to narratives of mythic nation building, for the heroes in the archives of *gene*lycology need the ferocity of a wolf for the future founding and ruling of their nations; like Marie's king, they require wolfness at their side or within them to mete out punishment without weakness. And they need to eradicate their twin brother first to do it—like Freud's primal murder of the father that binds men into a pact of guilt ensuring their cooperation, in nation-founding tales, it is the brother who is sacrificed and incorporated or assimilated.

True to the feminist observation that masculinist cultural fantasies consign women to the roles of mother and whore, the other female wolf, the "she-wolf," is rapacious, a sexual and economic predator. The

medieval and early modern nickname for prostitutes was "lupae" (she-wolves), because they sapped their clients' wealth. The other side of the coin is something like Shakira's song "She Wolf," where a sexually rapacious and predatory savagery is experienced by the woman as a form of liberation from the excessive docility required of her by her workaday life and boyfriend.[32]

We can also think of Freud's argument that women incompletely sublimate and are thus more tied to the instinctual drives of animality than men; women are connected, like wolves, to the wild and to earth, and they are amoral. Anne Carson writes,

> The wolf is a conventional symbol of marginality in Greek poetry. The wolf is an outlaw. He lives beyond the boundary of usefully cultivated and inhabited space marked off as the polis, in that blank no man's land called to apeiron ("the unbounded"). Women, in the ancient view, share this territory spiritually and metaphorically in virtue of a "natural" female affinity for all that is lawless, formless and in need of the civilizing hand of man.[33]

This is perhaps what motivates some of the postmodern and feminist twists in tales of Red Riding Hood, from Angela Carter's agential representations to the television series *Once Upon a Time*'s "Red Handed." And, in an interestingly gender-twisted rendition, David Kaplan's 1997 *Little Red Riding Hood* assigns the wolf an indeterminate gender and accords a queer feminine sexual agency to Christina Ricci's far-from-completely-innocent Red Riding Hood.[34]

Do modern versions of the conjoining of human and wolf and the figuration of the human in wolfly terms forge alternatives to the traditional narratives that link wolves to primitive masculinity and tyrannical savagery? Can refigurations of the relationship between wolves and humans—perhaps especially human women—have an effect on, even transform, the species, race, and sex nexus in which this relationship is knotted up? Karen Russell's "St. Lucy's Home for Girls Raised by Wolves" presents a feminist way of valorizing the connection between women and wildness that critiques masculinist fantasies of culture and articulates wolf-women in their own terms.[35] Too often, even in their feminist incarnations, women and wolves coexist in mutual relation to a now positively valorized wildness and nature. Russell's story nevertheless offers a way to think about wolfliness that queers the stories of lone heroic or rapacious individualists. She

describes a devastating rite of passage whereby young girls raised by wolves are taken from their packs to convent schools to learn to become human, an allegory for the boarding schools to which Native American children were forcibly taken to "educate and civilize" them into Western Christian North American values (thus troping, again, the connection between wolves and indigenous Americans). The process involves unlearning collectivity and solidarity, unlearning the Deleuzian pack or swarm, in favor of the oedipalized individual.[36] Indeed, so many of the Western cultural fantasies of being-wolf exaggerate what is, in wolf land, an extreme exception: the lone wolf. Wolves live in packs, in collectivities, and a feminist Deleuzian becoming-wolf that refuses masculinist heroic or demonic individualism might offer a line of flight for both women and wolves. "We have never been individuals," as Scott Gilbert asserts, and we become so at our peril (see Gilbert, this volume).[37]

This essay began by discussing wolves and humans, their similarities, their proximities within the naturecultures where they coexist, their mutual relations, their difficult entanglements, their cultural histories. There is no longer—if there ever was—a way to address wolves "as such," just as there is also no way to address "the human" as such. What cultural analyses can do is to think through the material-semiotic—the meaningful enfleshment—of bodies, histories, and meanings called human and nonhuman animal. To forgo any representation at all in order to avoid the traps of anthropomorphism is, I think, to relinquish responsibility for (in the sense of responding, responding to) the coarticulations of lives, histories, and cultures called human and animal. Wolves, even, perhaps especially the ones with which humans now choose to repopulate the wilderness, are, for human culture, spectral, even as they are also living material beings. The spectral wolf, which includes a long line, a *gene*lycology, a multiplicity, of wolves brutally and deliberately exterminated over centuries and centuries of human history, haunts human myths, human stories, human psyches, and continues to haunt the figure of the human-as-animal in Western literature, political theory, and popular culture. This haunting and the monster that emerges in the figure of the wolf-human hybrid also show us the degree to which human and animal share not only a history of comparison and analogy—some humans are like animals, some humans have been animalized—but also a history of interdependence and indeed of coimplication. It is

for those who recognize the connection—not the analogy—of this relationship to forge alternatives to the story of competing rights and hierarchically differential valuations of "life." Wolves and humans are conjoined in their (our) fates. Old stories tell us something about the fantasies that overdetermine the relationship, and new ones offer alternative modes of relating that narrate, like Russell, shared woundings, while also giving us glimpses of better possibilities for "arts of living" as damaged beings on a damaged planet, arts that recognize mutual dependence and resemblance in what will surely be—if it is not already—a posthuman world.

CARLA FRECCERO is an innovative literary critic who juxtaposes diverse sources of insight, from posthumanism to Renaissance studies, in her writing on animals, popular culture, literature, and critical theory. She holds the position of Distinguished Professor in the departments of Literature, History of Consciousness, and Feminist Studies at the University of California, Santa Cruz. She is the author of *Queer/Early/Modern* and the editor, with Claire Jean Kim, of a recent special issue of *American Quarterly,* "Species, Race, Sex."

Notes

1. Carla Freccero, "Carnivorous Virility, or Becoming-Dog," *Social Text* 29, no. 1 (2011): 177–95.
2. Jacques Derrida, "Passages—From Traumatism to Promise," in *Points . . . Interviews, 1974–1994,* ed. Elisabeth Weber, trans. Peggy Kamuf (Stanford, Calif.: Stanford University Press, 1995), 386.
3. Jacob Metcalf, "Intimacy without Proximity: Encountering Grizzlies as a Companion Species," *Environmental Philosophy* 5, no. 2 (2008): 99–128.
4. Derrida, "Passages," 387. An example of the unrecognized and unanticipated form of future merging might be the "coywolf": a hybrid that does in fact have something to do with "us" but in ways "we" might not have anticipated. See Marissa Fessenden, "Coywolves Are Taking Over Eastern North America," http://www.smithsonianmag.com/smart-news/coywolves-are-taking-over-eastern-north-america-180957141/?no-ist.
5. Lee Edelman, *No Future: Queer Theory and the Death Drive* (Durham, N.C.: Duke University Press, 2004).

6. "Woolen Under Where," *Merrie Melodies,* dir. Phil Monroe and Richard Thompson, aired May 11, 1963 (Burbank, Calif.: Warner Brothers Pictures, 1963).
7. Garry Marvin, "Wolves in Sheep's (and Others') Clothing," in *Beastly Natures: Animals, Humans, and the Study of History,* ed. Dorothee Brantz (Charlottesville: University of Virginia Press, 2010), 66; Aleksander Pluskowski, *Wolves and the Wilderness in the Middle Ages* (Woodbridge: Boydell Press, 2006), 25.
8. "Gray Wolf," http://en.wikipedia.org/wiki/Gray_wolf.
9. John Webster, *The Duchess of Malfi,* ed. Leah S. Marcus (London: Methuen Drama, 2009).
10. Pluskowski, *Wolves,* 108.
11. Ibid.
12. Donna Haraway, *When Species Meet* (Minneapolis: University of Minnesota Press, 2007); also Haraway, *The Companion Species Manifesto: Dogs, People, and Significant Otherness* (Chicago: Prickly Paradigm Press, 2003).
13. This essay was written before the media blitz about the coywolf (see note 4). It seems to be the case that there is a strategy of becoming-coyote at work, for the becoming-coyote of wolves gives the latter access to more adaptability in habitats densely populated by humans.
14. Titus Maccius Plautus, *Asinaria,* act 2, scene 4, in *The Comedies of Plautus,* trans. Henry Thomas Riley (London: George Bell, 1912), http://data.perseus.org/catalog/urn:cts:latinLit:phi0119.phi002.opp-eng1.
15. Jacques Derrida, *The Beast and the Sovereign, Volume I,* ed. Michel Lisse, Marie-Louise Mallet, and Ginette Michaud, trans. Geoffrey Bennington (Chicago: University of Chicago Press, 2011).
16. Giorgio Agamben, *The Open: Man and Animal,* trans. Kevin Attel (Stanford, Calif.: Stanford University Press, 2003).
17. Derrida, *Beast and Sovereign,* 10–11.
18. Ibid., 10.
19. Carl Schmitt, *The Concept of the Political,* trans. George Schwab (Chicago: University of Chicago Press, 1996).
20. Marvin, "Wolves in Sheep's (and Others') Clothing," 70, 72.
21. Aesop, "The Dog and the Wolf," in *Aesop's Fables,* trans. Laura Gibbs (Oxford: Oxford University Press, 2002).
22. *The Lais of Marie de France,* trans. Glyn Burgess and Keith Busby (London: Penguin, 1986), 68, ll. 9–11.
23. Ibid., ll. 3–4.
24. Peggy McCracken, "Translation and Animals in Marie de France's *Lais,*" *Australian Journal of French Studies* 46, no. 3 (2009): 206–18.
25. Giorgio Agamben, *Homo Sacer: Sovereign Power and Bare Life,* trans. Daniel Heller-Roazen (Stanford, Calif.: Stanford University Press, 1998),

also discusses "wolf" and Marie de France; for him, the metamorphosis of the knight into wolf signals the "state of exception" (104–11).

26. Jean de La Fontaine, "Le Loup et l'agneau," *Fables*, ed. Antoine Adam (Paris: Garnier Flammarion, 1966), fable X, 59–60. See also "The Wolf and the Lamb," in *Selected Fables*, ed. Maya Slater, trans. Christopher Wood (Oxford: Oxford University Press, 1995), 18–20.
27. Sigmund Freud, *Three Case Histories*, ed. and trans. Philip Rieff (New York: Touchstone, 1996); Gilles Deleuze and Félix Guattari, "1730: Becoming-Intense, Becoming-Animal, Becoming Imperceptible . . . ," in *A Thousand Plateaus: Capitalism and Schizophrenia*, trans. Brian Massumi (Minneapolis: University of Minnesota Press, 1987), 232–309.
28. Bruno Bettleheim, "Little Red Riding Hood," in *The Uses of Enchantment: The Meaning and Importance of Fairy Tales*, 166–81 (New York: Random House, 1975); *The Woodsman*, dir. Nicole Kassell (United States: Dash Films, 2004).
29. Sylvia Plath, "Daddy," in *Collected Poems* (New York: HarperCollins, 1992), ll. 48–50.
30. Stephenie Meyer, *Twilight* (New York: Little, Brown, 2005–2008).
31. Christiane Klapisch-Zuber, "Blood Parents and Milk Parents," in *Women, Family, and Ritual in Renaissance Florence*, trans. Lydia G. Cochrane (Chicago: University of Chicago Press, 1985).
32. "Lupanar (Pompeii)," http://en.wikipedia.org/wiki/Lupanar_%28 Pompeii%29; Shakira, "She Wolf," on *She Wolf* (Epic, 2009). On the question of werewolf gendering, see Rosalind Sibielski, "Gendering the Monster Within: Biological Essentialism, Sexual Difference, and Changing Symbolic Functions of the Monster in Popular Werewolf Texts," in *Monster Culture in the 21st Century: A Reader*, ed. Marina Levina and Diem-My T. Bui, 115–29 (New York: Bloomsbury Academic, 2013).
33. Anne Carson, "The Gender of Sound," in *Glass, Irony, and God*, 119–42 (New York: New Directions, 1992).
34. Angela Carter, *The Bloody Chamber and Other Stories* (New York: Harper and Row, 1979); Kimberly Lau, "Erotic Infidelities: Angela Carter's Wolf Trilogy," *Marvels and Tales* 22, no. 1 (2008): 77–94; "Red-Handed," *Once Upon a Time*, dir. Ron Underwood, aired March 11, 2012 (Burbank, Calif.: ABC Studios, 2012); and *Little Red Riding Hood and Other Stories*, dir. David Kaplan (United States: Little Red Movie Productions, 1997).
35. Karen Russell, "St. Lucy's Home for Girls Raised by Wolves," in *St. Lucy's Home for Girls Raised by Wolves*, 225–46 (New York: Random House, 2006).
36. Deleuze and Guattari, "1730."
37. Scott Gilbert, Jan Sapp, and Alfred I. Tauber, "A Symbiotic View of Life: We Have Never Been Individuals," *The Quarterly Review of Biology* 87, no. 4 (2012): 325–41.

6

UNRULY APPETITES

SALMON DOMESTICATION "ALL THE WAY DOWN"

Marianne Elisabeth Lien

IMAGINE A SALMON FARM IN WEST NORWAY. If you approach from the steep slopes surrounding the fjord, you may spot it as a cluster of circular shapes on the water surface, neatly organized in two parallel rows. If you were a school of mackerel, or pollock, you might be attracted to the smell, and perhaps you would swim against the current and toward its source until you found yourselves near a netting barrier where there is nearly always plenty to eat. Small quantities of feed pellets drift out of the nets with the currents.

If you were wrasse, you would find plenty to eat too, especially if you fancied sea lice. But unlike the mackerel and the pollock, you would experience the netting from the inside out, destined to spend the rest of your lives in the pen together with tiny sea lice that you might like to nibble on—and giant salmon that might like to nibble on you.

No animal is an island. Neither is a fish. This holds true even when we try to simplify organisms in order to manage them. Humans have a long history of trying to extract organisms from their ecologies to eat them, work with them, or keep them as companions or pets. Sometimes this involves confinement, fencing in, rounding up, or holding in netted enclosures, as in the case of aquaculture. Often it involves feeding them, too, and interfering in their mating practices. Such human practices are likely to have implications for the animals' offspring, not only as they become habituated to a different set of surroundings but

also through hereditary effects. This shift at the level of genes, which is often also visible in their bones, their coating, or their color, is what biologists refer to when they talk about domestication. Anthropologists tend to describe domestication as a set of intentional and practical arrangements that are often involved when nonhuman species are singled out in order to "work for us," such as taming, confinement, and control.[1]

Such human attempts at ordering, confining, and keeping separate draw our attention to domestication as a set of boundaries that materialize in the shape of a fence, a cage, or a pen. It is almost as if a fence itself bears a promise of control. But appearances betray. The fact is that confinement in itself tells us very little about what is going on inside the fence. Animals and plants are as dynamic and flexible as we are, and as long as they are alive, they continue to surprise us.[2] Under the surface of a salmon farm in West Norway, for example, multispecies communities are constantly in the making—shaping as well as undermining the ordering attempts that triggered them in the first place.

This chapter is about multispecies entanglements as they unfold within a salmon aquaculture enclosure and in adjacent tanks. Exploring domestication as ongoing practice, it details a multispecies assemblage of a particular kind: that between Atlantic salmon, sea lice, wrasse, and shrimp, and their entangled appetites, fears, and desires as they are assembled in order to mitigate the problem of sea lice on salmon farms. In this drama of life and death, a digestive battleground unfolds, one in which the role of "eater" and "eaten" is notoriously unsettled, and unsettling too. The lead figure in our story is the wrasse, its appetite, and, in particular, its appetite for sea lice.

Multispecies Domestication

Narratives of domestication are often told species by species, as the enrollment of animals one by one, from the wild to the farm, to the barn or, more recently, to netted enclosures of aquaculture. Guided by a notion of the farmed and the wild as fundamentally opposed domains, such narratives typically fail to notice the subsequent changes that unfold as humans and a plurality of nonhumans go on together, accommodating and appropriating each other's reproductive destinies.[3] Hence the stories draw our attention to "the effect of

culture on nature" rather than on the fuzzy interface of mutual becoming that unfolds in our midst, or as in the case of aquaculture sites on Norwegian fjords, in the shimmering cold water beneath our feet.

Contemporary salmon aquaculture is among the most recent turns in the human history of domestication, and in this way, it offers a unique chance to study domestication-in-the-making. Most of what we know about domestication is based on studies of what happened when goats and cows became part of human households in the early Neolithic era.[4] As archeologists have demonstrated, their transition to a more protected life triggered a host of transformations, such as a decline in body size and robusticity.[5] Some of these can be traced through archeological remains as morphological changes, such as through the study of broken fragments of ancient animals' bones. Such fragments have helped patch together what archeologist Gordon Childe once coined the "Neolithic Revolution." Other changes associated with domestication left more subtle traces in the landscape, such as the growth and abandonment of chestnut trees on Tuscan hillsides (see Mathews, in *Ghosts*). And then there are the many instances of microbiological flourishings that are mostly invisible to the human eye but that had dramatic implications for human digestion as well as human immunity. As immunities were thus unevenly distributed between those peoples who domesticated animals and those who did not, the latter became more vulnerable to the spread of disease during the colonial era, and in this way domestication also impacted the conditions for ecological imperialism.[6] Such microbiological changes have given rise to ecological transformations of unknown proportions and underlie James Scott's apt description of these early sites and phases of domestication as the "late neo-lithic multispecies resettlement camp."[7] Scott alludes to the way in which the "unprecedented crowding of many species spawned, over time, a novel ecological niche that favored the selection of specific insects and pathogens" that would dramatically change human health and mortality patterns.

Salmon farms are sites of multispecies flourishing, too, and like in the "late neo-lithic multispecies resettlement camps," their long-term effects are unpredictable and largely unknown. But unlike those that unfolded in the late Neolithic, ours are organized to extract a particular kind of value named "profit," which tends to thrive on so-called economies of scale; and often involve massive changes over a short period of time. Finally, they tend to be organized as "mono-cultures,"

although—as we shall see—this intention is not always achieved in practice. Technoscientific tools are not always sufficient to contain the cascading effects of domestication practices inside as well as outside the pen, the fence, or the barn.

This chapter invites you to consider a marine "late-industrial multispecies resettling camp" from the inside out. On an ethnographic journey to the salmon farms along the west coast of Norway, we will examine aquaculture through the uncharted liveliness of salmon, sea lice, algae, and wrasse. Make no mistake: much (or most) of what goes on "down under" where fifty thousand salmon school within the confines of a netted enclosure is beyond any human radar or monitoring device. Salmon are elusive, and so are the sea lice, not to mention the unknown hosts of other tiny-tinies that we humans have not yet noticed or named. But we know *something*. Or rather, *they* do: the biologists who care about the fjord, the salmon workers who care about the salmon and their jobs, and the veterinarians who recently began to care about the wrasse and their welfare. Let us turn to the field.

Making Shelter

It is a bright summer day in 2012, and Morten, a local farmhand, squats on the metal platform that connects the salmon cages. Below his feet, the water is shimmering blue. We are at Vidarøy, home to six hundred thousand farmed Atlantic salmon on the Hardanger fjord, owned and operated by the medium-sized, locally owned salmon firm Sjølaks. Vidarøy became a key field site for John Law and me in our ethnographic study of salmon farming conducted through intermittent ethnographic fieldwork between 2009 and 2012.[8] Vidarøy is one of several hundred so-called grow-out sites in Norwegian fjords, where the farmed salmon grow to the preferred size for slaughter (four to five kilograms), and it became a place to return to, a place to learn about salmon aquaculture and a site from which to observe changing practices.

After three years of working in the field at Vidarøy and elsewhere, we suddenly notice something we have not seen before. It is a sunny morning, and we see Morten holding a pair of scissors and a black plastic garbage bag. He slides the scissors along the full length of the bag and repeats until each bag is transformed into strips of plastic. Then he ties each pile into a bundle with a blue nylon rope, which is coiled

up in front of him, four or five bundles for each meter of rope. It is beginning to look like a black plastic version of Tibetan prayer flags, but what is it for?

"I am making shelter for the wrasse," he explains.

It turns out that another wrasse delivery is expected tomorrow. Wrasse, or *leppefisk* in Norwegian, is a generic term for several species of fish that are abundant along the Norwegian coast.[9] They are among the most common and least prestigious prey for kids who fish for fun. Wrasse are also occasionally eaten and have commonly been used as bait for crab and lobster fishing. Now, they have taken on a new role in the marine industrial food chain as cohabitants with salmon at Norwegian salmon farms and an important mitigation against parasitic sea lice *(Lepeophtheirus salmonis).* This summer they are badly needed. Sea lice counts are on the rise in the region, and medical mitigation efforts have recently become less effective as the sea lice populations have developed resistance against the most commonly used remedies.

Where there are salmon, there are sea lice. These parasitic copepods feed on salmon mucus and tissue.[10] As a common parasite in fjords with migrating salmonids, it was not a significant cause for worry until the expansion of salmon farming. But since the 1980s, sea lice have been found in high concentrations in and around the densely populated salmon farms, reducing the overall welfare of salmon and resulting in significant economic losses for the industry. If this is a problem for the farmed salmon, it is also a serious challenge for the migrating wild salmon, whose paths to the Atlantic Ocean pass by salmon farming grow-out sites. High densities of sea lice on the salmon farms means an increased risk that they will attach themselves to migrating smolts. Sea lice pose a particular risk to young smolts, whose small bodies are more vulnerable to the parasite, and sea lice are now considered one of the two most important mechanisms through which salmon farming poses a threat to wild salmon populations.[11] The problem is particularly acute in Hardanger, where there is a high concentration of salmon aquaculture localities. The Norwegian Institute for Marine Research found in 2010 that about half of the wild sea trout caught in the fjord were negatively affected by sea lice, which indicates that the migrating salmon smolts are at risk as well.[12]

The connection between sea lice and wrasse is almost as old as the salmon farming enterprise itself: in 1976, the first report about the potential for wrasse to control sea lice was published in a national

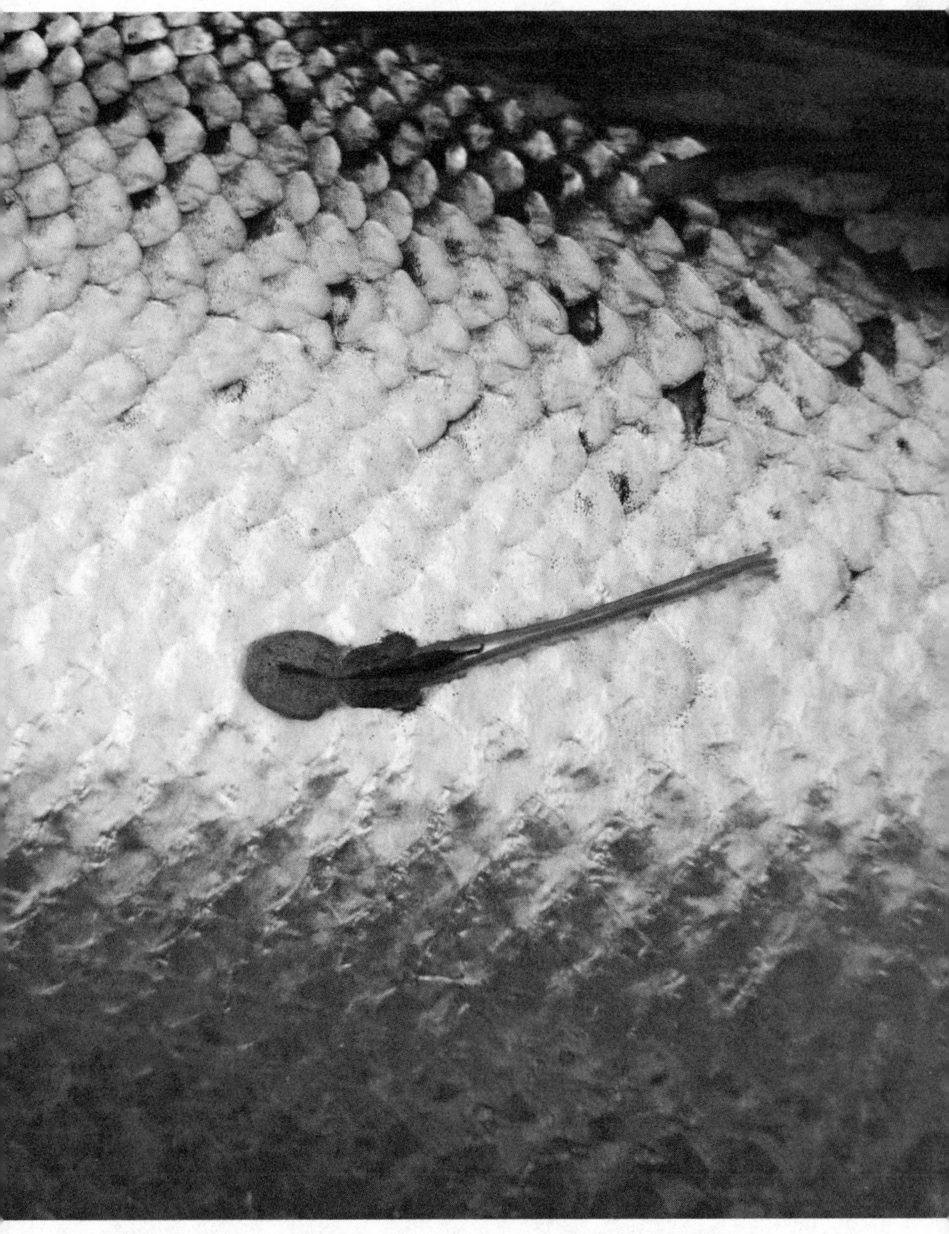

Figure M6.1. Sea louse *(L. salmonis)*. Photograph by John Law.

trade newspaper called *Fiskaren* (The fisherman).[13] According to the article, a salmon farmer in West Norway had solved his problems with sea lice by placing small wrasse inside his pens. The first controlled experiment took place in 1987, and an article about the "cleaning symbiosis" between wrasse and lice-infested salmon first appeared in 1988.[14] Experimental trials were conducted in the early 1990s, but it was not until more than fifteen years later that the wrasse caught broad attention in the industry as an alternative or supplemental remedy against sea lice infestation. This new interest in wrasse coincided with the increased prevalence of sea lice and with reports that sea lice were developing resistance against the common pharmaceuticals.

Bundles of Hope

During our fieldwork, scientific publications on the topic of wrasse appeared with increasing frequency and confirmed what salmon farmers had already noticed for some time: wrasse can be very effective at delousing salmon. Rumors had it that the stomach of a single ballan wrasse was once found to contain several hundred sea lice, but that may have been an exceptional case.[15] A more realistic estimate, as documented in a recent study published in *Aquaculture,* showed that ballan wrasse *(Labrus bergylta)* feeding off of salmon in aquaculture pens consumed a mean of twenty-three lice per wrasse per day, while their smaller cousin, the goldsinny wrasse *(Ctenolabrus rupestris),* consumed about twice that number.[16] No wonder, then, that the wrasse soon to arrive at Vidarøy are expected to keep the farmed salmon healthier and the surrounding waterways less affected by sea lice than would otherwise be the case. In this way, wrasse indirectly protect the migrating (wild) salmon smolts as well, or that—at least—is the intention. Such is the tinkering of contemporary fish farming: a rumor, an experiment, and then the hope that it might work, in what is nearly always a situation of significant uncertainty.

When we began fieldwork in 2009, wrasse were occasionally tried out at Vidarøy, on an experimental basis. During our first summer there, we saw wrasse delivered by the bucket directly to the grow-out sites. It was summer holiday in Norway, and the suppliers were young entrepreneurial teenage boys in their families' motorboats, who had caught them the same morning and sold them to the operating managers of salmon farms for a small profit.

In the beginning, wrasse were simply released into the netted enclosures at Vidarøy without any further ado. But gradually, they were enrolled as part of the salmon assemblage, numerically visible as an entity that could be counted and managed.[17] In 2011, a new standard inventory form was added to the weekly reports of salmon, with columns for the date, number, and specific species of the wrasse delivered to each pen; the number of mortalities each week; and their overall total number. Once unaccounted-for transitory migrants, they have now become registered salmon cohabitants and legal citizens in the "city of fish."

In the meantime, I noticed that their common Norwegian names, *berggylt* (wrasse) and *leppefisk* (literally, lip fish), were often replaced by the term *rensefisk* (cleaner fish), a term which also appears in the scientific literature. Their new name seals their purpose and signals their new position: from an abundant and insignificant pastime prey for kids with a fishing pole, they have become a scarce commodity and an active agent in sea lice mitigation. Removed from fjord currents to netted enclosures, they are about to become part of the salmon *domus* assemblage.

Like farmed salmon, their job revolves around their appetite. But while the salmon's appetite is geared toward their purpose of "putting on weight," the wrasse's appetite will relieve the salmon bodies of sea lice. Their own bodies matter less. In this sense, they could be seen as part of the salmon farming inventory, a technical solution, or a medicine comparable to the other medicines distributed to combat sea lice. But to do so would be to ignore their liveliness.

Their liveliness means that they can die, hence their deaths—as we have seen—must be counted. But their liveliness also means that they should "have a life." Their newly acquired status as companions to farmed salmon comes not only with duties ("eat sea lice," "clean salmon bodies") but with rights too. In the Norwegian world of aquaculture, this involves a legally enforced concern for their welfare, including the fitting of their new home.[18] Hence, when Morten lowers a rope with bundles of plastic, he is adapting the salmon *domus* to accommodate the wrasse and their alleged need for shelter.

Wrasse, I later learn, tend to gather near cliffs, rocks, and kelp forests.[19] Out in the fjord, you will find them resting where the seafloor offers some protection against predators. Strips of black plastic are an inexpensive tool that mimics a kelp forest. Can the wrasse tell

the difference between kelp and plastic? I suspect they might, but it doesn't really matter here: removed from their shallow sea floor habitat, destined to spend their lives in a world dense with salmon and sea lice, their choice of shelter is limited. Perhaps this is why they give the plastic the benefit of the doubt. Or, at least, that is how we humans are inclined to interpret their movements as they are represented on flat computer screens that occasionally transfer live images from underwater cameras: we can see them gathering at the bottom of the pens near the bundles of plastic. Seeking shelter is probably what they are up to.

The black plastic "prayer flags" guide my inquiry as bundles of hope. Quickly assembled on this bright summer morning, I think of them as a small gesture of human care for the wrasse that are given the lead role in this underwater drama of multispecies appetites. A question has been posed—indirectly—about "what they need to be

Figure M6.2. Making shelter. Photograph by John Law.

reasonably happy," and the wrasse have replied, collectively and with their body movements, gathering around those strings of blue nylon rope where their new habitat most resembles a kelp forest. With this move, they have appropriated their new home, or a *domus* that has now become a home for salmon and wrasse *together.* We may think of these artificial kelp forests as the first phase of wrasse domestication. But there is more to domestication than making shelter.

Unruly Appetites

Following the successful introduction of wrasse to salmon pens in 2009 and 2010, a number of experimental studies were set up. Questions were asked about their appetite, survival rates, eating behavior, welfare requirements, and interaction with salmon, such as the ideal ratio of wrasse to salmon in a salmon pen. Ballan wrasse, it turned out, are rather opportunistic eaters, and even if they have been found to eat an average of twenty-three lice per day in experimental studies, their appetite for sea lice is easily diverted and seems to depend on what else is available. (Whose appetite isn't?) It is well known, for example, from a study from Ireland, that when algae grow densely on the netting, this so-called biofouling can attract the wrasse's appetites and divert them from their task of consuming sea lice.[20] Hence keeping the nets fairly clean, and thus limiting alternative food sources, is essential for encouraging the so-called cleaning behavior of wrasse.[21]

Hunger is key to ensuring the wrasse's sea lice appetite, but too much hunger creates another problem, because ballan wrasse are indeed unruly eaters: according to Skiftesvik and colleagues, their experiments revealed that "low availability of food," due to the absence of fouling on the nets, "probably caused the ballan wrasse to nibble on the fins and opercula of the salmon."[22] When sea lice counts are low or the wrasse are "too effective," additional feeding for the wrasse is therefore recommended. But the nibbling goes both ways: salmon are a carnivorous species, and salmon smolts can also show aggressive behavior toward wrasse that have been introduced to the nets, especially if their introduction is sudden.[23] Hence salmon and wrasse are caught in a relation of mutuality that crucially depends on sea lice and wrasse feeding desires. But it is an unstable triangular relation, one in which their respective appetites can turn toward their cohabitants and make the arrangement hostile.

Vision matters too: experiments indicate that wrasse are most active as cleaner fish during the day, that ballan wrasse begin their cleaner-fish career as so-called visual eaters, and that they probably use vision to identify the lice.[24] Hence it is important to ensure enough light inside the pens for the wrasse to perform their cleaning behavior effectively.

In many (or most) of these scientific accounts of the behavior of wrasse, the practices of each experimental group of fish speak to the universal behavior of ballan, corkwing, or goldsinny wrasse as singular units. But occasionally, unexpected variation in wrasse behavior is exposed and the possibility of divergent wrasse behaviors is discussed in some detail. In the Irish experiment, for example, and based on observations and gut analysis, the scientists noted a tendency "for a few individuals to specialize in cleaning, while the others foraged on the net."[25] Such specialized cleaner behavior is also reported from other studies and discussed in relation to what the authors describe as "dominant behavior." Whether such specialized cleaning behavior is a lifelong career of individual wrasse or reflects a specific life phase that most wrasse go through is uncertain, and the authors conclude that "further research into the cleaning behavior of wild wrasse would provide important information that could be applied to their successful use on salmon farms."[26]

In the meantime, the wrasse have indeed demonstrated their usefulness as sea lice mitigators on salmon farms. The number of wrasse put to use as cleaner fish on Norwegian salmon farms increased every year during our fieldwork, thanks to opportunistic fishing practices among local recreational fishermen and what seemed to be an inexhaustible demand for wrasse among salmon farmers. Out at Vidarøy, they were quickly incorporated as part of the salmon assemblage. By 2010, the estimated use of wrasse in Norwegian salmon farms surpassed 10 million fish, but it was still insufficient to meet local demand. As a result, millions of wrasse were transported over long distances by trucks, from the southern coasts of Norway and Sweden to salmon aquaculture localities farther north.[27] In the Hardangerfjord area alone, the mean annual stock of farmed salmon had increased from 9.9 million individuals in 2002 to 25.5 million individuals in 2009. During approximately the same period, the proportion of wrasse in relation to salmon, the so-called labrid-to-salmon percentage, had increased from around 1 percent to around 4 percent.[28] The following

year, in 2011, considerably fewer wrasse were captured locally, and while the reasons for this decline are uncertain, overfishing is mentioned as a possibility.

With the successful application of wild-caught wrasse to combat sea lice, a range of potential unintended outcomes begin to emerge: the signs of uncontrolled and too high harvests, the unknown impact of long-distance transport of locally adapted wrasse from one coastal region to another, and an increasing concern that what appeared first as a "natural" remedy in relation to sea lice could have unforeseen ecological consequences for a range of relations that marine biologists currently know very little about.[29] As a precautionary measure to protect the wild stocks, but also as a response to the increased demand for and scarcity of cleaner fish, wrasse domestication trials have gained momentum. Gradually, some of these are beginning to yield results, and the most recent scientific papers address not only the effectiveness of wrasse as cleaner fish and the sustainability and management of the wrasse fishery but also the management of wrasse as a new and "uncharted" domesticated species. Another species is about to be domesticated, and its inclusion as husbandry animal is enacted at many different sites, in tanks and pens as well as scientific reports.

Picky Eaters and Good Bacteria

Caring for wrasse in a domesticated setting requires more than making shelter. As soon as one shifts from sourcing wild wrasse in fjords and transferring them from the outside to the inside of the salmon pens to raising them in aquaculture settings, a host of new questions emerge: What should their tanks look like? What color should they be?[30] How do you catch the broodstock, and what is the ideal ratio of females to males in each tank?[31] What is the ideal stocking density, and how do you collect eggs? Such questions are currently addressed in online reports and scientific articles and contribute to the ongoing domestication of wrasse in Norway. But the most pressing issue, and what causes most concern among the producers, is the question of feed. The challenge is to find affordable and palatable sources of protein, and this is difficult, because wrasse turn out to be picky eaters. According to Espen Grøtan, responsible for the first large-scale production of domesticated ballan wrasse in Norway in 2012, "ballan wrasse have a choosy palate [*kresen gane*]. They refuse to eat the feed options that

are commercially available. Their favorite food is ground shrimp. That is expensive."[32] Grøtan, who works for Marine Harvest Labrus outside Bergen, explains that they have tried mussels, krill, low-quality shrimp meal, squid, and fish meal, but nothing seems to beat the shrimp. Hence they are still looking for an affordable alternative.

The question of wrasse feed may be tricky in relation to the adult cohorts, but this is nothing compared to the challenges in feeding the young: ballan wrasse larvae need live feed from about four to five days after they hatch. First, they are fed enriched rotifers (*Brachionus* sp.) for about thirty days, then they shift to *Artemia* for another thirty days, and only then can they gradually shift to dry feed. Rotifers are a class of microscopic animals, less than 0.5 millimeters long, that form part of the zooplankton and are an important food source for other marine organisms in the ocean. *Artemia* is a genus of arthropods known as aquatic crustaceans, also called brine shrimp, that do not appear to have changed in 100 million years.[33] Up to fifteen millimeters long, they are thirty times as big as the rotifers and attractive to slightly older cohorts of wrasse. But the harvesting of these is problematic. According to a recent manual on wrasse production, such harvests rely on natural blooms of prey organisms, and this allows little control over species composition and the possible diseases and parasites that such tiny organisms might introduce to the tank environment. Hence the use of artificially cultivated organisms is recommended, and the most common one is a related copepod, known as *Acartia tonsa* Dana.[34] Copepods have been cultivated for more than ten years in Norway, triggered by the need to feed young farmed cod, which—like the wrasse—cannot tolerate dry feed during their first life phase. Recent papers recommend cultivated copepods as superior to wild-caught rotifers and *Artemia* as first feeding for ballan wrasse larvae, as the cultivated feed has been shown to "improve growth, survival and handling stress tolerance for the larvae."[35]

Correct feeding is important, but not sufficient to ensure the health and well-being of cultivated young wrasse. A recent article published in the Norwegian business newspaper *Dagens Næringsliv* goes a long way to explain the intricacies of wrasse cultivation. According to the article, which draws on experiments conducted at SINTEF (a research center for applied research, technology, and innovation), the move from the sterile egg incubation trays to the open sea involves a massive exposure to bacteria: as soon as the eggs hatch and the larvae

are exposed to water, they are colonized by bacteria. Hence the bacterial milieu surrounding the larvae when they open their mouths for the first time is crucial for survival and further development. As the bacterial flora in the gut is as important for wrasse larvae as it is for humans, they need to be exposed to a cocktail of "good" bacteria and protected from the harmful ones.[36] In other words, caring for wrasse means caring for arthropods and copepods as well as the bacterial composition of surrounding waters. In this way, the domestication of salmon is also the cultivation of particular formations of multispecies relations at multiple scales.

Relational "All the Way Down"

Aquaculture is all about relations. When salmon farming first began on the Norwegian coast more than forty years ago, it capitalized on an inexpensive local surplus of feed: fish scraps from the fish processing industry conveniently located near the salmon farms. As the industry grew, such feed became both scarce and cumbersome. The invention of the feed pellet—composed of fish meal and fish oil sourced at a global market—facilitated an exponential industrial growth that has continued ever since.[37] Zooming out from the fjords of Hardanger, we can imagine the expansion of salmon farming as the hardening of global trajectories of feed distribution channels, linking anchovies from the South Pacific to the North Atlantic coast. Salmon farming is this, but it is also more.

In this chapter I have zoomed in, scaling down to the level of sea lice, larvae, and bacteria. Through a focus downward, and inside the pens and tanks, I have traced a chain of relations as they unfold through the lively organic excess triggered by salmon farming. The relations are dynamic, their patterned form is shifting, and there is no stable resting point. Instead, we see a shifting pattern of relations that are parasitic as well as mutual, nurtured by hunger as well as fear. Appetite and profit are both part of the equation; loss and suffering are present too. The late-industrial multispecies resettlement camp is ripe with human intentions, procedures, and mechanisms for control. But in practice, what goes on underneath the water's surface is not easily captured by workplace protocols.

I have described a drama in which humans and nonhumans are deeply entangled and where any attempt to draw a line between the

domestic and the wild is bound to be arbitrary. As in the case of the late Neolithic multispecies settlement camp, we have seen how the bringing in of one new species to the "camp" triggers the flourishing of many others and creates a cascade of new relations. We are reminded that singling out and staying in control are constantly undermined by a multispecies world that turns out to be far more lively than humans imagined.

What we have is an ocean, a fjord, a pen, a tank, and a drop of water, each of which could be described both as a "cosmos unto itself" and as deeply entangled, teeming with lively and unexpected relations all the way down. This is our damaged planet, and this is our planet of hope.

In collaboration with science studies scholar John Law, anthropologist **MARIANNE ELISABETH LIEN** explores the possibilities of human-animal ethnography on salmon farms in West Norway. Her book *Becoming Salmon* inspires further queries into the histories and practices of salmon domestication in Norway and beyond. She is professor of social anthropology at the University of Oslo and led the project "Arctic Domestication in the Era of the Anthropocene" (2015–16) at the Norwegian Centre for Advanced Studies (CAS).

Notes

I am grateful to the anonymized "Sjølaks A/S" for kindly allowing John Law and me to locate our ethnography within the firm and for its additional generous practical support. We are grateful to all those who work there (they too are anonymized) for their warm welcome, their help, and their willingness to let us watch them at work. I am grateful to Kristin Asdal, John Law, Heather Swanson, and Gro Ween for extended salmon conversations and to John for being a collaborator in the field and beyond. This chapter draws on some observations previously published in the book *Becoming Salmon: Aquaculture and the Domestication of a Fish* (2015). Research for this chapter was funded through the project "Newcomers to the Farm" by the Norwegian Research Council (2009–2013). The chapter also benefited from the project "Anthropos and the Material" at the Department of Social Anthropology, University of Oslo. It was finalized at the Centre for Advanced Research (CAS) during the project "Arctic Domestication in the Era of the Anthropocene" (2015–2016).

1. A much-cited author has defined domesticated animals as those that are "bred in captivity for purposes of subsistence or profit, in a human community that maintains complete mastery of its breeding, organization of territory and food supply." See Juliet Clutton-Brock, "The Unnatural World: Behavioral Aspects of Humans and Animals in the Process of Domestication," in *Animals and Human Society,* ed. A. Manning and J. A. Serpell (London: Routledge, 1994), 24. See also Marianne Lien, *Becoming Salmon: Aquaculture and the Domestication of a Fish* (Berkeley: University of California Press, 2015), 11.
2. Anna Tsing, "Unruly Edges: Mushrooms as Companion Species," *Environmental Humanities* 1 (2012): 141–54.
3. Gifford-Gonzalez and Hanotte suggest that a preoccupation among archeologists with domestication as an invention has led to less interest in the ongoing changes that take place among husbandry animals. See Diane Gifford-Gonzalez and Olivier Hanotte, "Domesticating Animals in Africa: Implications of Genetic and Archaeological Findings," *Journal of World Prehistory* 24, no. 1 (2011): 2.
4. The Neolithic Revolution is often described as a historical watershed moment when humans conquered animals and plants. Hence domestication produced not only husbandry animals and cultivated plants but also a notion of nature as that which is not cultivated.
5. Helen Leach, "Selection and the Unforeseen Consequences of Domestication," in *Where the Wild Things Are Now: Domestication Reconsidered,* ed. Rebecca Cassidy and Molly Mullin (Oxford: Berg, 2007), 80.
6. Alfred W. Crosby, *Ecological Imperialism: The Biological Expansion of Europe, 900–1900* (Cambridge: Cambridge University Press, 1986).
7. James Scott, "Four Domestications: Fire, Plants, Animals, and . . . Us," address delivered at Harvard University, May 4–6, 2011, http://tannerlectures.utah.edu/_documents/a-to-z/s/Scott_11.pdf.
8. See Lien, *Becoming Salmon*; John Law and Marianne Lien, "Slippery: Field Notes on Empirical Ontology," *Social Studies of Science* 43, no. 3 (2012): 363–78, doi:10.1177/0306312712456947.
9. These include species known in Norway as *berggylt* (*Labrus bergylta,* or ballan wrasse), *bergnebb* (*Ctenolabrus rupestris,* or goldsinny wrasse), and *grønngylt* (*Symphodus melops,* or corkwing wrasse).
10. Anne Berit Skiftesvik, Anne Berit, Reidun M. Bjelland, Caroline M. F. Durif, Inger S. Johansen, and Howard I. Browman, "Delousing of Atlantic Salmon (*Salmo salar*) by Cultured vs. Wild Ballan Wrasse (*Labrus bergylta*)," *Aquaculture* 402–3 (2013): 113–18, doi:10.1016/j.aquaculture.2013.03.032.
11. The other threat is genetic hybridization, or breeding between farmed "escaped" salmon and their wilder cousins. See Lien, *Becoming Salmon,* 150.

12. Geir Lasse Taranger, Terje Svåsand, Abdullah S. Madhun, and Karin K. Boxaspen, et al., *Risikovurdering—Miljøvirkninger av Norsk Fiskeoppdrett* [Risk assessment—environmental impacts of Norwegian fish farming] (Bergen: Havforskningsinstituttet [Institute of Marine Research], 2011). This led to a new, concerted management regime in the region (called Lusalaus), including biweekly sea louse counts, coordinated treatments, and a near-zero tolerance for sea lice on salmon farms. In practice, it meant that salmon were treated with various forms of pharmaceuticals more often, and resistance to specific pharmaceuticals in the sea lice population soon became a significant problem. See also Lien, *Becoming Salmon,* 72–73.
13. *Fiskaren,* September 27, 1976, as cited in Per G. Kvenseth, "Berggylte som lusekontrollør" (Wrasse as lice controller), *Kyst og Havbruk* (2009): 185–86, http://www.imr.no/filarkiv/kyst_og_havbruk_2009/Kap_3.6.4.pdf/nb-no.
14. Sandra Deady, Sarah J. A. Varian, and Julie M. Fives, "The Use of Cleaner-Fish to Control Sea Lice on Two Irish Salmon *(Salmo salar)* Farms with Particular Reference to Wrasse Behaviour in Salmon Cages," *Aquaculture* 131, no. 1–2 (1995): 74.
15. Kvenseth, "Berggylte som lusekontrollør," 184.
16. Anne Berit Skiftesvik, Geir Blom, Ann-Lisbeth Agnalt, Caroline M. F. Durif, Howard I. Browman, Reidun M. Bjelland, Lisbeth S. Harkestad, et al., "Wrasse *(Labridae)* as Cleaner Fish in Salmonid Aquaculture—The Hardangerfjord Case Study," *Marine Biology Research* 10, no 3. (2014): 117.
17. Lien, *Becoming Salmon,* 53.
18. A new Norwegian animal welfare law came into effect in 2010, placing farmed salmon firmly within the category of farm animals, with specific requirements for farmed fish. See Lien, *Becoming Salmon,* 141–42.
19. Stine W. Dahle and Andreas Hagemann, "Experiment 4: Feeding Preference of Ballan Wrasse Larvae at First Feeding," in *Production of Ballan Wrasse—Science and Practice (a Production Manual),* ed. Synnøve Helland, Stine Wiborg, Courtney Hough, and Jørgen Borthen, 49–50 (Bergen, 2014), http://www.rensefisk.no/fileadmin/Opplaeringskontoret/cropped_image/publication-Production-of-ballan-wrasse.pdf, 104.
20. Deady et al., "Use of Cleaner-Fish," 87; see also Skiftesvik et al., "Delousing of Atlantic Salmon," 117.
21. Deady et al., "Use of Cleaner-Fish," 87.
22. Skiftesvik et al., "Delousing of Atlantic Salmon," 117.
23. Deady et al., "Use of Cleaner-Fish," 86.
24. Skiftesvik et al., "Delousing of Atlantic Salmon," 117.
25. Deady et al., "Use of Cleaner-Fish," 86.
26. Ibid.

27. Skiftesvik et al., "Wrasse *(Labridae)* as Cleaner Fish," 290.
28. The term *labrid* is used here to include all wrasse and is derived from the Latin name for the family, Labridae, of which wrasse are a part. See "Wrasse," https://en.wikipedia.org/wiki/Wrasse; Skiftesvik et al., "Wrasse *(Labridae)* as Cleaner Fish," 292.
29. See Skiftesvik et al., "Wrasse *(Labridae)* as Cleaner Fish."
30. Black, according to Synnøve Helland et al., "Best Practices for Farming of Ballan Wrasse," in *Production of Ballan Wrasse,* 127.
31. A higher ratio of females is recommended to avoid aggressive behavior in the males. See Helland et al., "Best Practices for Farming of Ballan Wrasse."
32. Torkil Marsdal Hansen, "Kresen yngel på rekefôr," http://forskning.no/fisk-havforskning-oppdrett/2012/06/kresen-yngel-pa-rekefor.
33. "*Artemia salina,*" https://en.wikipedia.org/wiki/Artemia_salina.
34. Andreas Hagemann, "Production Manual for *Acartia tonsa* Dana," in Helland et al., *Production of Ballan Wrasse,* 36.
35. Dahle and Hagemann, "Experiment 4," 50.
36. *Dagens Næringsliv,* "Cocktail mot lakselus," http://www.dn.no/meninger/debatt/2014/12/19/2156/Forskning-viser-at/cocktail-mot-lakselus.
37. Lien, *Becoming Salmon.*

7

WITHOUT PLANNING

THE EVOLUTION OF COLLECTIVE BEHAVIOR IN ANT COLONIES

Deborah M. Gordon

I STUDY ANTS BECAUSE I AM INTERESTED in collective behavior—in questions about how systems work together without central control to adjust to changing conditions. Collective behavior is ubiquitous in nature. It is found not only among animal groups but also in the movement and transformation of cells as an embryo develops and in the relations of neurons as brains spin. Ants provide many opportunities to learn about collective behavior. All ant species live in colonies consisting of one or more reproductive females, who lay eggs, and many sterile female workers, who do everything else, including caring for those who reproduce. No one is in charge, and no ant tells another what to do. Indeed, no ant understands globally what needs to be done. Instead, ants respond to what they detect right around them and to interactions with other ants nearby. This produces the patterns that we see as collective behavior.

Collective behavior, like any other phenotype, is embedded in ecology.[1] The entities in living systems—a cell, or an ant, or a molecule—act in relation to the world. This is true for behavior, what organisms do, as well as for morphology, how they are shaped. An ant acts as a forager when it searches for and retrieves food. A cell acts as a pancreatic cell when it produces and interacts with particular chemicals

that link it to certain other cells. Outcomes arise from natural forms because of the ways that form operates in relation to what is around it.

To consider how collective behavior evolves, the starting point is to consider how its form is related to the changing environment in which that behavior operates. The forms of collective behavior are the processes that combine local interactions to generate the behavior of the collective. These processes could be called rules, or algorithms, or mechanisms. For example, one class of forms of collective behavior stimulates or inhibits activity through interactions, creating feedback that regulates the activity of the whole. This is well known in developmental processes in which cells, through contact, stimulate each other to divide or to stop dividing. Another form of collective behavior uses local interactions to create a filter that collects something from outside the group and brings it in or prevents something outside the group from coming in, keeping it separate. For example, ants use interactions collectively to find and recruit to collect food, choose one nest rather than another, or allow nestmates inside the nest while keeping others out.[2]

The study of collective behavior so far has focused on identifying the algorithms, or rules, that connect individuals to produce outcomes for the group. There are unresolved philosophical questions about what it means to understand collective behavior. Biologists have struggled for centuries between two alternatives, a struggle that continues because both alternatives are incorrect.[3] One option is that each individual, ant or cell, is working independently off an internal program, currently envisaged as a sort of computer program contained in, created, and carried out by genes, and that all of these independent actions add up to make the organism, or colony, or tissue. The other is that there is another entity at the level of the whole system, such as the embryo, or superorganism, that somehow drives the relations among the individual entities. We need to develop new language and sets of metaphors that avoid both of these alternatives and instead describe collective behavior as a tangle of overlapping connections that is constantly being created, without any locus of control.

When the study of collective behavior expanded in the 1980s and 1990s, its main goal was to show that it is possible for a system to work without central control. Sometimes the notion of "emergence" resorts to a particular kind of conjuring trick: out of a collection of simple individuals, the behavior of the whole emerges in a puff of smoke. But

explanations that trace how local interactions, in the aggregate, produce complex outcomes show that collective behavior can occur without any conjuring. For example, a fish can sense how another nearby fish pushes the water and, using this cue, can change the way it moves. When a school of fish does this, the outcome is that the school turns. Examples of such behavior continue to proliferate, along with journals, conferences, books, and interdisciplinary centers in universities dedicated to the study of collective behavior.[4] These new insights mean that we are now in a position to move beyond merely showing that collective behavior is possible. Zooming out, we can look at collective behavior in different systems to consider systematically the links between form and outcome. I suggest that particular kinds or forms of rules, linking interactions to outcome, are likely to be used in particular ecological situations. The mechanisms by which collective behavior is organized are the result of evolution in a particular ecological context.

Although this thinking is new to the study of collective behavior in the natural world, it is familiar in evolutionary thinking about form in other areas of biology. It is easy to accept that polar bears have thick fur because they evolved in cold places; the form of thick fur keeps the bear warm in a particular environment where dealing with cold is important for bears. From there, it is also easy to imagine a set of adaptations related to temperature, keeping warm when it's cold and keeping cool when it is hot. But this kind of thinking is new to the study of collective behavior. The rules, or algorithms, that produce collective behavior have been treated as if they were independent of the world in which they function. Elegant models describe how a school of fish can turn but do not consider in what contexts the school turns, why it turns one way in one situation and another when conditions change, how often it turns, whether some schools turn differently from others, and so on.

It is interesting to consider briefly why ecology has so far been absent from the study of collective behavior. In part it is because the models of collective behavior are based on algorithms used in computer programs, and that led to an attribution in reverse, that if we can use a computer algorithm to describe something, then that something is a computer. A computer is different from a school of fish; a computer acts according to instructions, whereas the fish receive no instructions. Another reason, perhaps related to the first, for why the

study of collective behavior has developed this far without ecology is that the rules for collective behavior are considered to be packaged within, and carried by, the individuals, so that it seems that the system is functioning independently of its environment. We write down the rules for collective behavior by outlining how individuals interact and how they respond to those interactions. How an individual actually comes to respond in a particular way is a black box whose inscrutability encourages us to think of the individual's behavior as something it just does by itself and to forget that its behavior is the result of the collective behavior of its cells, and so on all the way down, and that each individual is swimming in a matrix of the collective, which is embedded in a changing world.

Forms of Collective Behavior in Ants

Ants provide a fascinating opportunity to begin the project of thinking ecologically about collective behavior. Ant species, in an astonishing range of places and conditions everywhere on earth, show an enormous diversity of things that ant colonies can do. All ant species nest somewhere, collect food, and reproduce. How they do this varies greatly, in stability, in specificity, and in the means of recovering after disruption. Ant colonies make nests by carving elaborate chambers out of dense soil, or by weaving together leaves using thread emitted by larvae, or in epiphytes, or in crevices left by beetles in trees, or under rocks, or in tiny acorns. Different species eat seeds, fungi that they grow themselves, rapidly moving insects that must be caught, nectar, pollen, the excretions of scale insects, dead insects. Each of these is based on collective behavior, regulated to respond to changing conditions.

The regulation of behavior in social insect colonies often depends on changing interaction networks.[5] Ants sense each other mostly through olfaction: either ants smell each other or chemicals that they emit into the air or onto a surface. Ants smell with their antennae, and antennal contacts are a common form of interaction in ants. When one ant touches another with its antennae, it decides whether the other ant is a nestmate.[6] Ants are covered with a layer of a greasy substance made up of hydrocarbons. These cuticular hydrocarbons vary among colonies and among task groups within colonies. For example,

in harvester ants, an ant's cuticular hydrocarbons change when it is exposed to the sun, so that ants that spend a lot of time outside foraging come to smell different from ants that work only inside the nest.[7] We did experiments in which we coated glass beads with hydrocarbons extracted from ants and showed that during each antennal contact, one ant assesses the task-specific hydrocarbon profile of another.[8] In many social insects, individuals respond to the rate at which they meet others.[9] The crucial information in each encounter is simply the fact of the encounter itself, not any further messages exchanged. An ant's recent rate of interaction influences its behavior.[10]

The diversity of ant species is also a diversity of processes, through which colonies nest, collect food, deal with their neighbors, and make more colonies. Of the many thousands of species of ants, only about fifty have been studied by scientists, but it is likely that in all of them, ants are carrying out their tasks using patterns of brief chemical encounters. Ants of different species are linked to where they move around, where they nest, and what they eat, and these are distributed in conditions that differ greatly in dynamics.

Different species have evolved to use networks of interactions in different conditions. As the evolution of the polar bear's fur is probably related to temperature, so the evolution of the forms of collective behavior used by different ant species is probably related to the conditions in which the ants work. A full ecological taxonomy of collective behavior would list the forms of collective behavior, analogous to fur, and the crucial features of the conditions that have shaped their collective behavior, analogous to temperature. Here I would like to discuss one form of collective behavior, how activity is regulated, and one feature of conditions, what it takes to keep activity going. I suggest that the form of collective behavior that regulates activity is linked to the relation of the flow of energy in and out of the system. When the flow of energy out is more rapid than the flow of energy in, regulation is needed to activate the system, keeping it at rest unless activity is worthwhile. When, by contrast, energy is coming in more quickly than it is being spent, regulation may be needed only when there is occasion to stop the system or inhibit activity.

Figure M7.1. Ants engaged in antennal contact. Photograph by Alex Wild.

The Rewards of Restraint

The foraging behavior of harvester ant colonies provides one example. The form of algorithm, or rules used to regulate the foraging activity of the colony, is linked to the ecology of this species. Lack of water is an important environmental challenge for harvester ant colonies in the desert. Ants lose water when they are out foraging in the hot sun and dry air, and colonies get water by metabolizing the fats from the seeds they eat. So a colony must spend water, as ants go out to search for seeds, to get water and food from the seeds that ants bring back.

I have been tracking a population of about three hundred ant colonies in the desert of southeastern Arizona since 1985. The colonies live for twenty to thirty years. By conducting a census of all the colonies in the population year after year, I have learned how a colony develops over its life cycle. A colony of *Pogonomyrmex barbatus* is founded by a single queen, who produces workers each year throughout her twenty- to thirty-year life span, using sperm from an original mating session when she is a few weeks old. Excavating colonies of known age and counting the ants showed how colony size changes as the queen matures. The colony begins with no worker ants, just the founding queen. It reaches a size of about ten thousand workers when the queen is five years old and then is reproductively mature; the colony produces reproductives to send to the population's annual mating flight.[11] Once a colony reaches the size of reproductive maturity at about ten thousand ants and five years, it continues at the same size for the rest of the queen's and the colony's life. The workers live about a year, so the queen must produce all of the ants every year; the ants in a colony one year are the younger sisters of those that were present the previous year. When the queen dies, the colony does not adopt a new queen, and so, once all the workers have died, the colony is dead.

A harvester ant colony regulates its foraging behavior using simple positive feedback based on interaction rate.[12] An outgoing forager's decision to leave the nest depends on its rate of interaction with foragers returning with food.[13] The regulation of foraging through interactions allows the colony to assess food availability collectively. The more food that is available, the shorter the search time is. Feedback from returning foragers adjusts colony foraging effort to food supply; the more food that is available, the more rapidly ants find it and return, and the more rapidly ants leave the nest to forage again. This

keeps foraging activity low unless there is enough food to make it worthwhile.

The regulation of foraging is a response to the trade-offs imposed by the hot, dry conditions in the desert environment. Colonies vary in the regulation of foraging activity to manage these trade-offs. Some colonies are more likely than others to reduce foraging on hot days. This behavior persists year after year in a particular colony, although the workers live only a year.[14] A single queen produces all of the workers in a colony. Consistent colony variation means that year after year, the queen produces successive cohorts of workers that behave similarly.

Colony foraging activity arises from interactions among foragers, so differences among colonies in foraging activity must arise from differences among colonies in how the ants respond to interactions. It seems that in some colonies, fewer interactions are needed to get a forager out, with the result that the colony tends to forage more, even when it is dry. We are currently investigating how ants of different colonies might differ in the neurophysiology of response to interactions.

By tracking colonies over many years, and by using genetic fingerprinting to match up parents and offspring, we were able to learn which colonies are the offspring of which parents.[15] We used microsatellite variation to identify which colonies were founded by the daughter queens of which parent colonies, measuring the female component of reproductive success.[16] Colonies reproduce in an annual mating aggregation. All of the colonies in a population send winged daughter queens and males to the mating aggregation. After mating, males die, and newly mated queens fly off to start new colonies. A queen continues to produce all of the workers, daughter queens, and males for more than twenty-five years, using the sperm stored in that original mating session. We identified parent–offspring pairs; an offspring colony is one produced by a queen that is a daughter of the queen in the parent colony.

Once we were able to match up parent and offspring colonies, it was possible to count which colonies produce more offspring colonies. This provided the first estimate of realized colony reproductive success in any social insect population. It appears that the regulation of foraging is heritable from parent to offspring colonies. Because we can identify which colonies are the offspring of other colonies, we can investigate whether the offspring colonies resemble parent colonies.

Offspring colonies do resemble parents in the regulation of foraging, in particular in the choice of days in which to reduce foraging.[17] A daughter queen apparently produces workers that regulate foraging in the same way as her sisters, her mother's daughters.

When I looked to see whether the colonies that forage more actively are the ones that have more offspring colonies, the results were surprising. It was the other way around. Colonies that forage less when it is hot and dry are more likely to have offspring colonies. A colony stores food for many months, so bringing in less food does not lead to starvation. In the severe and deepening drought of the past ten to fifteen years in the southwestern United States, conserving water is more important than getting food. Natural selection is favoring the colonies in which outgoing foragers are more reserved in their response to interactions with incoming foragers; I called the article I wrote about this "The Rewards of Restraint."[18] In my lab, we are currently working to understand the physiological process that leads colonies to differ in how ants respond to interactions.

This work showed how colonies are evolving to use interactions to solve an ecological problem. As far as I know, this is the first time it has been possible to learn how collective behavior is evolving in a natural system. The collective behavior that harvester ants use to regulate foraging activity is a form of simple feedback. It keeps foraging activity low unless food availability is high. In this form of collective behavior, the system is inactive unless stimulated by interactions. The default is not to forage. Abundant seeds in the sand around the nest lead to short, successful searches by foragers and increase the rate at which foragers return. Because the collective regulation of foraging is based on interactions of outgoing and returning foragers, abundant seeds detected by the ants move the system away from the default state of no activity and stimulate foraging.

This form of collective behavior—stop unless stimulated to go—makes sense in the desert environment of harvester ants, in which water can be spent by foragers much more quickly than it is obtained. In other systems in which collective behavior regulates activity, are we likely to see similar feedback from the environment—stop unless stimulated to go—when the flow of energy into the system is more rapid than the flow out?

Turtle Ants in the Tropics

Because collective behavior is an adaptation to a specific environment, species differ in how they use interactions to regulate behavior. The arboreal ant *Cephalotes goniodontus,* called the turtle ant, which I have recently begun to study in a tropical forest in western Mexico, suggests that a different form of collective behavior is used in a very different ecological situation.[19] A colony of this species occupies several nests in dead tree branches, linked by a system of trails through the trees that lead from one nest to another and to temporary food sources, mostly nectar. The ants move along the trail from stem to stem in the intricate web of trees, bushes, vines, and dead vegetation that creates the canopy of the forest.

In the turtle ant system, energy can be gained more quickly than it is spent. The humidity is high in the tropical forest, and an ant can spend long periods outside the nest without losing water. I do not know how a forager is stimulated to leave the nest, but it seems that a forager can travel many times around the trail network without needing to return to the nest. Because the ants consume liquids, a forager can travel from place to place, drinking more nectar at each site, without returning to the nest to deposit it. Ants transfer nectar through trophallaxis—one ant regurgitates to another—and sometimes foragers do this along the trails without returning to the nest.

The trail system changes slowly as the ants form new trails to food sources that may last for a day or two and abandon food sources that are depleted.[20] There are other, more abrupt changes, as wind and the movements of animals move the vegetation and break off the dead branches in which the ants nest. The trail is a network constantly being built and modified on the network of vegetation. Every time the ants pass from one branch or stem to another, there is a choice of paths. Each junction or choice point is a node in a network.

A colony regulates its foraging behavior by modifying this network, which determines where the ants go, how quickly they move, how resources to feed the larvae are allocated among nests, and how many ants can reach a particular place. The regulation of foraging depends on how the ants construct and modify the trail network. It seems that for the turtle ants, the default is to go, to remain active. This is crucial for the algorithm that regulates the constantly shifting network of trails. Saket Navlakha, a computer scientist, and I are currently investigating

this algorithm. It seems that ants constantly lay pheromone as they move along, and an ant continues going on the trail when it detects sufficient trail pheromone, put down on the trail by ants that have passed that point recently enough that the pheromone has not yet evaporated. The trail continues moving around, from one nest to another and from nest to food source and back. Experiments with marked ants show that an ant tends to continue going around the same part of the trail circuit unless the ant is recruited to a new food source. Of course, the ants will never find anything new unless they sometimes leave the existing trail. We are currently testing a model that seems to explain how the ants create, maintain, and prune their trails.

For the turtle ants, living in the tropics in conditions in which energy can be gained more quickly than it is spent, the regulation of foraging is based on a form of collective behavior that uses negative feedback to reduce activity. Negative events include the presence of predators or competing species. In response to encounters with some competing species, the turtle ants return to the nest or are less likely to use trails where the other ants are present. In this way, foraging activity retracts, allocating ants elsewhere, in response to competition. The algorithm that maintains the foraging trail makes it easy to prune away a part of the trail where a threat is present and just keep going. There is much more to learn about how the severity of the threat is calibrated with the extent of the foraging response.

Toward an Ecology of Collective Behavior

Traits and environments do not match up in any simple way, even fur and cold, because organisms and environments change each other and because what we see now does not necessarily reflect what happened in evolutionary history.[21] I am not suggesting that we can simply make a list of types of collective behavior, and another list of types of environments, and sort one list by the other. Instead, I am suggesting that the collective behavior of a particular group, such as ant colonies of a given species, evolves as a set of relations that links a colony with the rest of its world. It is not necessarily true that there must be trends that associate certain types of relations with certain forms of collective behavior. But I think it is at least possible, even likely, that there are. Although we have only just begun to understand collective behavior in a few systems, we already see analogies in the forms of collective

behavior used by neurons, other types of cells, and ant colonies. This suggests to me that the number of forms of collective behavior used in different systems is not infinite, and so there is some hope that if we look at ecologies associated with the forms of collective behavior, we will see trends. This is a big project, and I hope that others join in.

DEBORAH M. GORDON researches collective behavior, how it works, and how it evolves. A professor of biology at Stanford University, she studies how ant colonies regulate their behavior without central control and explores analogies with other systems that work collectively, such as the Internet, the immune system, and the brain. In addition to numerous scholarly articles, she has written two highly accessible books on ants, *Ants at Work: How an Insect Society Is Organized* and *Ant Encounters: Interaction Networks and Colony Behavior.*

Notes

1. Deborah M. Gordon, "The Ecology of Collective Behavior," *PLOS Biology* 12, no. 3 (2014): e1001805, doi:10.1371/journal.pbio.1001805.
2. Deborah M. Gordon, *Ant Encounters: Interaction Networks and Colony Behavior* (Princeton, N.J.: Princeton University Press, 2010).
3. See Donna Haraway, *Crystals, Fabrics, and Fields: Metaphors of Organicism in 20th Century Developmental Biology* (New Haven, Conn.: Yale University Press, 1976); Bruno Latour, *Reassembling the Social: An Introduction to Actor-Network-Theory* (New York: Oxford University Press, 2005).
4. David J. T. Sumpter, *Collective Animal Behavior* (Princeton, N.J.: Princeton University Press, 2010).
5. Gordon, *Ant Encounters*; Deborah M. Gordon, "From Division of Labor to Collective Behavior," *Behavioral Ecology and Sociobiology* 70 (2015): 1101–8, doi:10.1007/s00265-015-2045-3.
6. Fernando Esponda and Deborah M. Gordon, "Distributed Nestmate Recognition in Ants," *Proceedings of the Royal Society, Series B* 282, no. 1806 (2015): 20142838, doi:10.1098/rspb.2014.2838.
7. Diane Wagner, Madeleine Tissot, and Deborah M. Gordon, "Task-Related Environment Alters the Cuticular Hydrocarbon Composition of Harvester Ants," *Journal of Chemical Ecology* 27, no. 9 (2001): 1805–19.
8. Michael J. Greene and Deborah M. Gordon, "Cuticular Hydrocarbons Inform Task Decisions," *Nature* 423, no. 6935 (2003): 32, doi:10.1038/423032a.

9. Reviewed in Gordon, *Ant Encounters*. See also Deborah M. Gordon, Richard E. Paul, and Karen Thorpe, "What Is the Function of Encounter Patterns in Ant Colonies?," *Animal Behavior* 45, no. 6 (1993): 1083–1100; Deborah M. Gordon, "The Organization of Work in Social Insect Colonies," *Nature* 380, no. 6570 (1996), 121–24, doi:10.1038/380121a0; Sean O'Donnell, "Worker Biting Interactions and Task Performance in a Swarm-Founding Eusocial Wasp (*Polybia occidentalis*, Hymenoptera:Vespidae)," *Behavioral Ecology* 12, no. 3 (2001): 353–59, doi:10.1093/beheco/12.3.353; Martin Burd and Nuvan Aranwela, "Head-On Encounter Rates and Walking Speed of Foragers in Leaf-Cutting Ant Traffic," *Insectes Sociaux* 50, no. 1 (2003): 3–8, doi:10.1007/s000400300001.
10. Gordon, "Organization of Work in Social Insect Colonies"; Deborah M. Gordon and Natasha J. Mehdiabadi, "Encounter Rate and Task Allocation in Harvester Ants," *Behavioral Ecology and Sociobiology* 45, no. 5 (1999): 370–77; Michael J. Greene and Deborah M. Gordon, "How Patrollers Set Foraging Direction in Harvester Ants," *American Naturalist* 170, no. 6 (2007): 943–48, doi:10.1086/522843.
11. Deborah M. Gordon, "How Colony Growth Affects Forager Intrusion in Neighboring Harvester Ant Colonies," *Behavioral Ecology and Sociobiology* 31, no. 6 (1992): 417–27; Deborah M. Gordon, "The Development of an Ant Colony's Foraging Range," *Animal Behavior* 49, no. 3 (1995): 649–59, doi:10.1016/0003-3472(95)80198-7.
12. Balaji Prabhakar, Katherine N. Dektar, and Deborah M. Gordon, "The Regulation of Ant Colony Foraging Activity without Spatial Information," *PLOS Computational Biology* 8, no. 8 (2012): e1002670, doi:10.1371/journal.pcbi.1002670.
13. Noa Pinter-Wollman, Ashwin Bala, Andrew Merrell, Jovel Queirolo, Martin C. Stumpe, Susan Holmes, and Deborah M. Gordon, "Harvester Ants Use Interactions to Regulate Forager Activation and Availability," *Animal Behavior* 86, no. 1 (2013): 197–207, doi:10.1016/j.anbehav.2013.05.012.
14. Deborah M. Gordon and Bert Holldobler, "Worker Longevity in Harvester Ants," *Psyche* 94, no. 3–4 (1987): 341–46.
15. Krista K. Ingram, Anna Pilko, Jeffrey Heer, and Deborah M. Gordon, "Colony Life History and Lifetime Reproductive Success of Red Harvester Ant Colonies," *Journal of Animal Ecology* 82, no. 3 (2013): 540–50, doi:10.1111/1365-2656.12036.
16. Ibid. In this study, we measured only the female component of reproductive success; each queen mates with many males, and we did not track the males.
17. Deborah M. Gordon, "The Rewards of Restraint in the Collective Regulation of Foraging by Harvester Ant Colonies," *Nature* 498, no. 7452 (2013): 91–93, doi:10.1038/nature12137.

18. Ibid.
19. Deborah M. Gordon, "The Dynamics of Foraging Trails in the Tropical Arboreal Ant *Cephalotes goniodontus*," *PLOS ONE* 7, no. 11 (2012): e0050472, doi:10.1371/journal.pone.0050472.
20. Ibid.
21. F. John Odling-Smee, Kevin N. Laland, and Marcus W. Feldman, *Niche Construction: The Neglected Process in Evolution* (Princeton, N.J.: Princeton University Press, 2003); Stephen J. Gould and Richard C. Lewontin, "The Spandrels of San Marco and the Panglossian Paradigm: A Critique of the Adaptationist Programme," *Proceedings of the Royal Society, Series B* 205, no. 1161 (1979): 581–98, doi:10.1098/rspb.1979.0086.

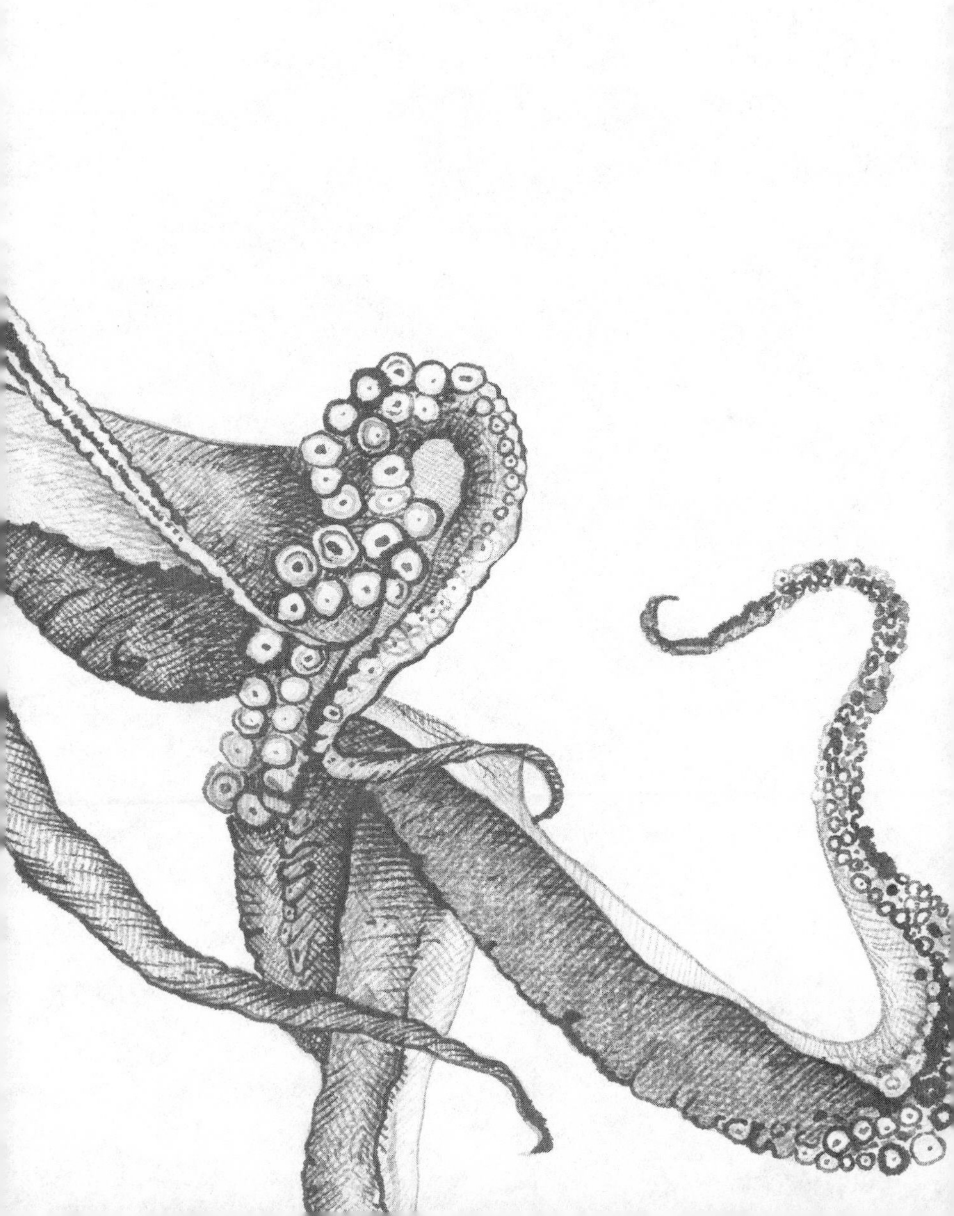

AT THE EDGE OF EXTINCTION

WE OFTEN TALLY THE PLANTS AND ANIMALS at risk of extinction one by one on lists of endangered species. But single species are not the best units through which to see extinction—because they are not the units of life.

Through the synchronized lives of horseshoe crabs and red knot birds, Peter Funch shows us how we both live and die in entanglements with others. Every spring, along the U.S. East Coast, horseshoe crabs crawl onto beaches and spawn en masse, laying billions of eggs in shallow, sandy nests. The lives of red knot birds depend on those of the crabs. During their epic spring migrations from South America to the Canadian Arctic, red knots have relied on the nutrients from horseshoe crab eggs to refuel. In recent years, however, horseshoe crab populations have declined, hit hard by habitat loss and intensive harvesting. Without this food source, the red knots cannot sustain their long flight, and they, too, slip toward extinction.

The plant communities of the Great Meadow, located on the University of California, Santa Cruz campus, have been completely reconfigured by species introduced by Spanish colonists, writes ecologist Ingrid Parker. The invading grasses did not displace just one native plant, or even a handful; they up-ended California coastal ecologies so significantly that these new species now make up "a full 90 percent of the vegetation growing in the Great Meadow." The transformation has been so rapid and widespread that it is difficult to know what was there before; we lose our ability to see and remember. How could an entire world be lost so quickly, in only three hundred years? Despite the limitations, Parker stresses the importance of trying to imagine what has been lost; more holistic forms of restoration depend on our ability to consider the structure and composition of past plant and animal ecologies, as well as the practices of the indigenous peoples who once cultivated these lands. •

8

SYNCHRONIES AT RISK

THE INTERTWINED LIVES OF HORSESHOE CRABS AND RED KNOT BIRDS

Peter Funch

SPECIES ARE BOUND UP WITH EACH OTHER; when one declines, those who depend on it often share a similar fate. Such is the case for those who live in Delaware Bay, a shallow estuary southwest of New Jersey, where salty Atlantic waters mix with freshwater from numerous rivers. This is a busy waterway for oceangoing ships surrounded by salt marshes and mudflats. Delaware Bay is one of thirty-eight sites in United States that has been designated as an internationally important wetland, because the area supports numerous shorebirds, marine turtles, and blue crabs along with all the other species with which they interact.[1] At the same time, however, the edges of the bay are far from pristine. In many areas, development encroaches: scores of low wooden houses line the dunes, an asphalt road protected by rock riprap hardens the shoreline. There used to be plenty of delicious oysters in the bay, but both overharvesting for human consumption and oyster diseases, partly introduced by the oyster culture industry, have reduced the populations markedly.[2]

The Delaware estuary is also the most important spawning place for the American horseshoe crab, *Limulus polyphemus,* a species belonging to a group of unusual marine creatures, the Xiphosura, named after their swordlike tails. Only four xiphosuran species exist today:

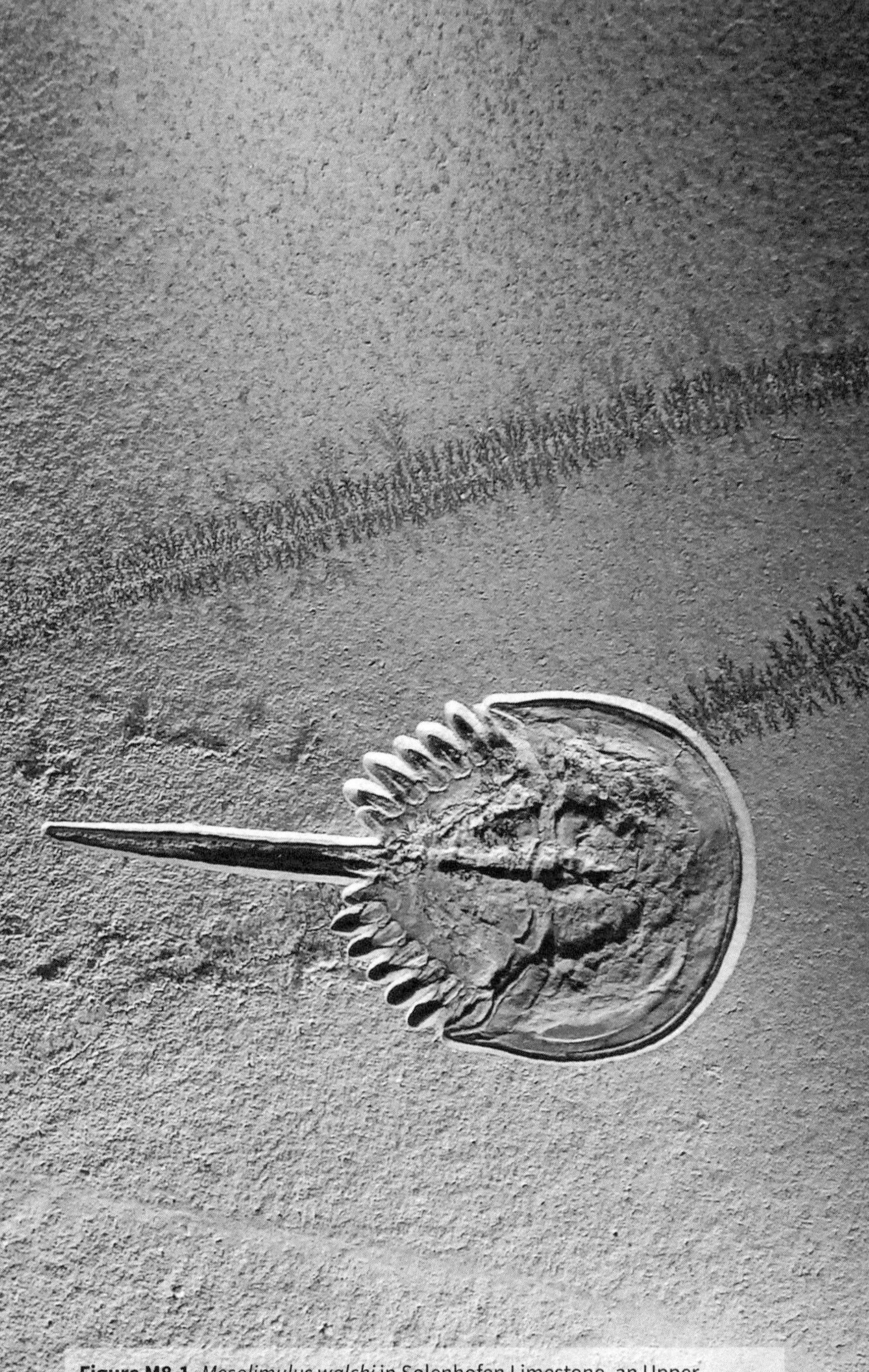

Figure M8.1. *Mesolimulus walchi* in Solenhofen Limestone, an Upper Jurassic (150 million years old) horseshoe crab, morphologically identical to modern horseshoe crabs. Photograph by Ghedoghedo, CC BY-SA 3.0 US.

one in North America and three in Southeast Asia. But in the fossil record, we can track horseshoe crabs back to the ancient Ordovician sea populated by ammonites, sea scorpions, and trilobites, some 450 million years ago.[3] Whereas the ammonites, sea scorpions, and trilobites have since gone extinct, horseshoe crabs have survived into the present-day world, while hardly changing their morphology. Indeed, comparison between Jurassic and recent horseshoe crabs illustrates how they are almost identical (Figure M8.1).

Synchronic Spawning

The horseshoe crabs in Delaware Bay, however, are noteworthy not only for their diachronic continuity; they also enact stunning synchronic patterns in their seasonal mating cycles. Each year, at full and new moons in the spring and summer, American horseshoe crabs emerge from the sea to engage in spectacular mass spawnings. As the high tide approaches its maximum, these strange sea creatures invade the shore. It looks like squadrons of squirming soldier helmets suddenly marching up the beach by the thousands. Before arriving on land, some of the horseshoe crab males have already located a female in the greenish water and grasped her with muscular legs called claspers. Male horseshoe crabs that have not managed to find a female to grasp follow the pairs, and they all aim for the beach, where they arrive at the exact time of the highest tide. They move with the tide to the upper beach, where each female is typically surrounded by numerous males, in addition to the one that is attached (Figure M8.2).

Within a few days of spawning, a single female horseshoe crab will lay approximately eighty thousand green eggs, two to three millimeters in diameter and so sticky that they attach to each other and to the sand grains. They are laid in clusters forming shallow nests, each some eight centimeters deep and typically containing two thousand to four thousand eggs. As the eggs are extruded, both the attached and unattached males release huge quantities of sperm over the eggs, and paternal analysis shows that unattached males close to the female often gain considerable paternity.[4]

Clustered in dugout nests, many eggs stay in place during early development. But the shallowness of the nests also means that a significant portion of the eggs is flushed back into the sea by wave action, especially if there is a storm. On a crowded beach, female horseshoe

Figure M8.2. Mass spawning of *L. polyphemus* in Delaware Bay. Photograph by Peter Funch.

crabs will also often unearth the eggs of other females when they dig their nests. These multiple forms of nest disturbance dislocate a sizable numbers of eggs to the upper centimeters of the sand, typically leaving some one hundred thousand eggs exposed for every square meter of a nesting beach during the peak of the horseshoe crab mating season.[5]

Red Knot Birds

The time for the massive land invasion by horseshoe crabs is predictable and determined by lunar and tidal rhythms. Because it can be predicted, other species have tuned in to benefit from the huge quantities of exposed eggs. More than 1 million birds utilize Delaware Bay in the spring. Among these are certain migratory shorebirds, such as the red

knot *(Calidris canutus rufa)*. Each May, red knots use Delaware Bay as a stopover during an unbelievably demanding migration. Red knots typically travel from South America to arctic Canada, and they do so with little rest. It has been shown that at least one red knot migrated from the border between Uruguay and Brazil to North Carolina in six days nonstop. It was capable of this stunning migration even while carrying a little extra weight in the form of a lightweight geolocator attached to its legs. This device revealed that the bird traveled a distance of almost five thousand miles (eight thousand kilometers) nonstop![6] Birds that migrate such distances rapidly deplete their energy reserves. But the red knots have developed an ingenious solution: the perfect timing of their trek to coincide with the land invasion of the American horseshoe crabs. They arrive in places like Delaware Bay precisely when horseshoe crabs perform their mass spawning. Here the red knots feast on the abundance of disturbed crab eggs to rapidly restore their protein and fat reserves.

Multispecies Vulnerabilities

Today, however, the red knot is at risk of extinction in the western Atlantic region, and the dramatic decline of the bird population has been ascribed to the overharvesting and habitat loss of the American horseshoe crab.[7]

Birds are not the only ones to take advantage of temporal and spatial synchronies to obtain resources. Humans have long done the same. Both Native Americans and colonial settlers made use of the predictable spawning patterns of horseshoe crabs, catching them by hand for use as fish bait, soil fertilizer, and livestock feed. But not all human attunement to the cyclical spawning of horseshoe crabs has been sustainable, not least when it has been part of industrial production. The mass spawning of horseshoe crabs in Delaware Bay has led to overexploitation by humans at least twice. During the nineteenth century, horseshoe crabs began to be used on an industrial scale. The crabs were caught in pounds, dried, and crushed, often to be used as fertilizer in commercial peach orchards.[8] By the late nineteenth century, the fertilizer industry had increased dramatically, harvesting up to 4 million crabs annually. This niche in the fertilizer industry lasted until 1970 and was the likely cause of a rapid decline in the horseshoe crab population in Delaware Bay.

In the past few decades, the horseshoe crab population in Delaware has partly rebounded, but populations in New England and New York have continued to decline. The continuing challenges for horseshoe crabs are many. One of the reasons for ongoing declines is the use of horseshoe crabs as bait in commercial fisheries for eels and whelks (called conchs by fishermen). Horseshoe crabs are easy to catch when they take to the beaches to mate, and the eel fishing industry prefers large females loaded with eggs to attract more eels to eel pots. This preference is problematic for the crabs, as these larger females are precisely the individuals who can potentially contribute the most to bolstering the numbers of future generations of horseshoe crabs. (Males, in contrast to females, can spawn multiple times in a season, so reductions in their numbers do not have as much of an impact on horseshoe crab reproductive success.)

According to the Atlantic States Marine Fisheries Commission, 2.7 million crabs were used as bait in 1998 on the U.S. East Coast.[9] More recently, these numbers have fallen, as states have enacted regulations to restrict bait harvesting. Of the states that border Delaware Bay, New Jersey has completely banned horseshoe crab harvesting for bait, and Delaware has limited bait harvests to males. It seems that the restrictions may be helping crabs in some areas, but it is difficult to know for certain. The overexploitation of horseshoe crabs, however, also has another source: the medical industry. Horseshoe crabs have unique blood. In contrast to human blood, made red by iron-based oxygen transport proteins, the blood of horseshoe crabs, as well as that of many crustaceans, mollusks, and arachnids, relies on copper. Colorless when inside the crab, this blood becomes blue when oxygenated. Because this respiratory pigment likely evolved from a protein that helped to fight pathogens, it retains some immune system functions. The blood also has remarkable coagulation properties when it encounters common ocean bacteria, a function thought to limit the spread of infectious organisms inside horseshoe crabs.

Medical companies have found such properties useful, and they now extract blood from live crabs to produce a medical product called LAL.[10] The bleeding companies harvest five hundred thousand individuals of the American horseshoe crab each year, mostly females. They extract about 30 percent of their total blood volume and return the crabs to the same area where they were caught. Mortality is between 10 and 30 percent, but the bleeding also negatively affects the

surviving crabs, by, for example, weakening their immune systems.[11] The medical companies extract substances from the blood cells that can be used to test for the presence of gram-negative bacteria, even tiny amounts. If gram-negative bacteria contact horseshoe crab blood, the blood rapidly coagulates, indicating the presence of pathogens. Testing for such bacteria is a routine part of ensuring that a variety of medical products are sterile. It is also sometimes used in the diagnosis of certain diseases and to examine drinking water for contamination. Today, artificial alternatives exist, but horseshoe crabs remain the cheapest choice.

Shoreline development is yet another danger threatening horseshoe crabs. The American horseshoe crab is challenged by the fact that it nests close to densely populated areas along the coast. Seaside property owners have built solid walls or laid down rock or other concrete erosion control structures on many nesting beaches to protect buildings, roads, and other forms of infrastructure. The walls act as a physical barrier that prevents horseshoe crabs from reaching the beaches where they breed and nest. When faced with rock barriers, horseshoe crabs will still try to nest, even though they often become trapped in the rocks and die. Along one section of the Delaware Bay shoreline, you can see horseshoe crabs crushed along a roadway, run over by pickup trucks.

When all of these factors compound to cause horseshoe crab declines, they do more than imperil a single species; they also make vulnerable all who depend on it. When American horseshoe crabs take a hit, red knot birds are among those who feel the loss most acutely. As crab populations dip, there are fewer crab eggs for red knot birds to feed upon during their migration.[12] Red knots are especially sensitive to the reduction in egg abundance, because Delaware Bay is the most important and sometimes the *only* stopover area for them during their spring migration. Without horseshoe crabs, red knots die.

Extinction and Conservation in an Entangled World

Biologists and conservationists are actively working to mitigate the damage. One approach that has yielded some benefits for the crabs has been the implementation of local community programs, such as Just Flip 'Em, that tries to reduce crab mortality by encouraging people to

return stranded horseshoe crabs to the sea.[13] Such an approach, however, does not necessarily mesh with other conservation methods. The U.S. Fish and Wildlife Service has recently proposed that the red knot be listed as a threatened species under the Endangered Species Act, but this raises some concerns for current approaches to horseshoe crab conservation. Red knot birds are sensitive to human disturbance, and they feed more efficiently if they are not bothered by beachgoers. But if the protection plan for the red knot restricts people's access to the shores during their migratory stopover to reduce human disturbance for them, it makes it difficult to continue horseshoe crab programs that rely on having people return crabs to the water.

These are the challenges of conservation in an already strained world of co-species coordinations—and they call out for more creative and more holistic approaches to living with other species. Single-species approaches to conservation are not sufficient for an entangled world. This is not a problem limited to Delaware Bay or even the U.S. East Coast. All four species of horseshoe crabs existing today are experiencing declines around the world due to pollution, overharvesting, and habitat destruction.[14] In no case do these declines affect horseshoe crabs alone. It is time that we give more attention not only to the crabs but also to those at risk of codecline and coextinction.

Ordovician fossils tell us that horseshoe crabs have survived several mass extinction events through their remarkable 450 million year history on earth. Human history on earth is a mere two hundred thousand years old. Nevertheless, scientists suspect that the activities of modern man are killing off the crabs, along with many others. The types of industrial development that are endangering the crabs are similar to those leading to disappearing species in many parts of the globe. Such individual extinctions are scary in part because they are never "individual": they cascade. If the American horseshoe crab disappears, the red knot likely will, too.

Because red knots are "global connectors" whose migrations link ecosystems in the arctic to those in South America, they themselves are potentially vulnerable to environmental changes in a number of different locales. But their movements may also make the decline of horseshoe crabs ripple outward in unexpected ways. The impact of global connectors on the stability of local ecological networks is still poorly understood, but changes in the lives of migrators like the red knot may potentially alter the dynamics of geographically distant

ecological webs. For example, research indicates that, as horseshoe crab populations shrink, red knots eat more bivalves, marine organisms with a markedly different suite of parasites. Questions of whether the birds transmit those parasites to other creatures near their breeding grounds in the arctic and, if so, to what effect remain to be answered. At this point, the effects of both red knot population declines and behavioral changes on the multiple ecosystems they connect are unknown.

A growing number of scientists now believe that we are in the midst of a sixth mass extinction event, one comparable to catastrophes like the famous Cretaceous-Tertiary extinction that knocked out the dinosaurs. The entanglements between species compound the dangers and seem likely to expand the death. If we are to do more than momentarily slow the decline, we will need to pay more attention to the relations among species.

Zoologist **PETER FUNCH** studies arthropods and microscopic organisms with a fascination for the diversity of life. One life-form he discovered lives on the mouthparts of lobsters; this phylum, Cycliophora, has a complicated life cycle with alternating generations, different free individuals, and distinct asexual and sexual phases. He is associate professor in the Department of Bioscience at Aarhus University and a member of the Aarhus University Research on the Anthropocene (AURA) group.

Notes

1. See "Delaware Bay Estuary," https://rsis.ramsar.org/ris/559.
2. K. D. Lafferty, C. D. Harvell, J. M. Conrad, C. S. Friedman, M. L. Kent, A. M. Kuris, E. N. Powell, D. Rondeau, and S. M. Saksida, "Infectious Diseases Affect Marine Fisheries and Aquaculture Economics," *Annual Review of Marine Science* 7, no. 7 (2015): 471–96, doi:10.1146/annurev-marine-010814-015646.
3. P. Van Roy, P. J. Orr, J. P. Botting, L. A. Muir, J. Vinther, B. Lefebvre, K. el Hariri, and D. E. G. Briggs, "Ordovician Faunas of Burgess Shale Type," *Nature* 465, no. 7295 (2010): 215–18, doi:10.1038/Nature09038.
4. H. J. Brockmann, "Nesting Behavior: A Shoreline Phenomenon," in *The American Horseshoe Crab,* ed. C. N. Shuster, R. B. Barlow, and

H. J. Brockmann, 33–49 (Cambridge, Mass.: Harvard University Press, 2003).

5. M. L. Botton, R. E. Loveland, and T. R. Jacobsen, "Site Selection by Migratory Shorebirds in Delaware Bay, and Its Relationship to Beach Characteristics and Abundance of Horseshoe Crab *(Limulus polyphemus)* Eggs," *Auk* 111, no. 3 (1994): 605–16.

6. L. J. Niles, J. Burger, R. R. Porter, A. D. Dey, C. D. T. Minton, P. M. Gonzalez, A. J. Baker, J. W. Fox, and C. Gordon, "First Results Using Light Level Geolocators to Track Red Knots in the Western Hemisphere Show Rapid and Long Intercontinental Flights and New Details of Migration Pathways," *Wader Study Group Bulletin* 117, no. 2 (2010): 123–30.

7. Lawrence J. Niles, Jonathan Bart, Humphrey P. Sitters, Amanda D. Dey, Kathleen E. Clark, Phillip W. Atkinson, Allan J. Baker, et al., "Effects of Horseshoe Crab Harvest in Delaware Bay on Red Knots: Are Harvest Restrictions Working?," *BioScience* 59, no. 2 (2009): 153–64, doi:10.1525/bio.2009.59.2.8.

8. C. N. Shuster, "Two Perspectives: Horseshoe Crabs during 420 Million Years, Worldwide, and the Past 150 Years in the Delaware Bay Area," in *Limulus in the Limelight,* ed. J. T. Tanacredi, 17–40 (New York: Kluwer Academic/Plenum, 2001).

9. Atlantic States Marine Fisheries Commission (ASMFC), "Addendum IV to the Interstate Fishery Management Plan for Horseshoe Crab," Fishery Management Report No. 32d, 2006, pp. 1–5, http://www.asmfc.org/uploads/file/addendumIV.pdf.

10. The abbreviation is derived from *Limulus,* "amoebocyte," and *lysis.* LAL is produced from the horseshoe crab blood cells called amoebocytes, which are lysed. Lysing is a way to treat the cells so they burst and release their content.

11. R. L. Anderson, W. H. Watson, and C. C. Chabot, "Sublethal Behavioral and Physiological Effects of the Biomedical Bleeding Process on the American Horseshoe Crab, *Limulus polyphemus,*" *Biological Bulletin* 225, no. 3 (2013): 137–51.

12. C. P. McGowan, J. E. Hines, J. D. Nichols, J. E. Lyons, D. R. Smith, K. S. Kalasz, L. J. Niles, et al., "Demographic Consequences of Migratory Stopover: Linking Red Knot Survival to Horseshoe Crab Spawning Abundance," *Ecosphere* 2, no. 6 (2011): 1–22, doi:10.1890/ES11-00106.1.

13. See "Just Flip 'Em Program," http://www.horseshoecrab.org/act/flipem.html.

14. B. Morton and G. Blackmore, "South China Sea," *Marine Pollution Bulletin* 42, no. 12 (2001): 1236–63; P. Hajeb, A. Christianus, A. Ismail, Sh. Shakiba Zadeh, and C. R. Saad, "Heavy Metal Concentration in Horseshoe Crab (*Carcinoscorpius rotundicauda* and *Tachypleus gigas*)

Eggs from Malaysian Coastline," in *Biology and Conservation of Horseshoe Crabs,* ed. J. T. Tanacredi, M. L. Botton, and D. Smith, 455–63 (Dordrecht, Netherlands: Springer, 2009); M. L. Botton and T. Itow, "The Effects of Water Quality on Horseshoe Crab Embryos and Larvae," in Tanacredi et al., *Biology and Conservation of Horseshoe Crabs,* 439–54.

9

REMEMBERING IN OUR AMNESIA, SEEING IN OUR BLINDNESS

Ingrid M. Parker

THE GREAT MEADOW IS AN ICONIC PLACE on the campus of the University of California, Santa Cruz (UCSC). Like other California grasslands, it turns golden early in the summer. The thick annual grasses look like fields of grain and contrast beautifully with the dark evergreen trees, huddled in crowds like cross-country runners ready to sprint off the starting line. For many, this landscape with its golden rolling hills is among the most beautiful sights in California. The Great Meadow is protected from campus development to preserve the viewshed for those on the campus as well as in the city below.

What is not known to many people is that this iconic landscape is almost entirely devoid of what ecologists would call "native plants." In a study of plant biodiversity in the Great Meadow, my colleagues and I quantified the relative abundance of all plant species.[1] We found that 84 percent were species introduced by Spanish colonists beginning in the eighteenth century, including familiar European plants like wild oat *(Avena fatua),* ripgut brome *(Bromus diandrus),* wild radish *(Raphanus raphanistrum),* Italian ryegrass *(Lolium multiflorum),* and wild mustards (*Brassica* spp.).[2] Because many of these introduced species are aggressive competitors, they also dominate the plant cover: in our study, introduced species made up a full 90 percent of the vegetation growing in the Great Meadow.

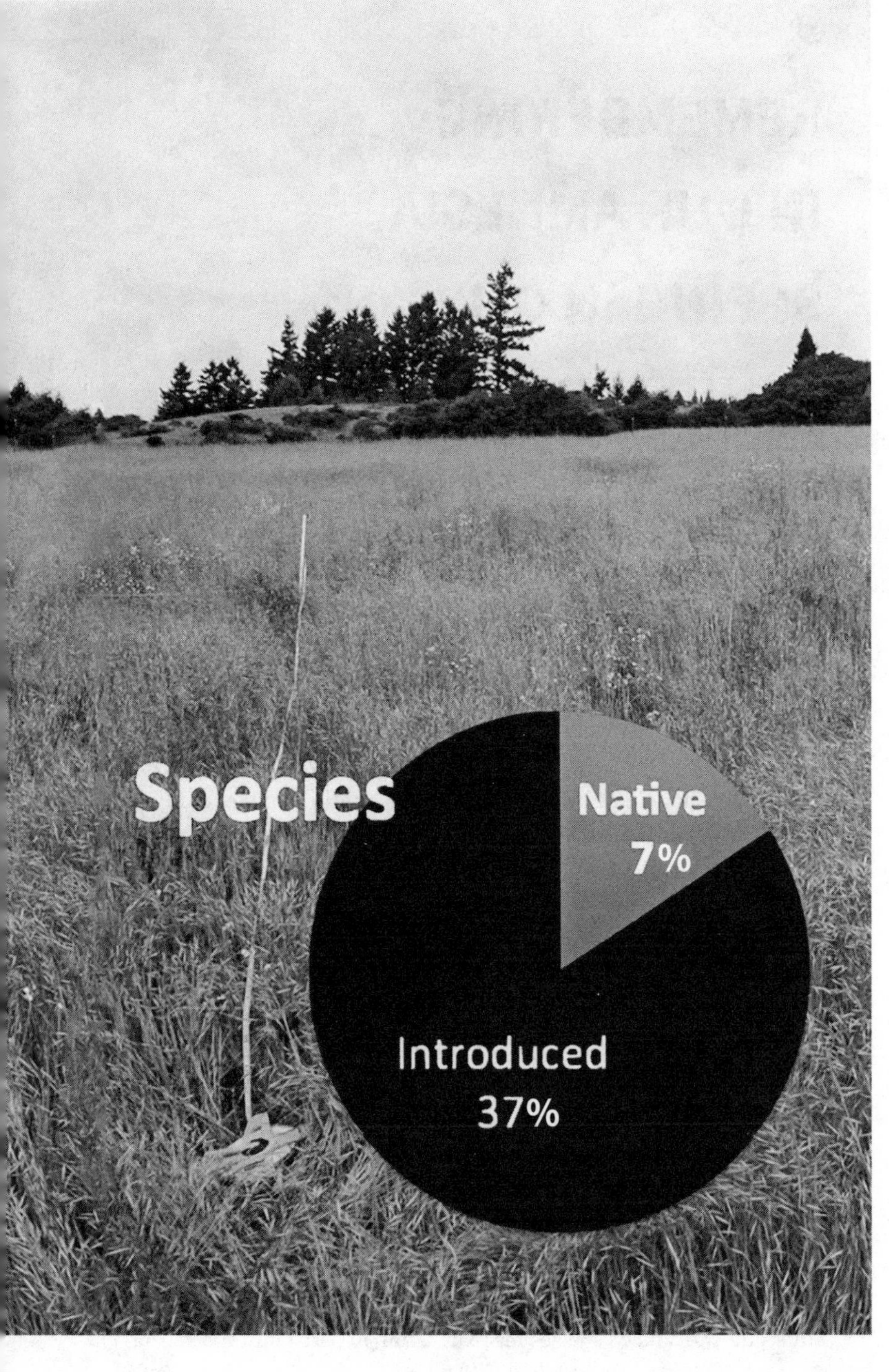
Species
Native
7%
Introduced
37%

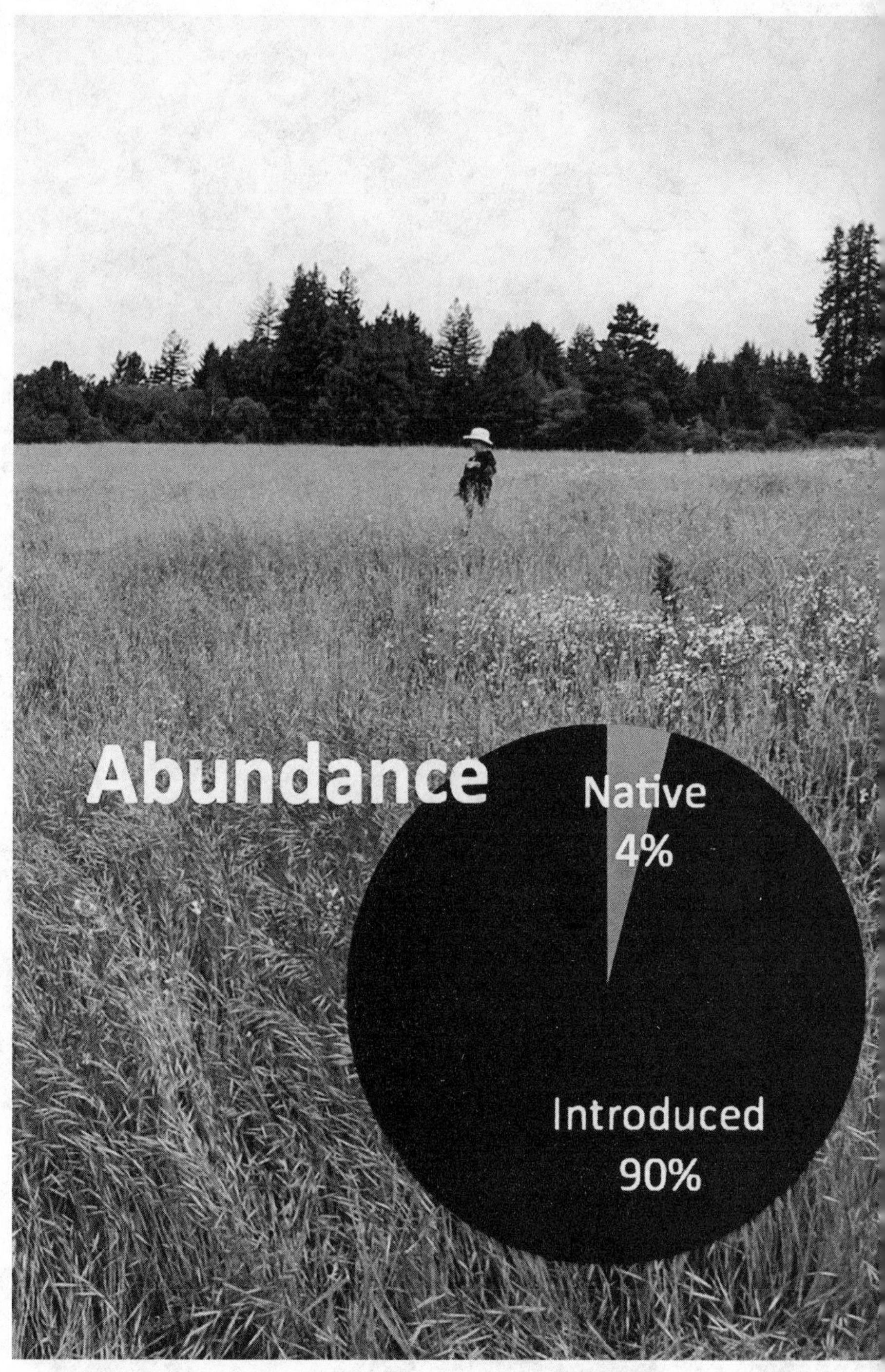

Figure M9.1. The Great Meadow of the campus of the University of California, Santa Cruz. Photograph copyright Gregory Gilbert.

Figure M9.2. California's grasslands were once dominated by multicolored expanses of wildflowers, such as this white fairy lantern *(Calochortus albus)*. Photograph by Miguel Vieira.

Amnesia

What we consider to be one of our most iconic landscapes would have been unrecognizable to a botanist three hundred years ago. But what did the Great Meadow look like before these plants arrived from Europe? We don't really know—an ignorance that is both disturbing and fascinating. A classic hypothesis argues that the coastal grasslands of California were originally dominated by native grasses, such as purple needle grass *(Stipa pulchra)* and creeping wild rye *(Leymus triticoides)*.[3] These species are perennials, some of which can live for hundreds of years. In contrast, the European grasses are annuals, and the golden color of the contemporary California hills in summer reflects that all of the plants are dead. Ecological studies in the past two decades have shown that ecosystem characteristics and functions can differ dramatically between grasslands dominated by European annual grasses and those dominated by native perennials. For example, these systems differ in the availability and timing of soil moisture and vegetation structure, influencing which species can coexist there.[4]

A captivating idea that has more recently grown to be the dominant theory is that the golden fields of European grasses were once filled not with native grasses but with wildflowers. In his book *California's Fading Wildflowers,* Richard Minnick draws on a range of historical and ecological evidence to make the argument that at the time of European contact, spring in California was a riot of color.[5] Spectacular fields of shimmering blue lupines and baby blue-eyes, pink owl's clover, prickly yellow fiddlenecks, scarlet paintbrushes, and uncountable poppies like orange dots in a pointillist painting would have stretched from San Diego to San Francisco and beyond. In addition to historical evidence, research on phytoliths (plant microfossils) now supports the theory that fields of wildflowers were the original California "grasslands."[6] Competition from European annual grasses in combination with the introduction of grazing animals like sheep and cattle wiped out the more vulnerable forms.

Those wildflowers are like ghosts to me. The vision of their exuberance haunts me as I gaze upon the rolling hills around my home.

I believe that to understand the relationship of humans to our landscape today, we have to come to terms with two challenges, which I will call amnesia and blindness. In the case of the wildflowers of California, we don't remember them. Even though the ecological

conversion happened in only a few hundred years—less than a blink of an eye in evolutionary terms—we don't even have formal records of their disappearance.

The first human inhabitants of the Great Meadow were the Amah Mutsun. Their role in shaping the landscape was profound, probably most importantly, but not exclusively, through their use of fire to maintain open meadows and forest edges.[7] Their mark on the history of UCSC is often forgotten, and the practices themselves are part of cultural knowledge that has been dormant and at risk of being lost.[8]

Our society's ecological amnesia is profound, and it limits us from understanding our current and past impacts on the species and ecosystems around us. Another example of this is the historical ecology work of Loren McClenachan, who used archived personal photos of recreational fishing expeditions in Key West, Florida, to see ecological change in the coral reef ecosystem.[9] Over a fifty-year period, trophy fish shrank almost an order of magnitude, from an average size of 19.9 kilograms to 2.3 kilograms. There were also dramatic changes in fish species in the photos, with large sharks and goliath groupers vanishing over time. What is considered a productive day of sport fishing in Florida today would have been considered a pathetic haul in 1956. This is the concept of the "shifting baseline," and it is recognized as a serious problem in fisheries management and in conservation generally.[10] We are hampered when we set conservation or restoration goals based on our knowledge of recent times alone, without an understanding of the structure and composition of plant and animal communities even a hundred years ago, nor of the practices of the peoples who interacted with the land before European colonization.

Blindness

A second challenge in our relationship to changing ecological assemblages is our limited ability to perceive. Here I am not even referring to our distractedness, our addiction to electronics and social media, or our children's lack of direct experience of nature.[11] I am referring to how ecological patterns—the diversity, composition, and overall structure of ecosystems—are often driven by factors that we cannot see. For example, microscopic fungi and bacteria are responsible for much of the fertility of soil and for regulating decomposition and mediating interactions among plants belowground, including

Figure M9.3. For the first time in more than 150 years, Pinzón giant tortoise hatchlings are successfully surviving on Pinzón Island, Galápagos, as a result of the eradication of invasive rats in 2012. The photograph shows the very first hatchling observed by humans as it emerged from its nest. Photograph by Francesca Cunninghame at Charles Darwin Foundation.

symbiotic mycorrhizal fungi and nitrogen-fixing rhizobia.[12] Other microbes, such as pathogenic fungi and oomycetes, bring balance to communities and increase biodiversity by causing disease on highly competitive plant species.[13]

The history of the scientific field of ecology is one of discovery through learning new ways of seeing. Microbes have become visible to us because we can now "see" their distinctive DNA and RNA in soil and inside of plants and animals.[14] This has allowed us to study how the diversity and composition of microbes influence everything from plant productivity to human digestive health. In another revolution in awareness, one of the most important developments in ecological science was the rise of the manipulative experiment as a tool for studying what could not be seen. In 1966, Robert T. Paine published the results of a two-year experiment on food web complexity and biodiversity.[15] In his experiment on Tatoosh Island, Paine removed ochre sea stars ("starfish," *Pisaster ochraceus*) from a stretch of Pacific Northwest rocky intertidal by flinging them out to sea. That study, which now has more than twenty-seven hundred citations, demonstrated a rapid loss of biodiversity as mussels took over in the absence of their predator.[16] The effects of the sea stars were entirely invisible until these unobtrusive yet voracious predators were experimentally removed.

The same principle was applied to discover the major driver of kelp forest ecology along the west coast of North America. In a large-scale and brutal hunting "experiment" from 1741 to 1911, Russian fur traders drove sea otters nearly to extinction.[17] By comparing sites with sea otters to areas where otters were missing, marine biologist Jim Estes found that sea otters have a dramatic effect on their ecosystem. The sea otters make it possible for lush and productive kelp forests to grow by eating urchins, keeping the marine herbivores in check.[18] Before Estes began his grueling work in the cold waters of Alaska, we were blind to the importance of the sea otter. Now where "urchin barrens" flaunt their bare ground without a scrap of living kelp, the marine ecologist sees the ghost of the otter.

Sometimes it is possible to use our imagination together with our knowledge to take action that leads to the conservation and restoration of species and their ecosystems, to, in effect, "bring back a ghost." The conservation nonprofit Island Conservation takes as its mission to prevent extinctions by removing invasive species from islands. Although islands represent only 5 percent of the world's landmass,

they are home to 45 percent of the world's critically endangered species globally. Most of the animal extinctions on islands are caused by invasive species such as cats and rats. For example, the Pinzón giant tortoise (*Chelonoidis nigra* subsp. *duncanensis*) was unable to breed in the wild for 150 years, because invasive rats ate their eggs and hatchlings. While the species struggled on from 1965 in a captive breeding colony, the Pinzón giant tortoise went extinct in the wild. In 2012, Island Conservation partnered with Galápagos National Park and the Charles Darwin Foundation to remove rats from Pinzón. In June 2013, baby tortoises were seen on Pinzón for the first time in more than 150 years.

The Pinzón giant tortoise, the sea otter, the wildflowers of California, the goliath grouper—a conservationist must learn how to transcend amnesia to remember these ghosts, to transcend blindness to a new kind of sight. To accomplish this, we need to bring all types of research to the table, from anthropology and archaeology to historical archival research, as well as experimental ecology and molecular ecology. Underlying it all is natural history, that slow art of getting to know a place in space and all of its creatures. Perhaps with a combination of imagination, scientific inquiry, and conservation inspiration, we will see more ghosts come back to life.

Combining natural history, experiments, and mathematical models, **INGRID M. PARKER** studies how plants interact with microbes and other organisms in ways that influence their distributions, traits, and ecological dynamics. She is known for her work on the biology of invasive species and on the conservation of rare plant species in California. She is professor of ecology and evolutionary biology at the University of California, Santa Cruz.

Notes

1. Ingrid M. Parker, Megan Saunders, Megan Bontrager, Andrew P. Weitz, Rebecca Hendricks, Roger Magarey, Karl Suiter, and Gregory S. Gilbert, "Phylogenetic Structure and Host Abundance Drive Disease Pressure in Communities," *Nature* 520 (2015): 542–44.
2. Richard Minnich, *California's Fading Wildflowers: Lost Legacy and Biological Invasions* (Berkeley: University of California Press, 2008).

3. Frederic E. Clements, "The Relict Method in Dynamic Ecology," *Journal of Ecology* 22 (1934): 39–68.
4. T. H. Holmes and K. J. Rice, "Patterns of Growth and Soil-Water Utilization in Some Exotic Annuals and Native Perennial Bunchgrasses of California," *Annals of Botany* 78, no. 2 (1996): 233–43; Nicole A. Molinari and Carla M. D'Antonio, "Structural, Compositional and Trait Differences between Native- and Non-Native-Dominated Grassland Patches," *Functional Ecology* 28, no. 3 (2014): 745–54.
5. Minnick, *California's Fading Wildflowers.*
6. Rand R. Evett and James W. Bartolome, "Phytolith Evidence for the Extent and Nature of Prehistoric California Grasslands," *Holocene* 23, no. 11 (2013): 1644–49.
7. Jon E. Keeley, "Native American Impacts on Fire Regimes of the California Coastal Ranges," *Journal of Biogeography* 29 (2002): 303–20; M. Kat Anderson, *Tending the Wild: Native American Knowledge and the Management of California's Natural Resources* (Berkeley: University of California Press, 2005).
8. Rick Flores, "The Amah Mutsun Relearning Program," http://arboretum.ucsc.edu/education/relearning-program/.
9. Loren McClenachan, "Documenting Loss of Large Trophy Fish from the Florida Keys with Historical Photographs," *Conservation Biology* 23, no. 3 (2009): 636–43.
10. Daniel Pauly, "Anecdotes and the Shifting Baseline Syndrome of Fisheries," *Trends in Ecology and Evolution* 10, no. 10 (1995): 430; S. K. Papworth, J. Rist, L. Coad, and E. J. Milner-Gulland, "Evidence for Shifting Baseline Syndrome in Conservation," *Conservation Letters* 2, no. 2 (2008): 93–100.
11. Richard Louv, *Last Child in the Woods: Saving Our Children from Nature-Deficit Disorder* (Chapel Hill, N.C.: Algonquin Books, 2005).
12. Yitzhak Hada and Kalliope K. Papadopoulou, "Suppressive Composts: Microbial Ecology Links between Abiotic Environments and Healthy Plants," *Annual Review of Phytopathology* 50 (2012): 133–53; Uffe N. Nielsen, Diana H. Wall, and Johan Six, "Soil Biodiversity and the Environment," *Annual Review of Environment and Resources* 40 (2015): 63–90.
13. James D. Bever, Scott A. Mangan, and Helen M. Alexander, "Maintenance of Species Diversity by Pathogens," *Annual Review of Ecology, Evolution, and Systematics* 46 (2015): 305–25.
14. Norman R. Pace, "A Molecular View of Microbial Diversity and the Biosphere," *Science* 276 (1997): 734–40; Erko Stackebrandt, *Molecular Identification, Systematics, and Population Structure of Prokaryotes* (Berlin: Springer, 2006).

15. Robert T. Paine, "Food Web Complexity and Species Diversity," *The American Naturalist* 100, no. 910 (1966): 65–75.

16. Ibid.

17. Marianne L. Riedman and James A. Estes, *The Sea Otter (Enhydra lutris): Behavior, Ecology, and Natural History,* Biological Report 90, no. 14 (Washington, D.C.: U.S. Fish and Wildlife Service, U.S. Department of Labor, 1990).

18. James A. Estes and John F. Palmisano, "Sea Otters, Their Role in Structuring Nearshore Communities," *Science* 185 (1974): 1058–60.

CODA

BEAUTIFUL MONSTERS

TERRA IN THE CYANOCENE

Dorion Sagan

THE WORLD IS A BEAUTIFUL, FRIGHTENING MONSTER. The internal combustion engine is a monster, and I myself often notice my heart jump at the sound of cars and chainsaws. But nature's monstrosity is not only concrete-pollution-machine gray and murderous red. It is also, instructively and paradoxically, considering the associations of the green movement, verdant in hue—for green is the color of the cyanobacteria that mutated 2 billion years ago, causing the greatest pollution crisis in planetary history. Green in another sense are we, newcomers to the biospheric sequelae of mass reproduction. Genetic analysis shows that cyanobacteria were ancestral to the green chloroplasts of plants and the browner plastids of kelp. They use hydrogen from water as their electron donor. Life had been living in water, but then it evolved to split water's molecular bonds. This made a monstrous fire. The green beings spread across the surface of the planet in a kind of green fire that has still not stopped burning. We—human civilization—are one of the latter-day flames.

The release of oxygen (O_2) from water (H_2O) must have been a horror show for any beings that could feel. The killingly reactive gas accumulated in the oceans and atmosphere on a global scale. Many life-forms, beginning with the green bacteria that first released oxygen gas as waste from photosynthesis (before life evolved metabolic means and behavioral stratagems to tolerate and then exploit the reactive gas), must have perished. Importantly, this far-more-than-human, and ultimately human-enabling planetary toxic crisis was brought on not from the outside by meteorites or tectonic events

but endogenously, by life's own operations. We are right to be worried about contributing to the increase in CO_2 in the atmosphere a fraction of a percent—a few score parts per million. Instead of waging wars for oil, we should study what sorts of communities thrived in the Pliocene, the last time the temperature was three degrees Fahrenheit higher; we should find what sorts of plants build topsoil at higher altitudes and prepare to plant them as glaciers recede. But without any human help, or plants, fungi, or animals of any kind, life survived the Cyanocene—my name for this global pollution crisis, inaugurated by the cyanobacteria, whose spread led to an increase of free oxygen until it composed 20 percent of the atmosphere, where it has long since stabilized, an increase of some two hundred thousand times.

Scars of this monstrous reproductive excess exist as uranium oxides and vast rust beds—iron oxides—on the earth's surface. But of course we—human beings—are not worried about this sort of global change. Indeed, it made our lives possible. Global life reached new heights of biodiversity and ecological complexity.

Life is one tough cookie. And a scary monster. And perhaps more scarily, we are part of it. A big difference between the self-titled Anthropocene and the Cyanocene is that ancient microbes—with much more time at their disposal—have long since evolved means of recycling the wastes precipitated by their monstrous growth. Networks of metabolically diverse beings more moderately growing in situ have formed ecosystems that use incoming energy to recycle limited compounds without destroying those grown from them.

Life is an energy-transducing phenomenon. It is not alone. Chemical "clocks" such as Belousov-Zhabotinski reactions (which change colors and make fractallike spiral patterns), Bénard cells (which maintain hexagonal shapes, metastable flow-forms appearing in certain fluids to reduce temperature gradients), and storm systems (which reduce temperature and atmospheric pressure gradients) cycle matter in regions of energy flow. Such systems tap into available energy gradients and persist until the energy runs out. Life, however, with its genetic infrastructure, is able to rebuild and vary the systems accessing energy, keeping the energy expenditure game going. Its most efficient forms have found ways to use energy without using it up. So far, as a whole, it has done this for almost 4 billion years, about a third of the estimated age of the universe. Terra, Earth, always contains the possibility for some of its energy-feeding forms to grow rogue, to become

teratological, monstrous growers that threaten the wholes from which they've sprung. The more successful tend to be slower-growing, massively energy-transducing, ably recycling late-stage ecosystems that don't strip their material or energetic resources. Nor do they, as polycultures, pollute themselves to death or become devoured by the predators or pathogens that may devastate a monoculture.

Impressed by the movement of munitions in World War I, Soviet geochemist Vladimir Vernadsky understood living matter as an energetically infused moving mineral, an impure form of water, a "geological force." Part of an Earth-solar *process,* humankind and technology ride on the multi-billion-year transduction of solar energy. Captured solar energy even allows life to "defy" gravity: mountain-sized swarms of insects, migrating birds, and airplanes remain airborne using the energy of the sun. Even the Himalayas and other mountains, theoretically dependent on the subduction of calcium carbonate gathered at the ocean bottom from a continuous submarine "rain" deposition of (virally infected) calcifying algae, may rise courtesy of life and the sun. This is because the plates slide on the limestone, lubricating the tectonic crashing that leads to mountain building. Life, including all human life and technics, is essentially a solar energetic phenomenon.

Successful ecosystems produce entropy, spread energy, and reduce gradients but recycle their wastes and avoid producing too much heat near their sensitive surfaces so as not to impair their ongoing operations. The biosphere as a whole has buried carbon and transpired water, cooling itself, reducing the energetic gradient between 5,400 Kelvin sun and 2.7 Kelvin outer space. As a whole, life feeds on sunlight and emits heat into space. Satellite readings surprisingly show Borneo and Amazon ecosystems to be as cold as Siberia in the winter. The heat-transporting properties via evaporation from leaves and soil in the summer in mid-latitudes is the equivalent of some fifteen tons of dynamite per hectare. Solar-powered growth, and transpiration through pores on the undersides of leaves (called stomata), which take in carbon dioxide and release moisture and volatile compounds, exchanging gases with the atmosphere, are part of life's evolved cooling processes. Emitted from algae and plants, sulfur compounds wafting from the oceans, and in plants hydrocarbons such as terpenes, found in essential oils, can serve as condensation nuclei for raindrops, fomenting rainstorms, each of which releases the energetic equivalent

of many megatons of dynamite. So it is not energy itself, entropic heat or solid or liquid or gaseous waste, that is the problem for life so much as how (and how close to life's sensitive surfaces) this necessary pollution is disposed of. And here the relative monocultures of city and farmland rate poorly compared to the biodiverse forests and rainforests that are way ahead of us in evolving stably recycling energy-spreading ecosystems. When Nietzsche, in his posthumous writing, holds up a mirror for us and describes the world as a monster of force beyond good and evil, he is emphasizing life's monstrous and necessary connection to energy in an energy-steeped cosmos without beginning or end. But another aspect of Terra's "Terans" are the fractal monsters of cellular networks, multispecies symbiogenetic enclaves that make use of such energy *stably*.

The nested pulsing interwoven energy-sensing networks of life are not confined to individuality at the animal level (see Gilbert's and McFall-Ngai's chapters in this volume). The thermodynamically open complex systems of cells and organisms permit not just companionship and mating and live birth but permanent interpenetration, the merging of metabolism. The tendency for exponential reproduction on a limited planet drives organisms not only into each other's territories but into one another. And monsters merge not only membranes, genes, and metabolisms but sensoria—perceptual abilities.

Geographer Jim MacAllister wrote to me about this in 2014:

> I was musing about defining us not as individuals but as communities and [how it relates to] your concluding chapter of *Cosmic Apprentice* [where you address] the possibility that drug effects "drop the veil" on reality giving us a dazzling perception of reality. . . . This thought came up: what if our consciousness—my I-ness—is just the job that the consciousness of my 90% bacterial cells and their obligate 10% animal multicellular cells require "me" to do? I have no consciousness of what all my cells are doing. I am like the super of an apartment building, I hear the complaints (e.g., an ache), I have to feed the furnace, put out the garbage, sweep the hallways, but I have no experience of the lives of the apartment dwellers or what they are really up to. My cells have lives and sentience of their own just like the apartment dwellers. Isn't this disconnection of awareness between scales of "ecosystems" fascinating? Maybe not disconnection but limited connection—sort of like our normal perception of reality that simplifies or filters out the dazzling truth of reality that is too complicated, messy or distracting for us to deal with all the time. It makes my consciousness sort of like our corporate media whose job it is to misinform, distract, dumb

> down, manipulate. That is a very different notion than that we are the Universe observing itself.
>
> I'm interested in your thoughts on this when next we break bread.

Life's dangerous, sometimes fatal tendency for exponential reproduction can be traced to nature's drive to find ways (including life's complex systems) to dissipate energy (a universal tendency, as described by thermodynamics's second law). The monstrous desire for "more"—exacerbated by shortsighted capitalism—has serious consequences for and even presently threatens global humanity. But life also makes use of the tendency over the long term to moderate its growth, store energy, recycle wastes, and thrive in resilient biodiverse ecosystems. Among the players in these multispecies enclaves are beings of literally fabulous complexity.

The late Oxford paleontologist Martin Brasier in his lectures would show the Great Sphinx of Giza, a lion's body with a human head rendered in yellow limestone. Brasier pointed out that microscopic examination of the yellow limestone from which the Sphinx was built revealed millions of fossils of real chimeras, amoeboid protists called foraminifera that had symbiotically merged. The Sphinx and pyramids of Egypt are composed almost entirely of benthic foraminifera, which annually generate an estimated 43 million tons of calcium carbonate per year. These foraminifera, moreover, are promiscuous protists, having merged with other species, including red algae, green algae, and diatoms, photosynthetic ray beings who make the translucent spicules of their bodies of silica, living glass. The yellow limestone depicting the mixed made-up animal monster is made of real symbiotic beings, forams and diatoms and dinoflagellates swimming in the Tethys Sea during the Eocene 50 million years ago. During the Eocene, our Ter(r)a was also sometimes all but ice-free, with crocodiles swimming at the North Pole—while the fabulous symbiotic forams used sun to precipitate calcium, their ocean-falling bodies burying carbon from which the Sphinx limestone was later formed.

Calcium ions inside marine cells, toxic if not pumped out, built up in ancestral forms. Adventitious stockpiling of calcium carbonate on the outside, of calcium phosphate on the inside, led, respectively, to shells and skeletons. As with the global oxygen dump of the Cyanocene, ancient calcium waste disposal is no longer a net negative for us. Indeed, without it, we would have no infrastructure to support

our microbially evolved flesh and blood, no skull or bone. We literally would not have a leg to stand on.

We should worry but not despair. The rock record shows that after each mass extinction, the organismically interweaving biosphere has regrown to form more species, cell types, metabolic skills, areas settled, networked intelligences, and complex sensory skills than before. Maybe this time, instead of hurting it, we can help it continue its multispecies energy-transducing recycling ways for billions of years more.

DORION SAGAN is an award-winning science writer and theorist who probes philosophical and empirical questions about life, evolution, and ecology. His publications include ten coauthored books with Lynn Margulis, his mother and the interdisciplinary biologist who pioneered modern theories of endosymbiosis. *Cosmic Apprentice: Dispatches from the Edges of Science* (Minnesota, 2013), his most recent collection of essays, illustrates his capacity to traverse a wide range of topics and genres.

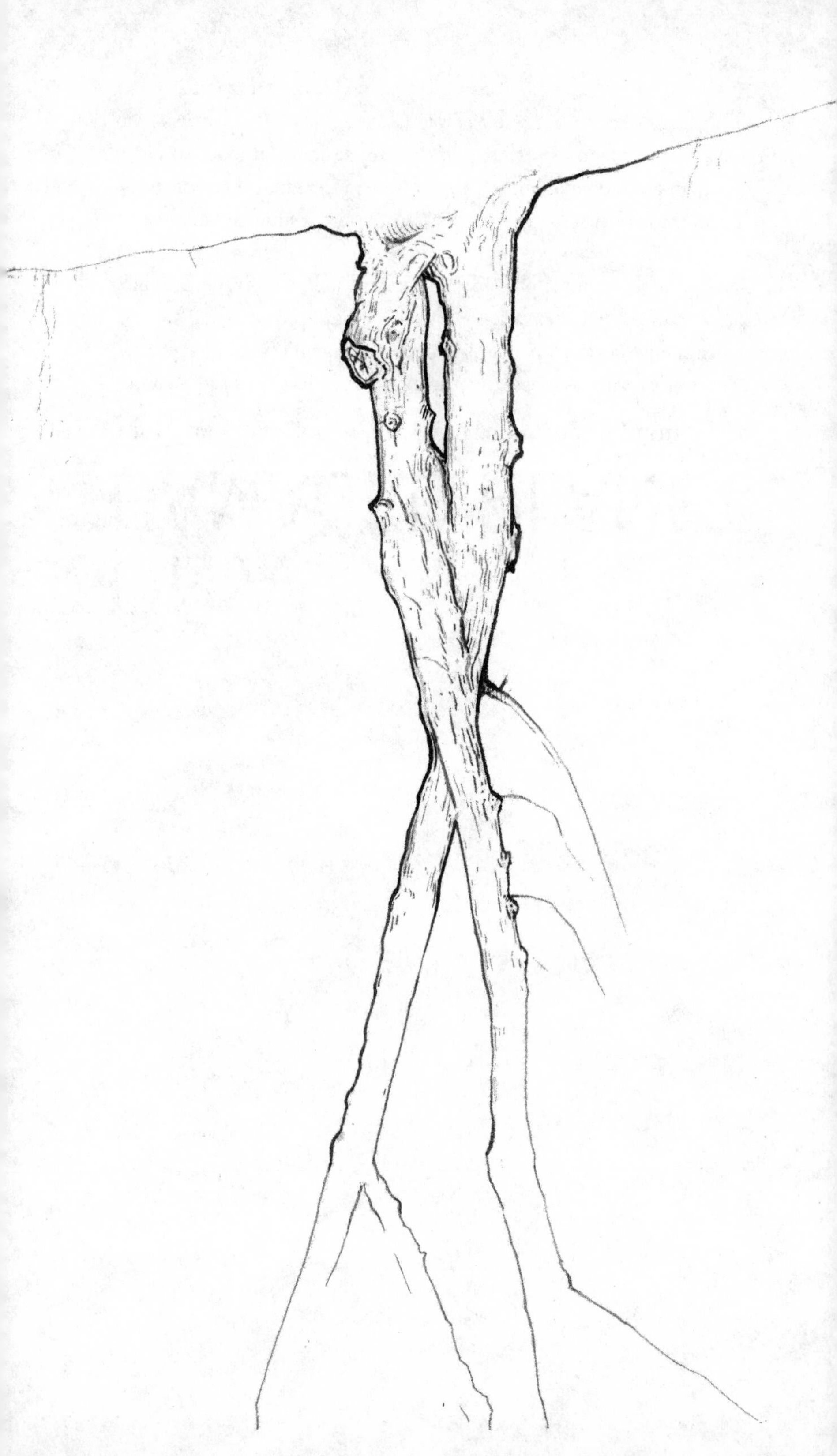

ON A DAMAGED PLANET

monsters and ghosts are figures hiding in plain sight.

They point us to forms of noticing that crosscut forms of knowledge, official and vernacular, science and storytelling. They show us co-species practices of living. If monsters are excess, ghosts are absence and invisibility. Monsters are entangled—and contaminated—bodies. Ghosts suffuse landscapes with many kinds of time. Following ghosts and following monsters are different ways to know the terrors of the Anthropocene. Ours is a playful invitation for collaborative methods, a call for new creativities for worlds that are possible and an urgent cry against the irreversibility of damage and the extinction of things that are not able to survive.

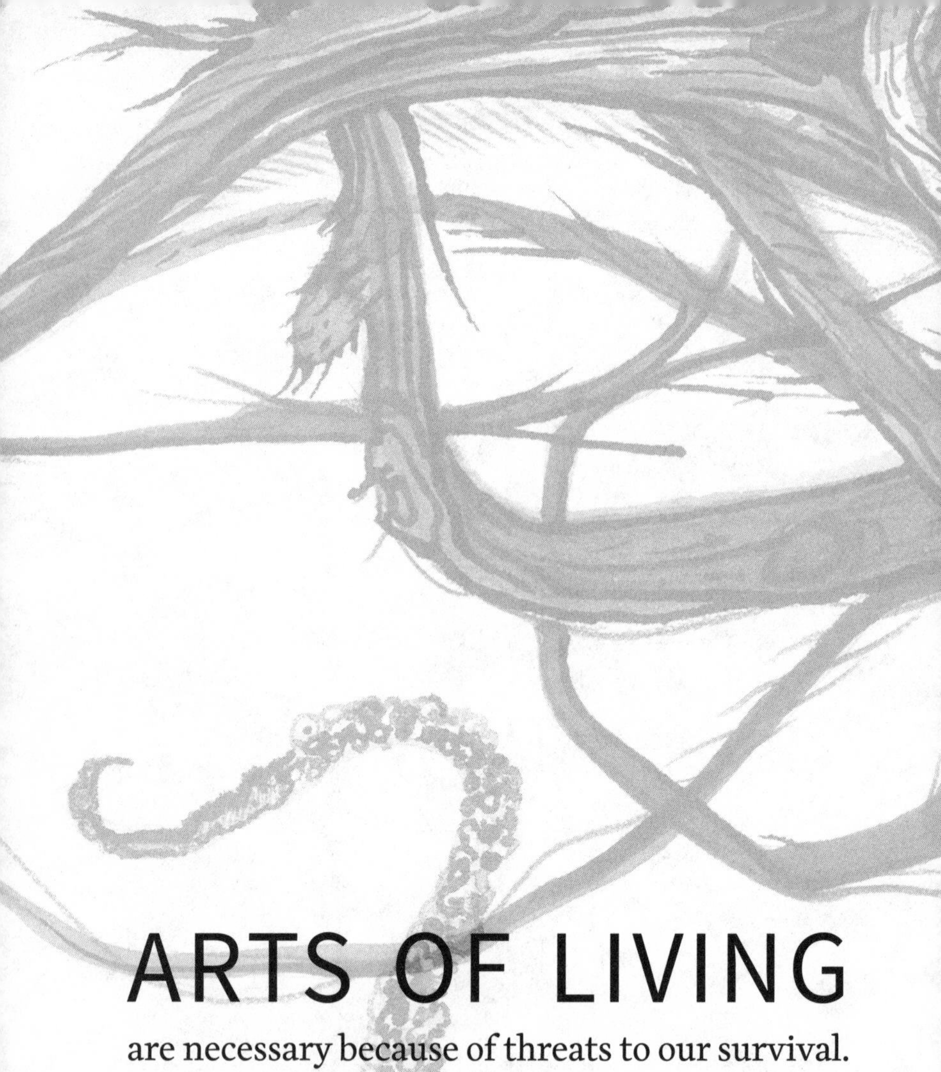

ARTS OF LIVING

are necessary because of threats to our survival.

The livability of the earth is at stake—perhaps not for extremophile bacteria but for the many forms of life that humans and our companion species have learned to love. Mounting crises from chemical contamination, land grabs, and biodiversity loss are prompting interdisciplinary dialogues and urgent calls to action. The sheer magnitude of disruption has pushed scientists, artists, and humanists to reconsider relationships between nature and culture, subjects and objects of knowledge, heroes and ghosts of progress. A major challenge is how to think geological, biological, chemical, and cultural activity together, as a network of interactions with shared histories and unstable futures. There is something mythlike about this task: we consider anew the living and the dead; the ability to speak with invisible and cosmic beings; and the possibility of the end of the world.